Fedja Hadzic, Henry Tan, and Tharam S. Dillon

Mining of Data with Complex Structures

Studies in Computational Intelligence, Volume 333

Editor-in-Chief
Prof. Janusz Kacprzyk
Systems Research Institute
Polish Academy of Sciences
ul. Newelska 6
01-447 Warsaw
Poland
E-mail: kacprzyk@ibspan.waw.pl

Further volumes of this series can be found on our homepage: springer.com

Vol. 311. Juan D. Velásquez and Lakhmi C. Jain (Eds.)
Advanced Techniques in Web Intelligence, 2010
ISBN 978-3-642-14460-8

Vol. 312. Patricia Melin, Janusz Kacprzyk, and Witold Pedrycz (Eds.)
Soft Computing for Recognition based on Biometrics, 2010
ISBN 978-3-642-15110-1

Vol. 313. Imre J. Rudas, János Fodor, and Janusz Kacprzyk (Eds.)
Computational Intelligence in Engineering, 2010
ISBN 978-3-642-15219-1

Vol. 314. Lorenzo Magnani, Walter Carnielli, and Claudio Pizzi (Eds.)
Model-Based Reasoning in Science and Technology, 2010
ISBN 978-3-642-15222-1

Vol. 315. Mohammad Essaaidi, Michele Malgeri, and Costin Badica (Eds.)
Intelligent Distributed Computing IV, 2010
ISBN 978-3-642-15210-8

Vol. 316. Philipp Wolfrum
Information Routing, Correspondence Finding, and Object Recognition in the Brain, 2010
ISBN 978-3-642-15253-5

Vol. 317. Roger Lee (Ed.)
Computer and Information Science 2010
ISBN 978-3-642-15404-1

Vol. 318. Oscar Castillo, Janusz Kacprzyk, and Witold Pedrycz (Eds.)
Soft Computing for Intelligent Control and Mobile Robotics, 2010
ISBN 978-3-642-15533-8

Vol. 319. Takayuki Ito, Minjie Zhang, Valentin Robu, Shaheen Fatima, Tokuro Matsuo, and Hirofumi Yamaki (Eds.)
Innovations in Agent-Based Complex Automated Negotiations, 2010
ISBN 978-3-642-15611-3

Vol. 320. xxx

Vol. 321. Dimitri Plemenos and Georgios Miaoulis (Eds.)
Intelligent Computer Graphics 2010
ISBN 978-3-642-15689-2

Vol. 322. Bruno Baruque and Emilio Corchado (Eds.)
Fusion Methods for Unsupervised Learning Ensembles, 2010
ISBN 978-3-642-16204-6

Vol. 323. Yingxu Wang, Du Zhang, and Witold Kinsner (Eds.)
Advances in Cognitive Informatics, 2010
ISBN 978-3-642-16082-0

Vol. 324. Alessandro Soro, Vargiu Eloisa, Giuliano Armano, and Gavino Paddeu (Eds.)
Information Retrieval and Mining in Distributed Environments, 2010
ISBN 978-3-642-16088-2

Vol. 325. Quan Bai and Naoki Fukuta (Eds.)
Advances in Practical Multi-Agent Systems, 2010
ISBN 978-3-642-16097-4

Vol. 326. Sheryl Brahnam and Lakhmi C. Jain (Eds.)
Advanced Computational Intelligence Paradigms in Healthcare 5, 2010
ISBN 978-3-642-16094-3

Vol. 327. Slawomir Wiak and Ewa Napieralska-Juszczak (Eds.)
Computational Methods for the Innovative Design of Electrical Devices, 2010
ISBN 978-3-642-16224-4

Vol. 328. Raoul Huys and Viktor K. Jirsa (Eds.)
Nonlinear Dynamics in Human Behavior, 2010
ISBN 978-3-642-16261-9

Vol. 329. Santi Caballé, Fatos Xhafa, and Ajith Abraham (Eds.)
Intelligent Networking, Collaborative Systems and Applications, 2010
ISBN 978-3-642-16792-8

Vol. 330. Steffen Rendle
Context-Aware Ranking with Factorization Models, 2010
ISBN 978-3-642-16897-0

Vol. 331. Athena Vakali and Lakhmi C. Jain (Eds.)
New Directions in Web Data Management 1, 2011
ISBN 978-3-642-17550-3

Vol. 332. Jianguo Zhang, Ling Shao, Lei Zhang, and Graeme A. Jones (Eds.)
Intelligent Video Event Analysis and Understanding, 2011
ISBN 978-3-642-17553-4

Vol. 333. Fedja Hadzic, Henry Tan, and Tharam S. Dillon
Mining of Data with Complex Structures, 2011
ISBN 978-3-642-17556-5

Fedja Hadzic, Henry Tan, and Tharam S. Dillon

Mining of Data with Complex Structures

Springer

Dr. Fedja Hadzic
Digital Ecosystems and Business
Intelligence Institute,
Curtin University
GPO Box U1987
Perth, Western Australia 6845
Australia

Prof. Tharam S. Dillon
Digital Ecosystems and Business
Intelligence Institute,
Curtin University
GPO Box U1987
Perth, Western Australia 6845
Australia

Dr. Henry Tan
183rd St. SE., 3514
98012 Bothel Washington
USA

ISBN 978-3-642-26703-1 ISBN 978-3-642-17557-2 (eBook)

DOI 10.1007/978-3-642-17557-2

Studies in Computational Intelligence ISSN 1860-949X

Softcover reprint of the hardcover 1st edition 2011

Typeset & Cover Design: Scientific Publishing Services Pvt. Ltd., Chennai, India.

Printed on acid-free paper

9 8 7 6 5 4 3 2 1

springer.com

I would like to equally dedicate this book to my neglected darling Mishele for her everlasting love, patience and understanding during my whole research career and to my mother, who has provided strong support to me throughout my life and inspiration to take a research path.

Fedja

I thank my wife Theresia and my two daughters Enrica & Eidee for consistently giving me courage and inspiration. Last but not least, I devote the book to my parents, who have been instrumental in guiding my life and encouraging me to succeed.

Henry

We would all like to thank Professor Elizabeth Chang for providing a unique environment within the Digital Ecosystems and Business Intelligence Institute that allowed us to concentrate on top level research.

Tharam, Fedja and Henry

Foreword

With the rapid development of computer technology and applications, data collected is mounting up both in size and in complexity on interconnections and structures. Thus, "*mining of data with complex structures*" becomes an increasingly important task in data mining. Although there are many books on data mining, this is a unique book dedicated to mining of data with complex structures, especially on tree structures. Tree structures have many important applications. XML documents, ontological structures, many semantic structures on the internet, structures in many social and economic organizations, Weblog structures, patient records in healthcare, and so on are tree-structured data.

Despite of the existence of a lot of general data mining algorithms and methods, tree-structure data mining deserves dedicated study and in-depth treatment because of its unique nature of structure and ordering, which leads to many interesting knowledge to be discovered, including simple subtree patterns, ordered subtree patterns, distance-constrained embedded subtrees, various kinds of application-oriented subtree patterns, and so on; and these kinds of patterns will naturally promote the development of new pattern analysis methods. From the discussions of various kinds of tree pattern mining methods, one can see that tree pattern mining contains many challenging research problems, and recent research has made good contributions to the understanding and solving the problems. Moreover, starting with tree pattern mining, this book also discusses methods for mining several other kinds of patterns with complex structures, including frequent subsequences and subgraphs.

This book, by Fedja Hadzic, Henry Tan, and Tharam S. Dillon provides a comprehensive coverage on these topics timely, with conciseness and clear organization. The authors of the book are active researchers on tree pattern mining and have made good contributions to the progress of this dynamic research theme. This ensures that the book is authoritative and reflects the current state of the art. Nevertheless, the book gives a balanced treatment on a wide spectrum of topics, well beyond the authors' own methodologies and research scopes.

Mining data with complex structures is still a fairly young and dynamic research field. This book may serve researcher and application developers a comprehensive overview of the general concepts, techniques, and applications on mining of data with complex structures and help them explore this exciting field and develop new methods and applications. It may also serve graduate students and other interested readers a general introduction to the state-of-the-art of this promising research theme.

I find the book is enjoyable to read. I hope you like it too.

Professor Jiawei Han
University of Illinois

Preface

For many practical applications in domains such as biology, chemistry, network analysis, Web Intelligence applications, the expressional power of relational data is not capable of effectively capturing the necessary relationships and semantics that need to be expressed in the domain. This gave rise to semi-structured data sources, capable of dealing with 2-dimensional relationships among data entities that are manifested through structural relationships among attribute nodes. Some examples are XML databases, RDF databases, molecular databases, graph databases, etc. Generally speaking, developing data mining methods for mining of data with complex structures is a non-trivial task and many issues exist that need to be carefully considered. The proper use of such methods needs to be explained to the practitioners in the application that may not be so familiar with the area. In the proposed book we intend to precisely define the existing problems associated with mining of data with complex structures, including trees, graphs and sequences, with a stronger focus on tree-structured data. The implications and possible applications for mining of different subpattern types under different constraints are discussed. We look at the current approaches for solving the different sub-problems in the area and discuss their advantages and disadvantages. A number of important application areas are discussed, where we explain how the described methods can effectively be used for the extraction of knowledge patterns useful for the domain.

The book overview is as follows. An introduction to the general aspects of the field of knowledge discovery and data mining is presented in **Chapter 1**, together with an overview of the sources of data with complex structures and the challenges of mining such data. **Chapter 2** is concerned with the problem of mining frequent patterns from data where the underlying information can be represented as a tree. We first discuss the motivation behind the problem together with an explanation of how a great deal of information can be effectively represented as a tree structure. We then show the importance of extracting frequent patterns from such data, known as the frequent subtree mining problem. This problem is the main focus of this book and detailed definitions of some general tree concepts are provided together with many specific terms necessary for understanding the problem in general. This involves the types of subtree patterns considered and existing frequency criteria definitions. In

Chapter 3 we look at the major issues that arise when developing algorithms for the frequent subtree mining problem. A number of different approaches are discussed and their advantages and disadvantages highlighted. At the end of the chapter, an overview of the existing frequent subtree mining algorithms is provided. The Tree Model Guided (TMG) framework for frequent subtree mining is discussed in **Chapter 4**. Here we discuss the underlying strategy of the TMG framework when approaching each of the implementation aspects discussed in Chapter 3. **Chapter 5** explains in detail the general mechanism of the TMG framework and its most generic implementation for mining of induced/embedded ordered subtrees. We also provide a mathematical model of the worst case analysis of a model-guided enumeration approach, and use this model to theoretically and practically show the differences in complexity between mining of induced and embedded subtrees. The approach is experimentally evaluated by comparing it with the current state-of-the-art techniques. A number of different real-world and synthetic datasets with varying tree characteristics are used to highlight the important differences and advantages/disadvantages of the different approaches. We then explain the necessary extensions to this general TMG framework to enable its use for mining of induced/embedded unordered, and ordered/unordered distance-constrained embedded subtrees, in **Chapters 6** and **Chapter 7**, respectively. Each extension is accompanied with a motivation of the problem, and a number of experiments with real world and synthetic datasets and comparisons to the current state-of-the-art approaches (when available). The problem of mining maximal and closed frequent subtrees is addressed in **Chapter 8**. A number of existing solution techniques are described, with one popular algorithm being explained in more detail. **Chapter 9** takes the frequent subtree mining problem and places it within the context of general knowledge analysis. The implications behind mining different subtree types using different support definitions and constraints are highlighted with the aid of motivating examples. A number of independent applications are then discussed related to analysis of health information, web data, knowledge structures for the purpose of matching, and protein structures. The problem of mining sequential data is addressed in **Chapter 10**. It starts with the necessary formulations of the problem, and looks at some existing sequence mining algorithms and applications. By considering a sequence as a special type of tree, we explain the way that the TMG framework can be used to mine sequential data. In **Chapter 11**, the graph mining problem is formally defined and existing approaches to the problem explained, with an overview of some existing algorithms. To conclude the book, in **Chapter 12**, we consider several future research directions in the area of frequent subtree mining, and discuss a number of emerging applications of the techniques to some important research areas.

Author Introduction

Dr Fedja Hadzic received his PhD from Curtin University of Technology in 2008. His PhD thesis is entitled: "Advances in Knowledge Learning Methodologies and their Applications". He is currently a Research Fellow at the Digital Ecosystems and Business Intelligence Institute of the Curtin University of Technology. He has contributed in a number of fields of data mining and knowledge discovery and published his work in a number of refereed conferences and journals. His research interests include data mining and AI in general with more focus on tree mining, graph mining, neural networks, knowledge matching and ontology learning.

Dr Henry Tan has received his PhD from University of Technology, Sydney in 2008. His PhD thesis is entitled: "Tree Model Guided (TMG) Enumeration as the Basis for Mining Frequent Patterns from XML Documents". He has become an expert in the field of data mining with a number of publications in refereed conferences and journals. Other research interests include AI, neural networks, game and software development

Professor Tharam Dillon is an expert in the field of software engineering, data mining XML based systems, ontologies, trust, security and component-oriented access control. Professor Dillon has published five authored books and four co-edited books. He has also published over 750 scientific papers in refereed journals and conferences. Over the last fifteen years, he has more than 4500 citations. Many of his research outcomes have been applied by Industry world wide. This indicates the high impact of his research work.

Acknowledgements

We would like to acknowledge Professor Elizabeth Chang and Professor Ling Feng who worked on aspects of mining of complex data structures under the Australian Research Council Discovery Grant. They were joint authors with us on several of the topics of this book and we are grateful to them for their permission to use some of the material of those papers in this book. We acknowledge the Australian Council for the Discovery Grant entitled “A Commercially Viable, Innovative XML - Enabled Association Rule Framework”, which made a lot of this work possible.

We also express our sincere thanks to Mr Christopher Jones for formatting the book and Ms Bruna Pomella for proofreading the book.

Contents

Chapter 1
Introduction

1.1 Introduction

Large amounts of data are collected and stored by different government, industrial, commercial or scientific organizations. As the complexity and volume of the data continue to increase, the task of classifying new unseen data and extracting useful knowledge from the data is becoming practically impossible for humans to do. This makes the automatic knowledge acquisition process not just advantageous over manual knowledge acquisition, but rather a necessity since the probability that a human observer will detect something new and useful is very low given the overwhelming complexity and volume of the information.

The process of discovering new and useful patterns from data that gives rise to the discovery of valuable domain knowledge is termed *knowledge discovery*. The step in the knowledge discovery process concerned with applying programs that are capable of learning and generalizing from the presented information is called *data mining*. Conventional quantitative analytical methods are increasingly being replaced by new and powerful data mining techniques that are generally applicable to a variety of domains. The traditional approaches are based mainly on statistical methods. In many domains, the systems under study are too complex to be mathematically formalized and efficiently analyzed. Data mining is a multi-disciplinary field that attracts work from many areas including database technology, machine learning, statistics, pattern recognition, information retrieval, neural networks, knowledge based systems, artificial intelligence, high-performance computing and data visualization (Han & Kamber 2006). Some features of data mining that make it superior to the traditional approaches include:

1. efficient processing of large and complex data (scalability)
2. automatically analyzing, detecting errors and inconsistencies, classifying, and summarizing the data with no human intervention (automation)
3. extracting novel and useful patterns which leads to new knowledge and discoveries (knowledge extraction)
4. combining the advantages of various disciplines (multi-disciplinary nature)

F. Hadzic, H. Tan, and T.S. Dillon: Mining of Data with Complex Structures, SCI 333, pp. 1–21.
springerlink.com

5. reducing the costs and time associated with the data analysis as a result of its automation (cost and time efficiency).

The purpose of this chapter is to provide a general introduction to the field of knowledge discovery and data mining. Further, it discusses the sources of data with complex structures and the difficulties that such data introduce to the task of knowledge discovery.

1.2 Data Mining Process

The core of the data mining process lies in applying an automated analysis algorithm to extract a knowledge model from the presented information. However, in reality, a number of steps need to occur so that the knowledge is extracted in an efficient and accurate manner. The first step is to define the objective and aim of the whole data mining process - in other words, to determine the kind of knowledge that is to be discovered and/or what questions are to be answered throughout the process. Generally speaking, the common steps include: data preparation, data mining, pattern evaluation and knowledge representation (Han & Kamber 2006). At the data preparation stage, any necessary pre-preprocessing of data is performed so that the application of the data mining algorithm is most effective. Knowledge patterns are then extracted using a suitable data mining algorithm. In the pattern evaluation step, the aim is to evaluate the available knowledge patterns and reduce and simplify these extracted patterns to only the essential knowledge patterns that capture all the useful knowledge for the application at hand. Knowledge representation is concerned with representing the acquired knowledge in a format suitable for the users' needs, which should be easy to comprehend, analyze and if necessary utilize in other applications. Each of these sub-tasks is discussed in more detail in the following subsections.

1.2.1 Data Preparation

Prior to the application of the data mining algorithms to the available data, a few processes need to occur in order for the application to be carried out in an effective manner. The data preparation step may in fact be one of the most time consuming and expensive processes in the whole data mining framework because it is very hard to obtain high quality data which can be effectively analyzed to serve its particular domain purposes. Data cleaning corresponds to the step where the noisy or inconsistent instances need to be detected, and corrected or removed from the dataset. Furthermore, data instances may contain a few missing attribute values. Removing these instances may lead to a loss of valuable information that may be provided only by that particular set of instances, and hence, this needs to be handled in an appropriate way. If the collected data has been derived from multiple sources, then data integration needs to take place. A data mining algorithm makes assumptions about the format in which the data will be presented and, if necessary, data transformation needs to occur so that the data is in the format expected by the algorithm. Furthermore, the attributes can be of different data types which need to be pre-processed in a specific way suitable for the application of the data mining algorithm in question. Another step in the data preparation stage, which may greatly affect the outcome of the data mining step, is *feature selection*. Depending on the aim of the application, not all attributes are relevant to the task at hand, and they can often interfere with the learning mechanism of the data mining algorithm applied. Furthermore, some of the attributes may be redundant in the sense that the information for which they are useful has already been provided by another attribute or sets of attributes. Hence, it is very important to detect such types of irrelevant or redundant attributes and remove them from the learning process which in general improves the data mining algorithm application in terms of efficiency, accuracy and generalization power.

1.2.2 Application of Data Mining Algorithms

After the data pre-processing stage, the data mining algorithms can be applied in order to extract interesting data patterns. There are a number of available data mining algorithms and the choice is application-dependent. Generally, there may not be a single machine learning algorithm which will achieve the desired results for all types of applications. In making the choice, one needs to consider the main aim of the application, the nature of data and the approach that would be the most useful and effective one for the domain needs. Even though this is the core of the data mining process, it may require the least effort and time, assuming that the data mining algorithms have already been developed and the data to be mined is in the appropriate format.

1.2.3 Pattern Evaluation

The patterns extracted from a dataset may be quite large in number and not all of them may be of interest to the domain application at hand. Hence, the patterns are commonly evaluated based on some interestingness measure as determined by the user. Furthermore, the patterns themselves may be subsets of other larger patterns and simplifying the pattern set will enable the discovery of more generalized knowledge.

1.2.4 Knowledge Representation

When considering the usefulness of a particular data mining technique in satisfying user needs, one of the main factors is the way in which the learned knowledge is represented. Without being able to clearly present to the user what has been learned, the user may not have great confidence in using the learned knowledge model. This is especially important when the knowledge model is to be used as a decision support tool in the domain at hand. Without a clear justification of the reasons behind the decisions predicted by the system, the user may be very reluctant to follow the advice provided by the system. In this step, it is common for visualization and various knowledge representation techniques to be used so that the learned knowledge is presented to a user in a form comprehensible to humans. In order to develop or extend a knowledge base, the domain experts may further analyze the knowledge representation. A knowledge base provides an organization with general domain knowledge and it provides the means for collection, organization and retrieval of knowledge. It is usually presented in machine readable form so that deductive reasoning methods can be automatically applied.

1.3 Different Types of Data Representations

Many data mining tools make assumptions about the nature of information being considered. The available forms of data can be generally categorized into relational, sequential, semi-structured or unstructured, each of which is discussed separately in this section.

1.3.1 Relational Data

Most commonly, a dataset used for training a data mining algorithm is comprised of a set of instances describing the occurring attribute values for a particular case in a domain. The information is represented in a two-dimensional table called a *relation* and this type of data is referred to as *relational*. The information is well structured and the schema or structure of the data is fixed and known beforehand. The nature of the entries in fields is specified for all the records. This type of data has traditionally been the most common way of representing related domain information. However, it has its limitations when representing more complex relationships between the attributes of the domain.

1.3.2 Sequential Data

A sequence corresponds to an ordered list of objects or events. A sequential database is used in domains where the order of data objects is important. It describes sequences of elements or events, and in many domains, a notion of time will also be recorded. Examples of sequential data are customer shopping sequences, web streams, biological sequences such as DNA sequences in DNA databases, and events that are ordered in time in temporal databases.

1.3.3 Semi-structured Data

For many practical applications in domains such as biology, chemistry, and network analysis, the expressional power of relational data is not capable of capturing the necessary relationships and semantics that need to be expressed in the domain. This gave rise to semi-structured data sources, capable of effectively dealing with 2-dimensional (2D) relationships among data entities that are manifested through structural relationships among attribute nodes. Some examples are XML databases, RDF databases, molecular databases, graph databases, etc. Semi-structured data might not have a fixed structure or schema for a precise description of concept attributes and their relationships. A semi-structured document can be composed of data from several heterogeneous sources each structured in a different way. With these characteristics, mining of semi-structured data poses additional challenges in the data mining field.

1.3.4 Unstructured Data

This type of data is referred to as unstructured because it has no schema that describes the underlying structure of the data, or the form of structure is not helpful for the desired processing task. Examples of unstructured data may include images, audio, video and text. Processing such unstructured data is very challenging using the currently available data mining methods, and the task has been referred to as multimedia data mining (Zaiane et al. 1998). Techniques that work so well on structured (or semi-structured) data have limited applications for unstructured data, and often require major pre-processing and reformatting of the available multimedia data. Since it is unknown what information a text document with multimedia contains, as well as where it is situated in the document, it is very difficult to formulate effective queries for analyzing and discovering useful knowledge from such data. Mining of unstructured data has become an increasingly important and popular area of research that calls for the development of new techniques specifically tailored for this task. A major subfield of unstructured data mining is mining of free form text in documents, which is often referred to as text mining.

Mining of unstructured data will not be discussed in this book as it requires a large amount of discussion of the specific issues and resolution methods which would require a book on their own.

1.4 Different Types of Knowledge Mined

The previous section has looked at some different types of representations used for representing domain data from which knowledge is to be extracted through a data mining technique. This section describes the type of information that is mined from such representations and the common format that is used for representing the extracted knowledge.

1.4.1 Type of Information Mined

Depending on the type of data representation that is to be mined, different types of information are present and may be of interest to the user. The data is represented in a particular way depending on the aim of the application and certain representations contain more information about data object relationships and properties.

Relational data is the simplest representation and one of the most common means of storing collected data. The task of data mining for relational data is often related to the detection of data patterns that can benefit the user in some way. These patterns often indicate the combinations of attribute values for which a particular outcome occurs. For example, patient records may be analyzed in order to detect the combination of possible causative factors that led to a particular illness. In these cases, the knowledge sought is the model describing the set of preconditions (i.e. attribute values) that must occur in order for a particular condition to occur. Another type of knowledge that may be mined relates to the deviations or abnormalities in the relational dataset. Such items may indicate exceptional knowledge or an irregular event which, after further investigation, can be turned into valuable domain knowledge.

Sequential data stores sequences of ordered events and therefore the ordering in which the attribute values occur in the dataset is considered important. By mining sequential data, one can detect the characteristics responsible for a change occurring in an object. Finding the characteristics for object evolution and the trend of object changes can aid in strategic planning and decision making (Han & Kamber 2006).

When semi-structured data sources are used, the attributes of the domain are organized in a hierarchical (tree) or graph structure to enable a more semantic representation of (complex) properties and relationships of data objects. These relationships and properties need to be captured in the data patterns extracted through a data mining application. Hence, rather than mining flat relational patterns, one needs to mine graph-structured patterns where structural relationships from the database are preserved. This information is necessary in fields such as web mining, bioinformatics, chemistry etc. The structural organization of a document often indicates the context in which an attribute value occurs and extracting substructures from the original database will keep the context information. When mining graph-structured data, the discovered patterns relate graph-structured entities. This knowledge which describes relationships and patterns between graph-structured entities will capture the complex structural relationships. Generally speaking, graphs have many undesirable theoretical properties with respect to algorithmic complexity. One of the common data mining tasks is the systematic enumeration of sub-graphs from a given graph, and to date no efficient algorithm exists to facilitate this task. The inherent complexity in graphs is caused by the existence of cycles. However, the number of cycles in real world databases is limited, and the core information can often be represented in a hierarchy.

If the graph contains no cycles, then we are talking about a special type of graph, called a tree. Tree-structured representations are very popular (e.g. XML) and the inherent complexity of mining a tree-structured database is much less in comparison to graphs. The task of enumerating all the subtrees from a given tree database is more feasible and a number of algorithms have been developed to efficiently solve this problem. A subtree pattern will capture the hierarchical relationships between data entities, and depending on the application, different types of tree patterns may be sought. For example, in some applications the order of the children of the sibling nodes may be irrelevant, in which case one is talking about unordered subtrees, while in an ordered subtree, this order needs to be preserved. A hierarchical relationship captured in a subtree pattern is often referred to as a parent-child or ancestor-descendant relationship. Fig. 1.1 depicts a tree showing an example of these relationships. If only parent-child relationships are allowed, then one would be interested in mining induced subtrees. On the other hand, if the captured relationships are to be generalized to those of ancestor-descendant nodes in the trees, then embedded subtrees are mined. With respect to semi-structured data, in this book the focus will be on the mining of tree-structured databases. A detailed explanation of various aspects of tree mining will be provided in Chapter 2.

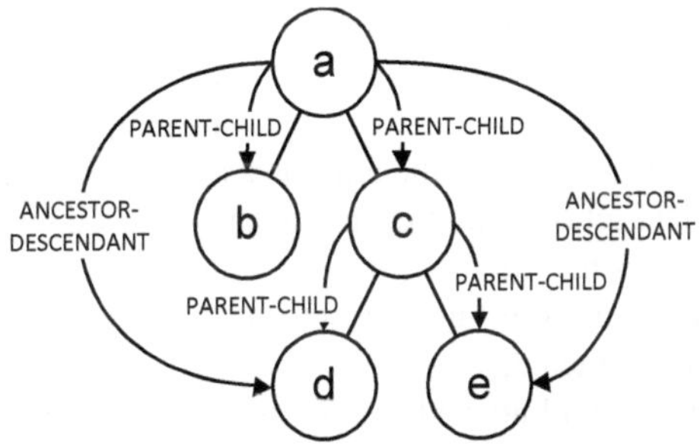

Fig. 1.1 Example tree and node hierarchical relationships

Mining of unstructured data is the most difficult task since there is no underlying schema describing the characteristics of the data. There is no indication of the type of patterns present in the data source. In order to extract useful information from unstructured data, the data mining techniques need to be integrated with other information retrieval and search techniques. For example, in order to effectively mine a text document, data mining needs to be combined with information retrieval, construction or use of hierarchies specific for text data (such as dictionaries and thesauruses), and a domain-oriented term classifications system (Han & Kamber 2006).

1.4.2 Representing the Mined Knowledge

Once a data mining technique has been applied for extraction of useful and meaningful patterns from the dataset, the knowledge learned needs to be presented in an appropriate form. The form chosen is often dependent on the type of machine learning algorithm used. This section looks at some common knowledge representations used to represent the outcome of a data mining or machine learning application.

One of the most common ways of representing the learned knowledge is in the form of standard *production rules*. A production rule consists of antecedent-consequence pairs. The antecedent part indicates a number of premises (attribute values or constraints) joined by logical 'AND' or 'OR' connectives. The consequent part corresponds to the value of the target or class attribute, about which the application aims to discover knowledge. In a conjunctive rule, the attribute values in the antecedent part are joined by logical AND connectives, whereas disjunctive rules have antecedents joined by OR connectives. Generally speaking, these rules correspond to mappings between the input space and the output space. In the case of a disjunctive rule, the information represented corresponds to a number of mappings between different regions in input space and one particular region of output space. As another rule-based knowledge representation, *association rules* are commonly used. They are in the form of $(A_1, A_2, \ldots, A_n) \Rightarrow B$,. In contrast to production rules, there may not be an independently defined target attribute that is to be the consequent part of the rule. Any attributes can be in the antecedent or consequent part of the rule and the rules describe associations between two or more attributes. The interestingness of an association rule is indicated by the support and confidence measures. The support corresponds to the number (or percentage) of examples from the dataset that contain the found association among the attributes (i.e. $(A_1, A_2, \ldots, A_n)$ and B). If X_A indicates the complete set of examples in the dataset that contain $(A_1, A_2, \ldots, A_n)$, the confidence indicates the number (or percentage) of examples from X_A that contain B. The user or domain expert usually sets these thresholds, so that only interesting or useful rules are extracted.

Another popular representation of learned knowledge is in the form of *decision trees*. A decision tree is easily constructed to represent the acquired knowledge and can be processed by both machines and humans. It is a directed graph consisting of labeled nodes and directed arcs. The root and the internal nodes are labeled with a question. The nodes correspond to the attributes of the training set, and the arcs emanating from the nodes are mutually exclusive values for that attribute. The leaf nodes represent the value of the target attribute. An instance is classified by starting from the root node, testing all the attribute values and moving down those branches that satisfy the test conditions for the instance. Eventually, a leaf node is reached that indicates the class of this instance. Generally, decision trees represent a disjunction of conjunctions of constraints on the attribute values of instances. Conjunctions occur at the nodes of the tree where the attribute testing occurs, and the tree itself is a disjunction of these conjunctions (Mitchell 1997).

In general, production rules and decision trees are appropriate for modelling empirically derived sets for heuristic rules. However, they lack the capability of capturing other types of knowledge such as abstractions and semantic relationships among the attributes. Other forms of knowledge representations are more suitable and these are discussed next.

A *concept hierarchy* is a suitable representation for describing inheritance relationships between the concepts of the domain. A concept is an abstraction for representing a collection of domain attributes. The necessary and sufficient conditions of a concept are defined by the set of permissible attribute values. Like a decision tree, a concept hierarchy can be modeled as a directed acyclic graph. The main difference is that the nodes in the graph correspond to concepts rather than attribute tests. The concepts situated at the higher level represent generalizations, while the concepts that are lower in the hierarchy inherit the properties of the higher-level concepts, and represent specializations in the form of additional attribute constraints. An instance is classified by starting from the root concept (i.e. most general concept) and moving down only those branches where the additional attribute constraints defining the lower concept are equal to the corresponding attribute values in the instance.

A *semantic net* is a directed graph consisting of a set of nodes representing the attribute objects or concepts and a set of edges describing the associated semantics. Hence, the main difference is that while a concept hierarchy can have a set of attributes associated with each concept node, in a semantic net the attributes themselves can be nodes with relationships to other nodes which may be concepts or attributes. A semantic net must label the nodes and the links in a meaningful way so that the attribute objects and the relationships between them are semantically described. Since the underlying structure is a graph, the relationships captured are not necessarily limited to the inheritance relationships as is the case in a concept hierarchy. Generally speaking, a semantic net allows for meaningful representations of the various aspects of the domain knowledge at hand.

A *frame* has been defined as a structure for capturing a collection of interrelated knowledge about a concept, a system state or an event (Dillon & Tan 1993). Each frame can represent a set or collection of instances (class frame), or it can refer to a specific item (instance frame). The 'is-a' relationship between the frames defines an inheritance by subclasses of the properties of superclasses. Frames are a very effective representation of a stereotypical object and are characterized by names, a number of slots and information about how the slots can be filled (Sestito & Dillon 1994). In addition, frames have a number of unique properties not possessed by the nodes representing concepts in a semantic net or a concept hierarchy. These slots have "facets" which determine how one obtains the result when a slot is accessed. These facets include:

1. a value facet – a value is attached to a slot;
2. a default facet – a default value is returned if there is no value defined for the slot;
3. an if-needed facet – which activates a piece of code known as a "daemon", that calculates a value and returns it if the slot is accessed.

1.5 Common Data Mining Tasks

Data mining tasks can be roughly categorized as being either descriptive or predictive in nature. The aim of the descriptive data mining tasks is to extract patterns and general characteristics of data objects in the data and present them in a form understandable to humans. In the predictive data mining task, the aim is to make inferences from the data objects from the available data set in order to predict future or unknown values of other data objects of interest. There are a number of ways in which the available information can be analyzed and the influencing factors include the aim of a particular application, the type of domain knowledge sought, and the characteristics and completeness of the data available. As a result, there are a number of different tasks that can be distinguished within the current data mining field, the most common of which will be discussed in the sub-sections that follow.

1.5.1 Association Mining

Association rule mining is one of the major data mining techniques used to discover novel and interesting relationships amongst data objects present in a database. Many businesses are interested in mining association rules from their databases to discover interesting relationships among the data objects present in large numbers of transaction records. Such discoveries can help businesses to make the right strategic decisions to boost their sales by marketing their products and services in the most effective manner (Han & Kamber 2006). Generally speaking, it aids the decision making process and provides the domain with new and invaluable knowledge. In this setting, there is no target concept defined about which knowledge is to be learned, but rather the organization is interested in discovering the items frequently associated together. Association mining consists of two important steps, namely frequent patterns discovery and rule construction (Agrawal, Imielinski, & Swami 1993; Agrawal & Srikant 1994, Agrawal et al. 1996). Frequent pattern discovery is considered as the major step in the process and the one that presents the greatest difficulty in terms of space and time complexity. A frequent itemset usually corresponds to a set of items that occur frequently together within a transactional database. The problem of frequent mining is to find all the frequent sub-patterns that occur at least as many times as the user-supplied minimum occurrence threshold. Because of the existence of different data types, a sub-pattern in this case can correspond to an itemset, sequence or a substructure. A motivating example of association mining use is the market basket analysis, the aim of which is to capture association relationships within customer transactions, which indicates that if a customer purchases a certain item(s), then he/she is also more or less likely to buy another item(s) (Han & Kamber 2006). Besides these applications, frequent pattern mining of tree-structured data can have many useful applications for general knowledge analysis and querying purposes since knowledge is often presented using a tree or graph structure. The extracted frequent substructures contain information that could be very useful for efficient document matching, information search, knowledge merging and other knowledge management related tasks.

1.5.2 Classification and Prediction

Classification is described as the process of discovering a model or a function that describes and distinguishes the data classes or concepts. The model discovery is achieved by analyzing a supervised data set, i.e. a dataset where all instances are labeled according to their class or target value. Hence, with respect to machine learning algorithms used, the type of learning that occurs is supervised. The learned model can then be used to predict the classes of future data instances for which the class label is unknown, and this is often referred to as the predictive task (Han & Kamber 2006). The data mining algorithms developed for this task are evaluated based upon the classification and predictive accuracy of their learned knowledge model. Classification accuracy is determined as the percentage of the correctly classified instances from the training set (the data set used for learning the model), whereas the predictive accuracy is calculated as the percentage of correctly classified instances from the test set or unseen data set. The learned knowledge model provides an organization with the basic knowledge of the roles and relationships of domain concepts and attributes, and is often used as a general decision support tool. For example, in a medical domain the supervised dataset would contain past patient information together with a disease diagnosis and the aim is to learn the underlying knowledge about that disease based upon the patient's symptoms. The learned knowledge can then be used to aid the medical expert in assessing the risk of future patients having that disease according to their symptoms. In addition to the meaning so far assigned to the term *prediction*, in some literature it is made more specific to include the prediction of unknown numerical data values rather than class labels in general (Han & Kamber 2006).

1.5.3 Cluster Analysis

As opposed to the classification task, when we carry out cluster analysis, the class labels of data objects or instances are not present in the training set, and hence the type of learning that needs to occur is of an unsupervised type. The main aim of the clustering methods is to group the data objects according to some measure of association, so that the data objects in one group or cluster are highly similar to one another while being very different from the data objects belonging to the remaining clusters. In other words, the aim is to maximize the intra-class similarity and minimize the inter-class similarity (Han & Kamber 2006). Clustering methods are particularly useful for exploring the interrelationships among the data objects presented in the dataset and are one way of assigning class labels to the data objects when no such information is available or known.

1.5.4 Outlier Detection

Depending on the aim of the application, the term 'outlier detection' has been commonly substituted by terms such as: anomaly, exception, novelty, deviation or noise detection. The definitions are quite similar throughout the literature and the general intent is captured by the definition given by (Hawkins 1980) 'an outlier is an observation that deviates so much from other observations as to arouse suspicion that it was generated by a different mechanism". Outlier detection is useful for applications where the intention is to find exceptional or rare events that occur in a particular set of observations. These rare events are mostly of greater interest than the common events in applications such as: fraud detection, network intrusion detection, security threats, credit risk assessment, pharmaceutical research, terrorist attacks, and some financial and marketing applications. In general, outlier detection is used in applications where the analysis of uncommon events is important and can provide extensional knowledge for the domain. Another application of outlier detection is to detect and remove the anomalous or noisy data entries so that the quality of data is increased which results in more accurate knowledge models.

1.6 Sources of Data with Complex Structures

Many organizations these days want to embed information about the complex relationships between the data objects in the domain data being collected. It is being realized that the relational or structured data has limited capabilities in presenting the domain specific information other than as just a collection of data object values. One often needs to express the specific relationship that occurs among the values as it is important information for that domain. Data where the information is organized using complex structures such as trees, graphs or lattices is of important use to model information such as: social networks, Web structure and content, circuits, protein structures, molecules, phylogenies, workflows. Generally speaking, complex structures occur in data coming from domains where complex relationships exist between the data objects (e.g. semantically enriched business and scientific data) or the data itself is inseparable from the complex structure in which it was found (e.g. chemical compounds). We next discuss examples of domains where one can expect to find data with complex structures, and these include: online information, chemical databases, bioinformatics and ontologies.

1.6.1 Online Information

In general, the information stored on the internet is often found in HTML or XML format. In fact, organizations often use the XML format to enrich their data with organization-specific meanings by adding descriptive tags and organizing it into a hierarchical structure that best describes the particular aspect of the domain. Many online documents such as XML, RDF allow for the representation of complex relationships between the data objects. One might be interested not only in mining the content on the web, but also the structural organization of the content of online documents. Furthermore, some web applications are interested in information regarding user access and browsing patterns. This information is collected using server logging facilities, and by representing it using complex structures, such as sequences or trees, information regarding user navigational patterns and Web site structure is made available for analysis. This will be discussed in detail in Chapter 9, where we discuss a number of different applications of tree mining. Analysis of the Web structure focuses mainly on hyperlink information where one can find useful information about the relatedness of a collection of interconnected Web documents. Generally speaking, any networks are effectively represented as graphs, as the entities (documents, people, communities, etc.) are represented as nodes whose relationships are expressed through links. Other online information is unstructured and it involves text documents and multimedia data. Mining of unstructured data is very challenging due to the freedom with which the data is organized, and often one attempts to organize this data into some form of structure or find the structure of the information inherent in it.

1.6.2 Chemical Databases

In these databases, one can expect to find information about stable molecules. These are traditionally represented using graph structures where edges indicate the chemical bonds between atoms which are represented as the nodes in the graph. Chemists search these databases using parts of molecules IUPAC (International Union of Pure and Applied Chemistry) names, structure components as well as constraints and properties. Chemical databases are also unique in the sense that they need to support sub-structure search. Other molecular properties include their physicochemical or pharmacological attributes referred to as descriptors. All these aspects can be used to define the similarity measure between molecules, and the finding of similar molecules or sub-structures is an important task for the drug design and discovery process.

1.6.3 Bioinformatics

In this domain, the practitioners collect large amounts of information about RNA structures, which are essentially structured as trees. A newly entered RNA structure is compared with known RNA structures looking for common topological patterns, which provide important clues to the function of the RNA. Existing information is distributed across a large number of information resources and is heterogeneous in its content, format and structure. The distributive and heterogeneous nature of the information is a main factor hindering efficient and effective information retrieval. This has motivated (Sidhu et al. 2005) to develop the ontology for proteins so as to enable easier information exchange, integration and querying. Examples of the sorts of instances that populate a protein ontology are Prions. Prion (short for proteinaceous infectious particle) is a type of infectious agent. Prions are abnormally structured forms of a host protein, which are able to convert normal molecules of protein into an abnormally structured form. The Human Prion Protein instance set and ontology was initially developed using the XML format. Here the underlying information is represented using a hierarchy of attributes and their values. One will be interested in discovering frequent or infrequent protein structures for comparison of protein datasets taken across protein families and species to help discover interesting similarities and differences.

1.6.4 Ontologies

In the Artificial Intelligence field, ontology is defined as a formal, explicit specification of a shared conceptualization (Gruber 1993a). 'Formal' corresponds to the fact that the ontology should be machine readable, 'explicit' means that the concepts and their constraints should be explicitly defined, and 'conceptualization' refers to the definition of concepts and their relationships that occur in a particular domain (Gruber 1993a, Gruber 1993b). The important factor that distinguishes an ontology from other formal and explicit domain knowledge conceptualizations is that it captures consensual knowledge, i.e. the conceptualization is accepted by a large community of users. This enables sharing and reuse of the consensual knowledge across applications and among groups of people. The set of concept terms and their relationships in an ontology is organized in a representational structure so that the inheritance relationships can be applied. It is often referred to as a 'concept hierarchy' or 'taxonomy' and the underlying structures that are commonly used are graphs or trees. Graphs are commonly used where there is a need to connect concepts situated in different parts of the hierarchy, thereby introducing a cycle; hence, the basic tree hierarchy becomes a graph. Except for very complex domains where many concepts are interrelated, there might not be that many cycles in the graph representation of the ontology and the core knowledge can still be represented using a hierarchy.

1.7 Complex Structures

This section provides high level definitions of the different types of structures that can be considered as complex. Graphs, trees and sequences fall into this category. A *graph* is a set of objects called points or vertices connected by links called lines or edges. A *tree* is an acyclic graph with one special node defined as the root, and the rest of the nodes are directed away from the root. A *sequence* describes an ordered set of events with or without the notion of time. These complex structures will be defined more formally and discussed in greater detail later in the book.

1.8 Emergence of Semi-structured Data Sources

Many electronic document formats such as HTML, Latex, SGML, BibTex, etc are semi-structured (Wang & Liu 2000). Unlike structured data, semi-structured data does not have a fixed structure and it can consist of data with different structures from several heterogeneous sources. It is different from unstructured data in that semi-structured data does have certain structures which are governed by schema.

XML adoption has spread to various domains including banking, finance, media, bioinformatics, biomedical science, information technology and science in general (Braga et al. 2002; Chi, Yang, Muntz 2004a; Davidson et al. 2001; Feng et al. 2003; Feng & Dillon 2004; Halverson et al. 2004; Hampton & vun Kannon 2001; Lewis et al. 2004; Shabo et al. 2006; Hadzic et al. 2006, 2007; Sidhu et al. 2005; Tan et al. 2005a, 2005b, 2006, 2008; Yang, Lee, Hsu 2003; Zaki & Aggarwal 2003; Zhang et al. 2005). In addition, XML has become the main data transport medium of AJAX technology. AJAX, Asynchronous Javascript and XML, is one of the pillars of Web 2.0 technology (Gehtland et al. 2006) and Web 2.0 is the next generation of the internet. MacManus & Porter (2005) even suggested that XML is the currency of choice in Web 2.0. Unlike traditional well-structured data whose schema is known in advance, XML data might not have a fixed schema, and the structure of data might be incomplete or irregular. This is why XML data is referred to as semi-structured data (Suciu 2000).

Several works have been proposed for mining XML documents (AliMohammadzadeh, Soltan, & Rahgozar 2006; Feng et al. 2003; Feng & Dillon 2004; Wang & Liu 2000; Yang, Lee, Hsu 2003; Zaki & Aggarwal 2003; Zhang et al. 2004; Zhang et al. 2005). If the focus is purely on values associated with the tags, this is by and large no different from traditional association rule mining. One interesting work is to discover similar structures among a collection of semi-structured objects (Feng et al. 2003; Feng & Dillon 2004). Recently, Zhang et al. (2005) proposed a hybrid approach, XAR-Miner, to transform XML documents into IX-Tree and Multi-DB depending on the size of the XML documents. Much research has emphasized performance but placed less emphasis on the semantics.

An XML document has a hierarchical document structure, where an XML element may contain further embedded elements, and each element can have a number of attributes attached to it. It is therefore frequently modeled using a labeled ordered tree. In Chapter 2, we will show how all the information in an XML document is modeled and preserved in a tree structure. Feng et al. (2003) initiated an XML-enabled association rule framework. It extends the notion of associated items to XML fragments to present associations among trees rather than simple-structured items of atomic values. Unlike classical association rules where associated items are usually denoted using simple structured data from the domains of basic data types, the items in XML-enabled association rules can have a hierarchical tree structure. However, it is worth noting that when each of the tree-structured items contains only one basic node, the XML-enabled association rules will degrade to traditional association rules. The adaptation of association mining to the XML document as is shown in (Feng et al. 2003) results in a more flexible and powerful representation of both simple and complex structured association relationships inherent in XML documents. Mining association rules from XML documents poses a greater challenge due to: 1) ordered data context, 2) rendered more complex hierarchy relations, and 3) greater size. The bigger data size arises from two sources: a) an XML record is more annotated through the tags than a relational record, and b) the actual amount of semi-structured (or unstructured) data (documents) exceeds the amount of relational data (Luk et al. 2002). The desire to focus on certain interesting rules has led Feng & Dillon (2004) to use a template approach to focus the search on the interestingness of the rule. An extension of this approach to define language constructs that allow one to carry out rule mining for a language such as XQuery is put forward in (Feng & Dillon 2005). However, despite the strong foundation established in (Feng & Dillon 2003), an efficient way to implement the framework is not discussed. Finding frequent patterns from XML documents without actually performing transformation to an intermediate representation as shown in (Tan et al. 2005a), can be the root of the poor performance of an algorithm. In general, XQuery-based approaches (Braga et al. 2002; Feng & Dillon 2004, 2005) suffer from a slow performance if used to mine association rules by exhaustively searching the large search space. The focus of this book is on the approaches developed to enable association rule mining from XML documents, which are known as tree mining or frequent subtree mining algorithms. All the aspects of the tree mining problem are discussed in detail in Chapter 2.

1.9 Challenges of Mining Data with Complex Structures

To date, most of the works done in association mining and other types of mining tasks were tailored for relational data and research into this type of data is relatively mature (Agrawal et al. 1993; Agrawal & Srikant 1994, Bayardo 1998; Brin, Motwani, Silverstein 1997; El-Haji & Zaiane 2003; Gouda & Zaki 2001; Han et al. 2002; Hidber 1999; Liu, Hsu, Ma 1999; Liu et al. 2002; Park, Chen, & Yu 1997; Pasquier et al. 1999; Pavn, Viana, and Gmez 2006; Sarawagi, Thomas and Agrawal 2000; Toivonen 1996; Wang, He, and Han 2000; Zaki 1997, 1998, 2000, 2002, 2003; Zheng, Kohavi and Mason 2001). Relatively less work has gone into mining of semi-structured, tree-structured and graph data (Asai et al. 2002, 2003; Chi, Yang, Muntz 2004a, 2004b; Chi, Yang, Xia & Muntz 2004, Chi et al. 2005; Dong 2004; Feng et al. 2003; Feng & Dillon 2004; Kuramochi & Karypis 2001; Inokuchi et al. 2001a, 2001b, 2003; Nijssen & Kok 2003; Wang and Liu 2000; Yang, Lee, Hsu 2003; Zaki 2003, 2005). The pillar for association mining is the first phase, i.e. frequent pattern search. This phase poses significant research challenges and has become the focus of many studies.

Relational data is limited to the ability to express only 1-dimensional (1D) relationships among entities, whereas tree-structured data is capable of dealing with 2-dimensional (2D) relationships among data entities that are manifested as hierarchical relationships among nodes. On the other hand, graph data can encode 3 dimensions of relationships among data entities in the form of network relationships. Generally, new research explorations of more complex data are sought due to the increasing complexity which is present in the collected data as well as an increasing demand for the ability to describe more complex semantics in data. Many real-world applications need patterns that are more complicated than data with linear relationships (sets and sequences) and are better represented through trees, lattices, graphs, networks, and other more complex structures (Han & Kamber 2006).

The challenges of frequent pattern mining for more complex data in the form of graph, tree and sequence structures are in general associated with an increasing complexity of structure and thus involve higher processing costs. Moreover, an ability to express more complex semantics, information encoded in tree-structured or graph form, can also translate to a larger data size, since additional hierarchical information is present and needs to be considered. With a more complex structure, utilization of a novel data structure for processing is becoming a more natural requirement for efficient processing. The computational intricacy of such complex structures can become a real issue. For example, a problem of computing if a subgraph is a subset of another graph (subgraph isomorphism) is in general NP complete (Garey & Johnson 1979). Fortunately, many problems can be modeled using a tree structure without losing too much semantics, such as XML data (Feng et al. 2003; Chi et al. 2005). Furthermore, it is generally assumed that for NP complete type problems, there is no known algorithm for performing efficient computation. In the context of frequent subgraph mining, there is no known algorithm for efficient systematic enumeration of subgraphs from a given graph. This poses serious efficiency problems. The existence of cycles in the graph is often the source of processing complexity.

In most cases, the number of cycles in graph instances in a database is limited, or the graphs may even be acyclic (Chi et al. 2005). An acyclic graph is called a tree and there are many efficient algorithms and techniques developed to deal with tree-structured data and the complexity of processing tree structures tends to be more manageable than processing graph-structured data. This makes tree mining relatively more attractive to researchers than graph mining. However, processing graph patterns is still an important task and will have broad applications including chemical informatics, bioinformatics, computer vision, video indexing, text retrieval, and Web analysis (Han & Kamber 2006).

1.10 Conclusion

This chapter has introduced the reader to some general aspects of the field of knowledge discovery and data mining. We have considered the different types of data mining tasks as well as different types of data representations and knowledge mined. The focus was then narrowed to the data with complex structures and the distinguishing aspects of such data were discussed. The emergence of semi-structured data was then overviewed with XML documents being the main case in point, and some challenges of mining such data were discussed.

References

1. Agrawal, R., Imieliski, T., Swami, A.: Mining Association Rules between Sets of Items in Large Databases. In: Proceedings of the ACM SIGMOD International Conference on Management of Data, Washington D.C., USA, May 26-28, pp. 207–216. ACM, New York (1993)
2. Agrawal, R., Srikant, R.: Fast Algorithms for Mining Association Rules. In: Proceedings of the 20th International Conference on Very Large Data Bases (VLDB), Santiago de Chile, Chile, Septemebr 12-15, pp. 487-499 (1994)
3. Agrawal, R., Mannila, H., Srikant, R., Toivonen, H., Verkamo, A.I.: Fast discovery of association rules. In: Usama, M.F., Piatetsky-Shapiro, G., Smyth, P., Uthurusamy, R. (eds.) Advances in Knowledge Discovery and Data Mining. American Association for Artificial Intelligence, pp. 307–328 (1996)
4. AliMohammadzadeh, R., Soltan, S., Rahgozar, M.: Template Guided Association Rule Mining from XML Documents. In: Proceedings of the 15th International Conference on World Wide Web, Edinburgh, Scotland, pp. 963–964. ACM, New York (2006)
5. Asai, T., Abe, K., Kawasoe, S., Arimura, H., Sakamato, H., Arikawa, S.: Efficient substructure discovery from large semi-structured data. Paper presented at the Proceedings of the 2nd SIAM International Conference on Data Mining (SIAM 2002), Arlington, VA, USA, April 11-13 (2002)
6. Asai, T., Arimura, H., Uno, T., Nakano, S.-i.: Discovering Frequent Substructures in Large Unordered Trees. In: Grieser, G., Tanaka, Y., Yamamoto, A. (eds.) DS 2003. LNCS (LNAI), vol. 2843, pp. 47–61. Springer, Heidelberg (2003)
7. Bayardo, R.J.: Efficiently mining long patterns from databases. Paper presented at the Proceedings of the ACM SIGMOD Conference on Management of Data, Seattle, USA, June2-4 (1998)

8. Braga, D., Campi, A., Ceri, S., Klemettinen, M., Lanzi, P.L.: A Tool for Extracting XML Association Rules. In: Proceedings of the 14th IEEE International Conference on Tools with Artificial Intelligence (ICTAI 2002), Washington, DC, USA, pp. 57–64 (2002)
9. Brin, S., Motwani, R., Silverstein, C.: Beyond Market Baskets: Generalizing Association Rules to Correlations. In: Proceedings of the ACM SIGMOD International Conference on Management of Data, Tucson, Arizona, USA, May 13-15, pp. 265–276. ACM, New York (1997)
10. Chi, Y., Yang, Y., Muntz, R.R.: HybridTreeMiner: An Efficient Algorithm for Mining Frequent Rooted Trees and Free Trees Using Canonical Forms. Paper presented at the Proceedings of the 16th International Conference on Scientific and Statistical Database Management (SSDBM 2004), Santorini Island, Greece, June 21-23 (2004a)
11. Chi, Y., Yang, Y., Muntz, R.R.: Canonical forms for labeled trees and their applications in frequent subtree mining. Knowledge and Information Systems 8(2), 203–234 (2004b)
12. Chi, Y., Yang, Y., Xia, Y., Muntz, R.R.: CMTreeMiner: Mining both closed and maximal frequent subtrees. In: Dai, H., Srikant, R., Zhang, C. (eds.) PAKDD 2004. LNCS (LNAI), vol. 3056, pp. 63–73. Springer, Heidelberg (2004)
13. Chi, Y., Muntz, R.R., Nijssen, S., Kok, J.N.: Frequent Subtree Mining - An Overview. Fundamenta Informaticae, Special Issue on Graph and Tree Mining 66(1-2), 161–198 (2005)
14. Davidson, S.B., Crabtree, J., Brunk, B.P., Schug, J., Tannen, V., Overton, G.C., Stoeckert Jr., C.J.: K2/Kleisli and GUS: Experiments in Integrated Access to Genomic Data Sources. IBM Systems Journal 40(2), 512–531 (2001)
15. Dillon, T., Tan, P.L.: Object Oriented Conceptual modelling. Prentice-Hall of Australia Pty Ltd. (1993)
16. Dong, A.: Treefinder: a First Step towards XML Data Mining. Paper presented at the Proceedings of the IEEE International Conference on Data Mining (ICDM 2002), Maebashi City, Japan, December 9-12 (2004)
17. El-Haji, M., Zaiane, O.R.: COFI-tree Mining: A New Approach to Pattern Growth with Reduced Candidacy Generation. Paper presented at the Workshop on Frequent Itemset Mining Implementations (FIMI 2003), in Conjunction with IEEE-ICDM, Melbourne, Florida, USA, November 19-22 (2003)
18. Feng, L., Dillon, T.S., Weigand, H., Chang, E.: An XML-Enabled Association Rule Framework. In: Mařík, V., Štěpánková, O., Retschitzegger, W. (eds.) DEXA 2003. LNCS, vol. 2736, pp. 88–97. Springer, Heidelberg (2003)
19. Feng, L., Dillon, T.S.: Mining XML-Enabled Association Rules with Templates. In: Goethals, B., Siebes, A. (eds.) KDID 2004. LNCS, vol. 3377, pp. 66–88. Springer, Heidelberg (2005)
20. Feng, L., Dillon, T.S.: An XML-Enabled Data Mining Query Language XML-DMQL. International Journal of Business Intelligence and Data Mining 1(1), 22–41 (2005)
21. Garey, M.R., Johnson, D.S.: Computers and Intractability, A Guide to the Theory of NP-Completeness. W.H. Freeman and Company, New York (1979)
22. Gehtland, J., Almaer, D., Galbraith, B.: Pragmatic Ajax: A Web 2.0 Primer. Pragmatic Bookshelf (2006)
23. Gouda, K., Zaki, M.J.: Efficiently Mining Maximal Frequent Itemsets. Paper presented at the Proceedings of the 1st IEEE International Conference on Data Mining, San Jose, USA, November 29 - December 2 (2001)
24. Gruber, T.R.: A translation approach to portable ontology specifications. Knowledge Acquisition 5(2), 199–220 (1993a)
25. Gruber, T.R.: Towards Principles for the Design of Ontologies Used for Knowledge Sharing. International Journal of Human and Computer Studies 43(5/6), 907–928 (1993b)

26. Hadzic, F., Dillon, T.S., Sidhu, A.S., Chang, E., Tan, H.: Mining Substructures in Protein Data. Paper presented at the IEEE Workshop on Data Mining in Bioinformatics DMB 2006, in conjunction with IEEE ICDM 2006, Hong Kong, December 18-22 (2006)
27. Hadzic, F., Tan, H., Dillon, T.S.: UNI3 - Efficient Algorithm for Mining Unordered Induced Subtrees using TMG Candidate Generation. In: Proceedings of IEEE Symposium on Computational Intelligence and Data Mining (CIDM), Honolulu, Hawaii, USA, April 1-5, pp. 568–575. IEEE, Los Alamitos (2007)
28. Hadzic, F.: Advances in knowledge learning methodologies and their applications. Curtin University of Technology, Perth (2008)
29. Halverson, A., Josifovski, V., Lohman, G., Pirahesh, H., Morschel, M.: ROX: Relational Over XML. In: Proceedings of the 30th Conference on Very Large Databases, Toronto, Canada, pp. 264–275 (2004)
30. Hampton, L., vun Kannon, D.: Extensible Business Reporting Language (XBRL) 2.0 Specification (December 14, 2001); XBRL.org
31. Han, J., Wang, J., Lu, Y., Tzvetkov, P.: Mining Top-K Frequent Closed Patterns without Minimum Support. Paper presented at the Proceedings of the 2002 IEEE International Conference on Data Mining, Illinois, USA (2002)
32. Han, J., Kamber, M.: Data Mining: Concepts and Techniques, 2nd edn. Elsevier, Morgan Kaufmann Publishers, San Francisco, CA, USA (2006)
33. Hawkins, D.: Identification of outliers. Chapman & Hall, London (1980)
34. Hidber, C.: Online Association Rule Mining. ACM Sigmod Record 28(2), 145–156 (1999)
35. Inokuchi, A., Washio, T., Nishimura, K., Motoda, H.: A Fast Algorithm for Mining Frequent Connected Subgraphs. IBM Research, Tokyo Research Laboratory, IBM Japan, Ltd., Tokyo, Japan (2001)
36. Inokuchi, A., Washio, T., Motoda, H.: Complete Mining of Frequent Patterns from Graphs: Mining Graph Data. Machine Learning 50(3), 321–354 (2003)
37. Inokuchi, A., Washio, T., Motoda, H.: A General Framework for Mining Frequent Subgraphs from labeled Graphs. Fundamenta Informaticae, Advances in Mining Graphs, Trees and Sequences 66(1-2) (2004)
38. Kuramochi, M. and Karypic, G, Frequent Subgraph Discovery. Paper presented at the Proceedings of the IEEE International Conference on Data Mining, ICDM 2001, San Jose, California, USA, November 29 - December 2 (2001)
39. Lewis, K.N., Robinson, M.D., Hughes, T.R., Hogue, C.W.V.: MyMed: A database system for biomedical research on MEDLINE data. IBM Systems Journal 43(4), 756–767 (2004)
40. Liu, B., Hsu, W., Ma, Y.: Mining Association Rules with Multiple Minimum Supports. In: Proceedings of the 5th ACM SIGKDD International Conference on Knowledge Discovery and Data Mining, San Diego, CA, USA, August 15-18, pp. 337–341. ACM, New York (1999)
41. Liu, J., Pan, Y., Wang, K., Han, J.: Mining Frequent Item Sets by Opportunistic Projection. Paper presented at the Proceedings of the 8th ACM SIGKDD International Conference on Knowledge Discovery and Data Mining, Edmonton, Alberta, Canada (2002)
42. Luk, R.W., Leong, H., Dillon, T.S., Chan, A.T., Croft, W.B., Allen, J.: A Survey in Indexing and Searching XML Documents. Journal of the American Society for Information Science and Technology 53(6), 415–438 (2002)
43. MacManus, R., Porter, J.: Web 2.0 Design: Bootstrapping the Social Web. Digital Web Magazine (2005)
44. Mitchell, T.M.: Machine Learning. McGraw-Hill Companies, Inc., Boston (1997)

45. Nijssen, S., Kok, J.N.: Efficient discovery of frequent unordered trees. In: Proceedings of the 1st International Workshop on Mining Graphs, Trees, and Sequences, Dubrovnik, Croatia (2003)
46. Park, J.S., Chen, M.-S., Yu, P.S.: Using a Hash-Based Method with Transaction Trimming for Mining Association Rules. IEEE Transactions on Knowledge and Data Engineering 9(5), 813–825 (1997)
47. Pasquier, N., Bastide, Y., Taouil, R., Lakhal, L.: Discovering Frequent Closed Itemsets for Association Rules. Paper presented at the Proceedings of the 7th International Conference on Database Theory, Jerusalem, Israel, January 10-12 (1999)
48. Pavon, J., Viana, S., Gomez, S.: Matrix apriori: speeding up the search for frequent patterns. Paper presented at the Proceedings of the 24th IASTED International Conference on Database and Applications, Innsbruck, Austria (2006)
49. Sarawagi, S., Thomas, S., Agrawal, R.: Integrating association rule mining with relational database systems: Alternatives and implications. Data Mining and Knowledge Discovery 4(2-3), 89–125 (2000)
50. Sestito, S., Dillon, T.S.: Automated Knowledge Acquisition. Prentice Hall, Sydney (1994)
51. Shabo, A., Rabinovic-Cohen, S., Vortman, P.: Revolutionary Impact of XML on Biomedical Information Interoperability. IBM Systems Journal 45(2), 361–372 (2006)
52. Sidhu, A.S., Dillon, T.S., Chang, E., Sidhu, B.S.: Protein ontology: vocabulary for protein data. Paper presented at the Proceedings of the 3rd International Conference on Information Technology and Applications (ICITA 2005), Sydney, Australia, July 4-7 (2005)
53. Suciu, D.: Semistructured data and XML Information. In: Tanaka, K., Ghandeharizadeh, S., Kambayashi, Y. (eds.) Information Organization and Databases: Foundations of Data Organization. Kluwer International Series In Engineering And Computer Science Series, pp. 9–30. Kluwer Academic Publishers, Dordrecht (2000)
54. Tan, H., Dillon, T.S., Feng, L., Chang, E., Hadzic, F.: X3-Miner: Mining Patterns from XML Database. Paper presented at the Proceedings of the 6th International Conference on Data Mining, Text Mining and their Business Applications, Skiathos, Greece, May 25 (2005)
55. Tan, H., Hadzic, F., Feng, L., Chang, E.: MB3-Miner: mining eMBedded subTREEs using tree model guided candidate generation. In: Proceedings of the 1st International Workshop on Mining Complex Data in Conjunction with ICDM 2005, Houston, Texas, USA, November 27-30, pp. 103–110 (2005)
56. Tan, H., Dillon, T.S., Hadzic, F., Chang, E., Feng, L.: IMB3-Miner: Mining Induced/Embedded Subtrees by Constraining the Level of Embedding. In: Ng, W.-K., Kitsuregawa, M., Li, J., Chang, K. (eds.) PAKDD 2006. LNCS (LNAI), vol. 3918, pp. 450–461. Springer, Heidelberg (2006)
57. Tan, H.: Tree Model Guided (TMG) enumeration as the basis for mining frequent patterns from XML documents. University of Technology Sydney, Sydney (2008)
58. Tan, H., Hadzic, F., Dillon, T.S., Feng, L., Chang, E.: Tree Model Guided Candidate Generation for Mining Frequent Subtrees from XML. ACM Transactions on Knowledge Discovery from Data 2(2) (2008)
59. Toivonen, H.: Sampling Large Databases for Association Rules. Paper presented at the Proceedings of the 22nd International Conference on Very Large Data Bases (VLDB 1996), Mumbai (Bombay), India (1996)
60. Wang, K., He, Y., Han, J.: Mining Frequent Itemsets Using Support Constraints. In: Proceedings of the 26th International Conference on Very Large Data Bases (VLDB), Cairo, Egypt, pp. 43–52 (2000)

61. Wang, K., Liu, H.: Discovering Structural Association of Semistructured Data. IEEE Transactions on Knowledge and Data Engineering 12(3), 353–371 (2000)
62. Yang, L.H., Lee, M.L., Hsu, W.: Efficient Mining of XML Query Patterns for Caching. In: Proceedings of the 29th International Conference on Very Large Data Bases (VLDB), Berlin, Germany, September 9-12, pp. 69–80 (2003)
63. Zaiane, O.R., Han, J., Li, Z.-N., Chee, S.H., Chiang, J.: Multimediaminer: a system prototype for multimedia data mining. Paper presented at the Proceedings of the 1998 ACM SIGMOD International Conference on Management of Data, Seattle, Washington, USA, June 2-4 (1998)
64. Zaki, M.J., Parthasarathy, S., Ogihara, M., Li, W.: New Algorithms for Fast Discovery of Association Rules. New York (1997)
65. Zaki, M.J., Ogihara, M.: Theoretical Foundations of Association Rules. Paper presented at the Proceedings of the 3rd ACM SIGMOD Workshop on Research Issues in Data Mining and Knowledge Discovery, Seattle, Washington, USA, June 2-4 (1998)
66. Zaki, M.J.: Scalable Algorithms for Association Mining. IEEE Transactions on Knowledge and Data Engineering 12(3), 372–390 (2000)
67. Zaki, M.J., Aggarwal, C.C.: XRules: An Effective Structural Classifier for XML Data. Paper presented at the Proceedings of the 9th ACM SIGKDD International Conference on Knowledge Discovery and Data Mining, Washington D.C., USA, August 24-27 (2003)
68. Zaki, M.J., Gouda, K.: Fast vertical mining using diffsets. Paper presented at the Proceedings of the 9th ACM SIGKDD International Conference on Knowledge Discovery and Data Mining, Washington D.C., USA, August 24-27 (2003)
69. Zaki, M.J.: Efficiently Mining Frequent Trees in a Forest: Algorithms and Applications. IEEE Transactions on Knowledge and Data Engineering 17(8), 1021–1035 (2005)
70. Zhang, J., Ling, T.W., Bruckner, R.M., Tjoa, A.M., Liu, H.: On Efficient and Effective Association Rule Mining from XML Data. In: Galindo, F., Takizawa, M., Traunmüller, R. (eds.) DEXA 2004. LNCS, vol. 3180, pp. 497–507. Springer, Heidelberg (2004)
71. Zhang, S., Zhang, J., Liu, H., Wang, W.: XAR-Miner: Efficient Association Rules Mining for XML Data. Paper presented at the Proceedings of the 14th International World Wide Web Conference Shiba, Japan, May 10-14 (2005)
72. Zheng, Z., Kohavi, R., Mason, L.: Real World Performance of Association Rule Algorithms. Paper presented at the Proceedings of the 7th ACM SIGKDD International Conference on Knowledge Discovery and Data Mining, San Fancisco, California, USA (2001)

Chapter 2
Tree Mining Problem

2.1 Introduction

As mentioned in Chapter 1, an important category of complex data is tree-structured data. It occurs in a variety of different domains and applications such as Web Intelligence applications, bioinformatics, natural language processing, programming compilation, scientific knowledge management and querying, etc. (Wang et al. 1994). Mining of tree-structured data introduces significant new challenges which are the subject of this chapter.

Association rule mining is a popular data mining technique used for discovering interesting associations among data objects in a wide range of domains. Mining frequent itemsets or patterns is a fundamental and essential step in the discovery of association rules. The aim of this chapter is to define the aspects necessary for understanding the problem, which is also often referred to as the frequent subtree mining problem. This is the first and the most difficult problem to consider when one wishes to perform association rule mining from tree-structured documents such as XML. Section 2.2 provides an overview of some general tasks within the traditional association rule mining framework. It is important to consider this so that issues related to tree mining and differences between mining relational and tree-structured data can be properly appreciated. The way that information contained in an XML document can be modeled using a tree-structure, and the challenges of mining such data are discussed in Section 2.3. Section 2.4 starts by providing definitions of some general tree concepts. The problem of frequent subtree mining is then discussed in detail, taking into account the different subtree types and support definitions according to which subtrees are considered as frequent.

2.2 Problem of Association Rule Mining

One form of knowledge is the association among entities. Association rule mining is directed towards finding interesting association relationships among items in a given data set. The first step in discovering association rules is to discover frequent

F. Hadzic, H. Tan, and T.S. Dillon: Mining of Data with Complex Structures, SCI 333, pp. 23–40.
springerlink.com

itemsets. From the discovered frequent itemsets, association rules in the form of $A \Rightarrow B$ are then generated. It is important that the generated rules meet certain criteria. A rule is said to be interesting if it meets the *support* and *confidence* thresholds. In the following sections, general concepts and terms used in association rule mining will be described.

2.2.1 Association Rule Framework

The databases considered in association rule mining are often comprised of transactional records. The term 'transaction' has extended its original definition in the data management field where it refers to an atomic interaction with a database management system. When the term is used in the data mining context, the following definition provided by (Bayardo, Agrawal & Gunopulos 1999) clarifies its meaning: *A transaction is a set of one or more items obtained from a finite item domain, and a dataset is a collection of transactions.* Let $I = \{i_1, i_2, \ldots, i_n\}$ be a set of items. Let D, be a transactions database for which each T is a transaction that contains a set of items such that $T \subseteq I$. Each transaction has a unique identifier denoted as *tid*. A set of items is referred to as an *itemset*. An itemset consists of k items and is commonly denoted as k-itemset. For example a set $\{A\}$ is a 1-itemset, a set $\{A,B\}$ is 2-itemset, and so on.

2.2.2 Support

An association rule is an implication of the form $A \Rightarrow B$, where $A \subseteq I$, and $B \subseteq I$, and $A \cap B = \phi$. Let the *support* (or *support ratio*) of rule $A \Rightarrow B$ be $s\%$. This implies that there are $s\%$ transactions in D that contain items A and B. In other words, the probability $P(A \cup B) = s\%$. Sometimes it is expressed as *support count* or *frequency*, that is, it reflects the actual frequency count of the number of transactions in D that contain the items that are in the rules. If the support ratio is expressed as a percentage, then its value ranges from 0% to 100%, otherwise it can be expressed as a ratio between 0.0 and 1.0. We say that an itemset is *frequent* if it satisfies the user-specified *minimum support* threshold.

2.2.3 Confidence

The rule $A \Rightarrow B$ has a *confidence* $c\%$ if $c\%$ of the transactions in D containing A, also contain B. In other words, the *confidence* of a rule $A \Rightarrow B$ is the conditional probability of a transaction containing the *consequent* (B) if the transaction contains the *antecendent* (A). Please note that antecedent and consequent can be of itemset too, such as $\{A,B,C\} \Rightarrow \{D,E\}$, etc. We say the rule is strong if it satisfies both *minimum support* (σ) and *minimum confidence* (τ). Often, an interesting rule is a rule with low support and high confidence. There are other rule interestingness measures such as *importance* (Chen, Han & Yu 1996), *leverage* (Piatetsky-Shapiro 1991), *gain* (Fukuda et al. 1996), *variance* and *chi-squared value* (Morishita 1998; Nakaya & Morishita 2004), *entropy gain* (Morimoto et al. 1998; Morishita 1998), *gini* (Morimoto et al. 1998), *laplace* (Clark & Boswel 1991; Webb 1995), *lift* (IBM, 1996) (a.k.a. *interest* (Brin, Motwani & Silverstein 1997) or *strength* (Dhar & Tuzhilin, 1993)), and *conviction* (Brin et al. 1997). Thus, support count, support and confidence can be expressed by equations 2.1 below.

$$\text{Support count}\,(A \Rightarrow B) : |\gamma|, \gamma : \{\exists t \subseteq T | A \cup B \subseteq t; |T| = N \tag{2.1}$$

$$\text{Support}\,(A \Rightarrow B) : \frac{|\gamma|}{N}, \gamma : \{\exists t \subseteq T | A \cup B \subseteq t; |T| = N \tag{2.2}$$

$$\text{Confidence}\,(A \Rightarrow B) : \frac{\text{Support}(A \cup B)}{\text{Support}(A)} \tag{2.3}$$

Association rule mining consists of two main processes: 1) frequent itemset discovery, and 2) association rule generation.

2.2.3.1 Frequent Itemset Discovery

Frequent pattern analysis is in itself an important data mining problem. It becomes the basis and pre-requisite for important data mining tasks such as: *association mining* (Agrawal, Imielinski & Swami 1993, Agrawal et al. 1996; Agrawal & Srikant 1994; Mannila et al. 1994), *correlations* (Brin, Motwani & Silverstein 1997), *causality* (Silverstein et al. 1998) and *sequential mining* (Agrawal and Srikant 1995; Ezeife & Lu 2005; Tan et al. 2006; Zaki 1997, 2001), *episodes* (Mannila et al., 1997), *multi-dimensional patterns* (Lent et al. 1997; Kamber et al. 1997), *maximal patterns* (Bayardo 1998), *partial periodicity* (Han et al. 1999), *emerging patterns* (Dong & Li 1999), and many other important data mining tasks. Moreover, frequent pattern analysis encompasses patterns from relational, sequential, tree-structured and graph data.

The problem of mining frequent itemsets can be formulated as: given a user specified minimum support threshold σ, find all frequent itemsets whose support $\geq \sigma$. This step is considered the most difficult and essential part of association rule mining. It is not only complex and expensive but also the overall performance of the association rule mining technique is determined by the frequent itemset discovery step. Therefore, many studies in the association rule mining field focus on frequent itemset discovery. Frequent itemset discovery, in general, comprises of candidate generation and candidate counting processes.

The problem of candidate generation is to generate all subsets of itemsets from the given dataset. The common approach is to generate candidate 1-itemset, 2-itemset, ..., *k*-itemset, where *k*-itemset is the set of data items which have *k* data items in each element *k* of the set. Thus, if (*A*, *B*, *C*, *D*, ..., *Z*) are items, an element in a 3-itemset could be (*B*, *C*, *D*). Another approach is to perform frequent itemset discovery without candidate generation (Han et al. 2004).

The second problem is to count the frequency of enumerated itemsets. Given that the enumerated itemsets can be numerous and very costly to compute, we are normally interested only in frequent itemsets, i.e. itemsets whose support is $\geq \sigma$. To avoid generating unnecessary candidates, pruning techniques are used. The most common one is to utilize the *downward closure* lemma (Agrawal & Srikant 1994) that states a k-itemset is frequent if and only if all of its subsets are also frequent. If any of its subsets is infrequent, then all of its superset would also be infrequent and there is no longer any need to consider such an itemset. Since this technique utilizes *apriori* knowledge, this pruning technique is commonly referred to as the *apriori* technique (Agrawal et al. 1993). In most cases, this technique helps to eliminate a large number of infrequent candidates during the candidate generation phase.

2.2.3.2 Rule Generation

The second phase of association rule mining is straightforward. Given that all frequent itemsets L_i are discovered, i.e. itemsets whose support $\geq \sigma$, and given a minimum confidence threshold τ, interesting association rules can be generated by computing the conditional probability $P(f_i|F_i - f_i) \geq \tau$, for $f_i \subseteq F_i$ and $F_i \subseteq L_i$ (Han & Kamber 2006; Zaki, 2000). Sometimes an additional rules pruning scheme might be necessary if the number of rules returned is very high, most of which are often bad rules. For example, Zaki, Lesh & Ogihara (2000) applied a sequence mining algorithm, SPADE (Zaki 1997, 2001), to the planning dataset to detect sequences which lead to plan failures. Since the domain of interest has a complicated structure with data redundancy, the algorithm returns a large number of highly frequent but non-predicting rules. To deal with this particular scenario, three pruning schemes were proposed to eliminate *normative patterns*, *redundant patterns*, and *dominated patterns*. The *normative patterns* were considered as patterns that are consistent with the background knowledge of the domain and reflect some normal or expected operation of the domain. The *redundant patterns* were viewed as those that have the same frequency as at least one of their proper subsequences, with the intuition that the smaller sequence is as predictive in the planning domain as its proper supersequence. The *dominated patterns* were considered as those patterns that are less predictive than any of their proper subsequences, and hence a pattern is eliminated if there is a smaller subsequence that has a higher confidence in predicting failure. This is, however, an example only and does not suggest that the same pruning schemes will always work when applied to problems in different domains without modification. The other research direction to address the too-many-rules (redundant rules) issue is to mine *closed* frequent itemsets (Zaki & Hsiao 2002; Zaki & Ogihara 1998) and *maximal* (Bayardo 1998) frequent itemsets.

2.3 Emerging Field of Tree Mining

As mentioned in the previous chapter semi-structured documents (e.g. HTML, Latex, SGML, BibTex, etc.) are increasingly being used to represent domain knowledge in a more meaningful and application oriented way. Many of semi-structured documents are hierarchical by nature, and the information can be effectively represented in a rooted ordered and labeled tree structure. In this section, we will take XML as a special case in point, and indicate how the information is modeled as a tree and discuss some general aspects that need to be considered when approaching the association rule mining problem from tree-structured documents.

2.3.1 XML and Association Mining

One of the tasks required to mine frequent patterns from XML is to find a representative structure with which XML can be modeled. Given that XML data contains a hierarchical relationship between nodes and there are no loops, XML can be modeled using tree structures.

```
<?xml version="1.0" encoding="utf-8"?>
<Bookstore>
  <Book isbn="123-013-1931">
    <Author>
      <Name>Michael Foster</Name>
      <Company>IT Professional</Company>
    </Author>
    <Title>Data Mining Concepts and Applications</Title>
    <Price>41.50</Price>
  </Book>
  <Book isbn="113-011-1911">
    <Author>
      <Name>Michael Foster</Name>
      <Company>IT Professional</Company>
    </Author>
    <Title>XML Bible</Title>
    <Price>49.90</Price>
  </Book>
</Bookstore>
```

Fig. 2.1 Example of an XML fragment

In the following sections, we explain how XML data can be viewed as a tree structure. Concepts and relationships that are developed for XML will be approached from the perspective of a tree structures domain.

2.3.1.1 XML Document Entities

In this section, we will describe XML constructs.

XML Nodes

Nodes can be categorized into *simple* and *complex* nodes. Simple or basic nodes have no edges emanating from them. In a tree structure, this type of node is called a leaf node. Complex nodes are also called internal nodes. Two important relationships can be constructed from complex nodes: parent-child and ancestor-descendant.

From Fig. 2.1, representatives of simple nodes would be <Name>, <Company>, <Title> and <Price>. They do not have any children and/or descendants. The complex nodes examples are <Bookstore>, <Book> and <Author>.

Element-Attribute Relationships

The relationship between element-attribute in XML is of significant value. When it comes to tree structure, this is more or less a depiction of a node with multi-labels and the level of relationships among them is of equal value. One is no more important than any other. When one needs to consider such a scenario, the next type of relationship, element-element, is more appropriate to be used.

Element-Element Relationships

Relationships between elements in XML are the basic construct for hierarchical relationships. The relationships of two elements are either parent-child relationships or ancestor-descendant relationships. The two elements that are connected by one edge are the basis for a parent-child relationship. The two elements that are connected by more than one edge are the basis for ancestor-descendant relationships. The element-element relationship must be constructed only by two elements from different levels. If two elements have the same levels and belong to the same parent, the relationship between them is a sibling relationship. There are no edges connecting sibling nodes, so it is more of a virtual relationship. Examples of both element-element and element-attribute relationships are shown in Fig. 2.2.

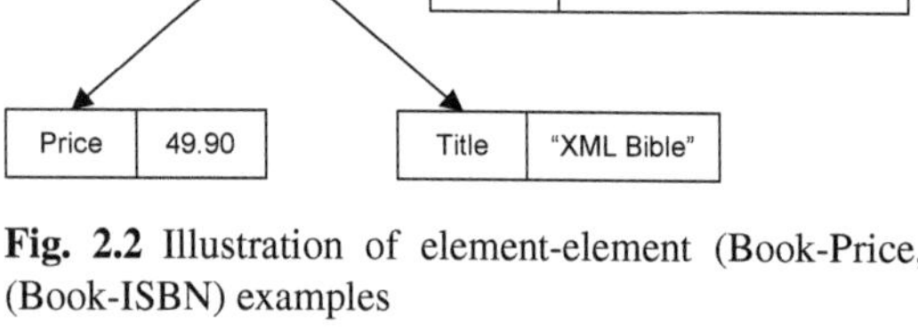

Fig. 2.2 Illustration of element-element (Book-Price, Book-Title) and element-attribute (Book-ISBN) examples

Tree-Structured Items

The basic construct of XML are tree-structured items. This is in contrast to relational data where the basic constructs are atomic items and contain only 1-1 relationships between items in itemsets (Agrawal & Srikant 1994; Han & Kamber 2006). XML, on the other hand, contains more complicated hierarchical relationships between tree-structured items, than there exist with relational data. A tree-structured item that does not have any children is known as an element. Examples of tree-structured items from Fig. 2.1 are shown below in Fig. 2.3.

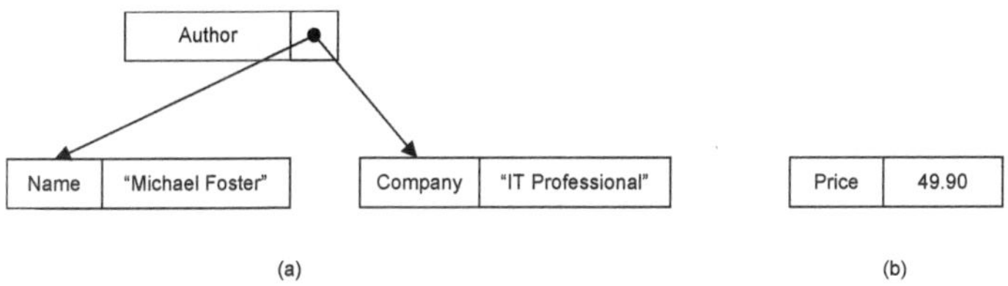

Fig. 2.3 Illustrations of tree-structured items with size 3 (a) and with size 1 (b)

2.3.2 The Parallel between XML and Tree Structure

A tree is an acyclic connected graph. XML data can be easily represented in a tree-like structure. A rooted tree is a tree in which one of the vertices is distinguished from others, and is called the root. All well-formed trees have only one unique root node. XML data is a rooted well-formed tree (Feng et al. 2003). We refer to a vertex of a rooted tree as a node of the tree. In case of XML data, a node refers to a tag or an element. A rooted ordered tree is a rooted tree in which the order of the children nodes of each node is important. Hence, if we have k nodes as children of say node A, a node at the left-most position would be at position 0 and the nodes at its right will have an incrementing position number up to the $(k-1)^{th}$ child. A rooted unordered tree is a rooted tree in which the order of the children nodes is not important. A labeled tree is a tree where each node of the tree is associated with a label. Two or more nodes can have the same label. An XML document has a hierarchical document structure, where an XML element may contain further embedded elements, and these can be attached with a number of attributes. Elements that form sibling relationships may have ordering imposed on them. Each element of an XML document has *name* and *value*. If only structure is to be considered, then only an element name is considered. Given such parallelisms, an XML document can therefore be modeled as a rooted labeled ordered (or unordered) tree.

Given that XML data is modeled through the above said constructs and there are strong parallelisms with tree structure concepts, from this point forward we concentrate on discussing tree structure. Thus, one can regard an XML document structure as a tree structure. Techniques developed for tree structure can be assumed to work as well for XML documents. If there is anything that needs to be done in a more specialized way for XML, it will be identified and discussed whenever necessary.

2.3.3 Problem of XML Document Association Mining

The problem of finding frequent patterns from XML data concerns tree-structured data and presenting associations among trees rather than simple items with atomic values. However, if each of the tree-structured items contains only root node, then mining XML data will be equivalent to mining traditional relational data. The task of mining frequent subtrees from XML documents also consists of a candidate generation phase and a frequency counting phase. The candidate generation for tree-structured data, however, is different and has to be formulated differently. It is more complex and the number of candidates generated from a tree with n nodes is generally much higher than for an itemset with n nodes. This is due to the hierarchical relationships inherent in tree-structured data. In general, candidate generation for tree-structured data involves enumerating all possible subtrees from a database of trees.

As a corollary to the problem of mining frequent itemsets in traditional association mining and modelling the database of XML documents as a database of trees, the problem of mining frequent tree-structured itemsets from a database of XML documents can be formulated as a problem of mining frequent subtrees from the database of trees.

The main problem in association mining from semi-structured documents such as XML still remains that of frequent pattern discovery, where a pattern corresponds to a subtree in this case, and a transaction to a fragment of the database tree whereby an independent instance is described. This problem is more complex than in traditional frequent pattern mining from relational data because structural relationships need to be taken into account. Furthermore, depending on the domain of interest and the task that is to be accomplished in a particular application, different types of subtrees can be mined using different support definitions. Throughout this book, these will be generally referred to as the tree mining parameters. In the next section, we provide some general definitions of tree-related concepts, as a prelude to subtree mining.

2.4 General Tree Concepts and Definitions

A *tree* is a special type of graph where no cycles are allowed. It consists of a set of *nodes* (or vertices) that are connected by *edges*. Each edge has two nodes associated with it. A *path* is defined as a finite sequence of edges and in a tree there is a single unique path between any two nodes. The *length of a path p* is the number of edges in p. A *rooted tree* has its top-most node defined as the *root* that has no incoming edges and for every other node there is a path between the root and that node. A node u is said to be a *parent* of node v, if there is a directed edge from u to v. Node v is then said to be a *child* of node u. Nodes with no children are referred to as *leaf nodes* and are also called internal nodes. Nodes with the same parent are called *siblings*. The *fan-out/degree* of a node is the number of children of that node. The *ancestors* of a node u are the nodes on the path between the root and u, excluding u itself. The *descendants* of a node v can then be defined as those nodes that have v as their ancestor. A tree is *ordered* if the children of each internal node are ordered from left to right. In an ordered tree, the last child of an internal node is referred to as the *rightmost child*. The *rightmost path* of a tree is the path connecting the rightmost leaf node with the root node. The *level/depth* of a node is the length of the path from root to that node. The *height* of a tree is the greatest level of its nodes. An *n-ary tree* is defined as a tree where every internal node has a degree no larger than n.

2.5 Frequent Subtree Mining Problem

As mentioned earlier, a tree-structured document can be modeled as a rooted ordered labeled tree. A *rooted ordered labeled tree* can be denoted as $T = (v_0, V, L, E)$, where (1) $v_0 \in V$ is the root vertex; (2) V is the set of vertices or nodes; (3) L is a labelling function that assigns a label $L(v)$ to every vertex $v \in V$; (4) $E = \{(v_1, v_2) | v_1, v_2 \in V$ AND $v_1 \neq v_2\}$ is the set of edges in the tree, and (5) for each internal node, the children are ordered form left to right.

While there exist trees with labeled edges, these are not considered in the definitions provided in this chapter, since the tree is used to model semi-structured documents, and this usually does not consider edge labels.

The problem of ***frequent subtree mining*** can be generally stated as: given a tree database T_{db} and minimum support threshold (σ), find all subtrees that occur at least σ times in T_{db}.

2.5.1 *Subtree Types*

Driven by different application needs, there are number of different subtrees that can be sought after in the general tree mining framework defined above. We reserve the explanation of instances where the different types of subtrees are the most appropriate ones to mine for Chapter 9, where we discuss some important tree mining applications. The tree mining problem is related to the isomorphism problem which determines whether two subtrees are equal to one another.

Given a tree $S = (vs_0, V_S, L_S, E_S)$ and tree $T = (vt_0, V_T, L_T, E_T)$, S and T are ***isomorphic*** to each other if there exists a structure preserving vertex *bijection* f: $V_T \rightarrow V_S$ such that $(v_1, v_2) \in E_S$ *iff* $(f(v_1), f(v_2)) \in E_T$ (Deistel 2000). Therefore, two labeled trees S and T are *isomorphic* to each other if there is a one-to-one mapping from V_S to V_T that preserves roots, vertex labels, and adjacency and non-adjacency of the vertices. An isomorphism of a tree T to itself is referred to as an ***automorphism***.

The following sections define the different types of subtrees in order of increasing complexity.

2.5.1.1 Bottom-Up Subtree

Given a tree $S = (vs_0, V_S, L_S, E_S)$ and tree $T = (vt_0, V_T, L_T, E_T)$, S is a ***bottom-up subtree*** of T, *iff* (1) $V_S \subseteq V_T$; (2) $L_S \subseteq L_T$, and $L_S(v) = L_T(v)$ (i.e. vertex labelling function is preserved); (3) $E_S \subseteq E_T$; and (4) $\forall v \in V_S$ then all the descendants of v in T must be in S. A bottom-up subtree S with root v can be obtained from a tree T by taking the vertex v of T together with v's descendants and all edges incident to those descendants.

2.5.1.2 Induced vs. Embedded Subtree

The two most commonly mined subtrees are induced and embedded. An induced subtree preserves the parent-child relationships of each node in the original tree. In addition to this, an embedded subtree allows a parent in the subtree to be an ancestor in the original tree, and hence, ancestor-descendant relationships are preserved over several levels. Therefore, an embedded subtree generalizes the definition of an induced subtree by preserving ancestor-descendant relationships.

Formal definitions follow:

Given a tree $S = (vs_0, V_S, L_S, E_S)$ and tree $T = (vt_0, V_T, L_T, E_T)$, S is an ***ordered induced subtree*** of T, *iff* (1) $V_S \subseteq V_T$; (2) $L_S \subseteq L_T$, and $L_S(v) = L_T(v)$; (3) $E_S \subseteq E_T$; and (4) the left to right ordering of sibling nodes in the original tree is preserved.

Given a tree $S = (vs_0, V_S, L_S, E_S)$ and tree $T = (vt_0, V_T, L_T, E_T)$, S is an ***ordered embedded subtree*** of T, *iff* (1) $V_S \subseteq V_T$; (2) $L_S \subseteq L_T$, and $L_S(v) = L_T(v)$; (3) $if(v_1, v_2) \in E_S$ then $parent(v_2) = v_1$ in S and v_1 is ancestor of v_2 in T; and (4) the left to right ordering of sibling nodes in the original tree is preserved.

If we consider a tree database T_{db} where all the subtree types are extracted, the following relationships hold: bottom-up subtree set $\subseteq$ induced subtree set $\subseteq$ embedded subtree set. Hence, the task of extracting all embedded subtrees is more complex than the rest, since ancestor-descendant relationships are allowed, thereby making the number of possible subtrees much larger. While we have provided the definition of a bottom-up subtree above, it is not the one that is often mined within the existing tree mining framework, and the existing algorithms focus mainly on induced or embedded subtrees as will be discussed in the following chapter.

2.5.1.3 Ordered vs. Unordered Subtrees

The subtrees defined above take into account the order among the child nodes of a vertex (siblings) and a requirement is that this ordering be preserved as detected in the original tree. This is what makes them ***ordered*** subtrees. In an ***unordered subtree***, the left-to-right ordering among the sibling nodes does not need to be preserved. Hence, if we were defining the induced and embedded subtrees for the unordered case, then condition 4 can be removed from the corresponding definitions. The information in most of the semi-structured documents (such as XML) is presented in a particular order. However, in many applications, the order among the sibling-nodes is considered irrelevant to the task and is often not available. If one is interested in comparing different document structures, it is very common that the order of sibling nodes may differ, but the information contained in the structure is essentially the same. In these cases, mining of unordered subtrees is much more suitable as a user can pose queries and does not have to worry about the order. All matching sub-structures will be returned, with the difference being that the order of sibling nodes is not used as an additional candidate grouping criterion. Hence, the main difference when it comes to the mining of unordered subtrees is that the order of sibling nodes of a subtree can be exchanged and the resulting tree is still considered the same. On the other hand, in an ordered subtree, the left-to-right ordering among the sibling nodes in the original tree has to be preserved and is used as an additional candidate subtree grouping criterion. Despite the fact that ordered subtrees have this additional requirement, mining of unordered subtrees is more difficult since each encountered subtree has to be ordered into a consistent and logical form so that candidates are grouped accordingly. This aspect is explained in detail in Section 2.7.

2.5.2 *Support Definitions*

Within the current frequent subtree mining framework to determine a support of a subtree t, generally denoted as $\sigma(t)$, two support definitions have been used namely transaction-based support and occurrence match support (also referred to as weighted support) (Zaki 2005; Chi et al. 2005; Tan et al. 2005). Another more recently proposed support definition is hybrid support (Hadzic et al. 2007) that combines both support definitions to extract more specialized subtree patterns with respect to the chosen support threshold.

2.5.2.1 Transaction-Based Support

When using the ***transaction-based support (TS)*** definition, the transaction-based support σ of a subtree t, denoted as $\sigma_{tr}(t)$ in a tree database T_{db} is equal to the number of transactions in T_{db} that contain at least one occurrence of subtree t.

Definition: Let the notation $t \prec k$, denote the support of subtree t by transaction k, then for T_S, $t \prec k = 1$ whenever k contains at least one occurrence of t, and 0 otherwise. Suppose that there are N transactions k_1 to k_N of tree in T_{db}, the $\sigma_{tr}(t)$ in T_{db} is defined as:

$$\sum_{N}^{i=1} t \prec k_i$$

In traditional frequent itemset mining from relational data, checking whether an item exists within a transaction is sufficient to determine the traditional support definition. Hence, using transaction-based support would appear to be the obvious choice when moving from relational to XML frequent pattern mining. Furthermore, it is common that in transferring from relational to XML data, an instance in relational data is described by one transaction in XML data. This has made transaction-based support the focus of many tree mining works and it is simpler than the occurrence-match support that is defined next.

2.5.2.2 Occurrence-Match Support

The ***occurrence-match support*** (*OC*) takes the repetition of items in a transaction into account and counts the subtree occurrences in the database as a whole. Hence, for *OC*, the support (σ) of a subtree t, denoted as $\sigma oc(t)$, in a tree database T_{db} is equal to the total number of occurrences of t in all transactions in T_{db}.

Definition: Let function $g(t,k)$ denote the total number of occurrences of subtree t in transaction k. Suppose that there are N transactions k_1 to k_N of tree in T_{db}, $\sigma oc(t)$ in T_{db} can be defined as:

$$\sum_{N}^{i=1} g(t,k_i)$$

2.5.2.3 Hybrid Support

The ***hybrid support (HS)*** definition can be seen as a transaction-based support that also takes into account the intra-transactional occurrence of a subtree. The support threshold is set as $x|y$, where x denotes the minimum number of transactions that must support subtree t, and y denotes the minimum number of times that t has to occur in those x transactions. HS provides extra information about the intra-transactional occurrences of a subtree. Hence, using HS threshold of $x|y$, a subtree is considered frequent if it occurs in x transactions and it occurs at least y times in each of the x transactions. Hence, while two values x and y are used as input, the support of a subtree is determined based upon the single value x that corresponds to the number of transactions in which that subtree has occurred at least y times. More formally, it can be defined as follows:

Definition: Given a hybrid support threshold of $x|y$, let function $g(t,k)$ denote the total number of occurrences of subtree t in transaction k, let the notation $t \prec k$, denote the support of subtree t by transaction k so that it is equal to 1 *iff* $g(t,k) \geq y$ and 0 otherwise. Suppose that there are N transactions k_1 to k_N of tree in T_{db}, the hybrid support of subtree t in T_{db}, denoted as $\sigma_{hs}(t)$ is defined as:

$$\sum_{N}^{i=1} t \prec k_i$$

2.6 Illustrative Example

As discussed in previous sections, there are a number of subtree types or support definitions used within the current tree mining framework. These tree mining parameters affect the resulting set of subtrees extracted. To illustrate the difference in these, please consider Fig. 2.4 which represents two subtrees (st_1 and st_2) from a tree database T_{db} consisting of two transactions T_1 and T_2.

The positions of the nodes in each of the transactions are numbered according to the pre-order traversal of that particular transaction tree. A ***pre-order traversal*** can be defined as follows:

If ordered tree T consists only of a root node r, then r is the pre-order traversal of T. Otherwise let T_1, T_2, ..., T_n be the subtrees occurring at r from left to right in T. The pre-order traversal begins by visiting r and then traversing all the remaining subtrees in pre-order starting from T_1 and finishing with T_n.

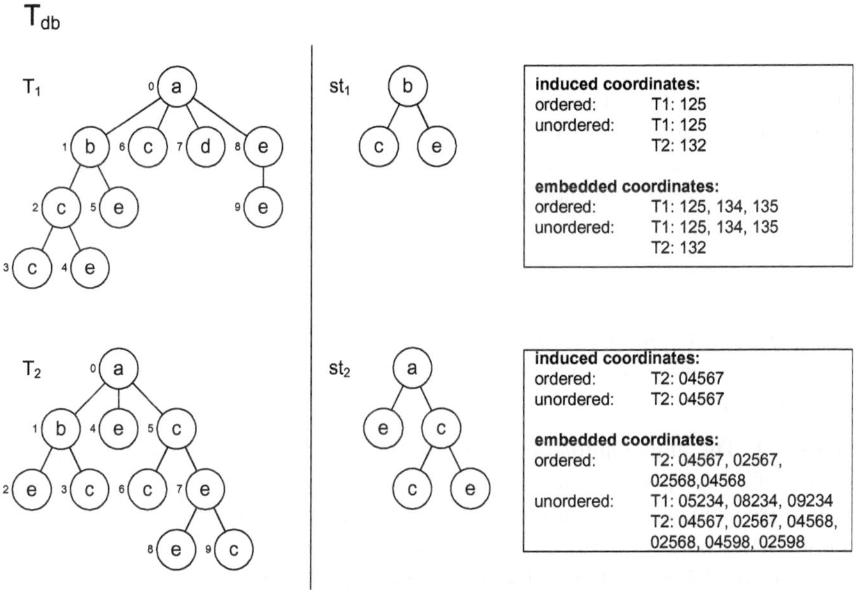

Fig. 2.4 Example Tree Database (T_{db}) consisting of two transactions (T_1 & T_2)

In Fig. 2.4, the label of each node is shown as a single-quoted symbol inside the circle, whereas its pre-order position is shown as an index on the left side of the circle. In this book, the term 'occurrence coordinate(s) *(oc)*' will be used to refer to the occurrence(s) of a particular node or a subtree in the tree database. In the case of a node, *oc* corresponds to the pre-order position of that node in the tree database, whereas for a subtree, *oc* is a sequence of *oc* from nodes that belong to that particular subtree. On the right hand side of Fig. 2.4, the *oc* of each subtree are presented with the corresponding transaction that they occur in. If ordered induced subtrees are mined, st_1 occurs only once in T_1 with *oc*:125; whereas, if unordered induced subtrees are mined, the order of node *'c'* and *'e'* can be exchanged, and hence, now st_1 occurs in T_2 as well with *oc*:132 (note: the order of occurrence coordinates is exchanged since the order of corresponding nodes is exchanged). If embedded subtrees are mined, a parent in st_1 subtree can be an ancestor in T_{db}, and hence, many more occurrences of st_1 are counted as can be seen in the top right corner of Fig. 2.4. Similarly, counting the occurrences of subtree st_2 in T_{db}, it occurs only once in T_2 with *oc*:04567 in the case of induced (ordered or unordered) subtree mining. If considering ordered embedded subtrees, then there is an additional occurrence of st_2 in T_2 with *oc*:02567 since we allow the extra ancestor-descendant relationship between node *'a'* (*oc*:0) and node 'e' (oc:2). Once the ordering among the siblings can be exchanged, there are many more instances of the st_2 subtree as shown on the bottom right corner of Fig. 2.4.

2.6.1 Issue with Pseudo Frequent Subtrees

Based on the downward-closure lemma (Agrawal & Srikant 1994), every subpattern of a frequent pattern is also frequent. In relational data, given a frequent itemset, all its subsets are also frequent. However, a question arises as to whether the same principle applies to tree-structured data when the occurrence-match support definition is used. To show that the same principle does not apply, we need to consider a counter-example.

Definition. Given a tree database T_{db}, if there exist candidate subtrees C_L and C'_L, where $C_L \subseteq C'_L$, such that C'_L is frequent and C_L is infrequent, we say that C'_L is a ***pseudo-frequent candidate subtree***. In the light of the downward closure lemma, each of these candidate subtrees is infrequent because one or more of its subtrees are infrequent.

Lemma 1. The antimonotone property of frequent patterns suggests that the frequency of a superpattern is less than, or equal to, the frequency of a subpattern. If pseudo-frequent candidate subtrees exist, then the antimonotone property does not automatically hold for frequent subtree mining.

From Fig. 2.4, suppose that the minimum support σ is set to 2, and we are focusing only on transaction T_1. Consider a candidate 2-subtree C_L that has a node with the label 'a' as the root and a node with label 'c' as the child of the root node. When an embedded subtree (ordered or unordered) is considered, there are three occurrences of C_L that occur at positions $\{(0, 2), (0, 3), (0, 6)\}$. Hence, the C_L subtree would be considered as frequent. The *(k-1)*-subtrees from C_L are singular nodes with labels 'a' and 'c'. The 1-subtree consisting of node 'a' is infrequent Therefore, in the light of the definition above, C'_L is a pseudo-frequent candidate subtree because it has an infrequent subtree 'a'.

Subsequently, since pseudo frequent candidate sub trees exist, according to Lemma 1, the anti-monotone property does not hold for frequent subtree mining when occurrence-match support is used. Hence, in the case where there exists a frequent subtree 's' with one or more of its subtrees being infrequent, then 's' also needs to be considered as infrequent in order for the anti-monotone property to hold. Tree-structured data has a hierarchical structure where 1-to-many relationships can occur, as opposed to relational data where only 1-to-1 relationships exist between the items in each transaction. This multiplication between one node to its many children/descendants makes the anti-monotone property not hold for tree-structured data. However, if transaction-based support is used no pseudo-frequent subtrees will be generated since the repetition of items is reported only once per transaction. This means that the 1-to-many relationship between a node and its children/descendants must be treated as a set of items similar to a relational database.

2.7 Canonical Form of a Subtree

In the process of generating candidate subtrees from a tree-structured database, the representation used for each subtree should uniquely map to only one subtree. This enables the use of traditional hashing methods for efficient subtree counting. The main problem that needs to be considered is to determine whether two trees are isomorphic to one another. A formal definition of the isomorphism problem for trees was given earlier. In general, two trees T_1 and T_2 are isomorphic, denoted as $T_1 \cong T_2$, if there is a bijective correspondence between their node sets which preserves and reflects the structure of the trees. Generally speaking, a canonical form (CF) of an entity is a representative form (or a function) for which many equivalent variations of an entity can be represented (mapped) into one standard, conventional, logical form in a consistent manner (Chi, Yang, & Muntz 2004; Valentine 2002). As mentioned earlier in Section 2.5.1.3, the main difference between ordered and unordered subtrees is that in an unordered subtree, the order of sibling nodes can be exchanged and the resulting subtree is considered to be the same candidate. Hence, the canonical forms for ordered and unordered subtrees are different. For ordered subtrees, the order of sibling nodes can remain the same as that in the original database, while for unordered subtrees, they may need to be ordered according to the chosen canonical form so that all candidate subtrees are grouped appropriately. Choosing an appropriate canonical form for representing a subtree is of extreme importance for unordered subtree mining. This is because in an unordered subtree, the order of sibling nodes can be exchanged and the resulting subtree is considered the same. This issue is explained in more detail in Chapter 3.

2.8 Conclusion

This chapter has presented essential concepts, definitions and problems occurring in the area of association rule mining. The focus was on problems associated with tree-structured documents such as XML. We have defined the subtree types and support definitions used within the general tree mining framework, as well as some necessary aspects that need to be considered. The purpose of this chapter was mainly to make the reader aware of the existing issues in the tree mining area as well as to provide formal definitions of the various aspects of the problem. The next chapter will address several major issues that need to be considered when developing tree mining algorithms and will also discuss some of the aspects defined in this chapter in more detail. It will provide some discussion at the implementation level and will give an overview of the existing tree mining algorithms.

References

1. Agrawal, R., Imieliski, T., Swami, A.: Mining Association Rules between Sets of Items in Large Databases. In: Proceedings of the ACM SIGMOD International Conference on Management of Data, Washington D.C., USA, May 26-28, pp. 207–216. ACM, New York (1993)
2. Agrawal, R., Srikant, R.: Fast Algorithms for Mining Association Rules. In: Proceedings of the 20th International Conference on Very Large Data Bases (VLDB), Santiago de Chile, Chile, Septemebr 12-15, pp. 487–499 (1994)

3. Agrawal, R., Srikant, R.: Mining sequential patterns. Paper presented at the Proceedings of the 11th International Conference on Data Engineering, Taipei, Taiwan, March 6-10 (1995)
4. Agrawal, R., Mannila, H., Srikant, R., Toivonen, H., Verkamo, A.I.: Fast discovery of association rules. In: Usama, M.F., Piatetsky-Shapiro, G., Smyth, P., Uthurusamy, R. (eds.) Advances in Knowledge Discovery and Data Mining. American Association for Artificial Intelligence, pp. 307–328 (1996)
5. Bayardo, R.J.: Efficiently mining long patterns from databases. Paper presented at the Proceedings of the ACM SIGMOD Conference on Management of Data, Seattle, USA, June 2-4 (1998)
6. Bayardo, R.J., Agrawal, R., Gunopulos, D.: Constraint-based rule mining on large, dense data sets. Paper presented at the Proceedings of the 15th International Conference on Data Engineering Sydeny, Australia, March 23-26 (1999)
7. Brin, S., Motwani, R., Silverstein, C.: Beyond Market Baskets: Generalizing Association Rules to Correlations. In: Proceedings of the ACM SIGMOD International Conference on Management of Data, Tucson, Arizona, USA, May 13-15, pp. 265–276. ACM, New York (1997)
8. Brin, S., Motwani, R., Ullman, J., Tsur, S.: Dynamic Itemset Counting and Implication Rules for Market Basket Data. Paper presented at the Proceedings of the, ACM SIGMOD International Conference on Management of Data, Tucson, Arizona, USA, May 13-15 (1997)
9. Chen, M.S., Han, J., Yu, P.S.: Data mining: An overview from a database perspective. IEEE Transactions on Knowledge and Data Engineering 8, 866–883 (1996)
10. Chi, Y., Yang, Y., Muntz, R.R.: Canonical forms for labeled trees and their applications in frequent subtree mining. Knowledge and Information Systems 8(2), 203–234 (2004)
11. Chi, Y., Muntz, R.R., Nijssen, S., Kok, J.N.: Frequent Subtree Mining - An Overview. Fundamenta Informaticae, Special Issue on Graph and Tree Mining 66(1-2), 161–198 (2005)
12. Clark, P., Boswell, P.: Rule Induction with CN2: Some Recent Improvements. Paper presented at the Proceedings of the 5th European Machine Learning Conference, Porto, Portugal, March 6-8 (1991)
13. Desitel, R.: Graph Theory, 3rd edn. Heidelberg Graduate Texts in Mathematics, vol. 173. Springer, New York (2000)
14. Dhar, V., Tuzhilin, A.: Abstract-Driven Pattern Discovery in Databases. IEEE Transactions on Knowledge and Data Engineering 5(6), 926–938 (1993)
15. Dong, G., Li, J.: Efficient mining of emerging patterns: Discovering trends and differences. Paper Presented at the Proceedings of the 5th ACM SIGKDD International Conference on Knowledge Discovery and Data Mining (KDD 1999), San Diego, CA, USA, August 15-18 (1999)
16. Ezeife, C.I., Lu, Y.: Mining Web Log sequential Patterns with Position Coded Pre-Order Linked WAP-tree. Data Mining and Knowledge Discovery 10(1), 5–38 (2005)
17. Feng, L., Dillon, T.S., Weigand, H., Chang, E.: An XML-Enabled Association Rule Framework. In: Mařík, V., Štěpánková, O., Retschitzegger, W. (eds.) DEXA 2003. LNCS, vol. 2736, pp. 88–97. Springer, Heidelberg (2003)
18. Fukuda, T., Morimoto, Y., Morishita, S., Tokuyama, T.: Data Mining using Two-Dimensional Optimized Association Rules: Scheme, Algorithms, and Visualization. Paper presented at the Proceedings of the 1996 ACM-SIGMOD International Conference on Management of Data, Motreal, Canada, June 4-6 (1996)

19. Han, J., Dong, G., Yin, Y.: Efficient mining of partial periodic patterns in time series database. Paper presented at the Proceedings of the 15th International Conference on Data Engineering sydeny, Australia, March 23-26 (1999)
20. Han, J., Pei, J., Yin, Y., Mao, R.: Mining Frequent Patterns without Candidate Generation: A Frequent-Pattern Tree Approach. Data Mining and Knowledge Discovery 8(1), 53–87 (2004)
21. Han, J., Kamber, M.: Data Mining: Concepts and Techniques, 2nd edn. Elsevier, Morgan Kaufmann Publishers, San Francisco, CA, USA (2006)
22. IBM, IBM Intelligent Miner User's Guide, Version 1, Release 1 (1996)
23. Kamber, M., Han, J., Chiang, J.Y.: Metarule-guided mining of multi-dimensional association rules using data cubes. Paper presented at the Proceedings of the 3rd International Conference on Knowledge Discovery and Data Mining, Newport Beach, CA, USA, August 14-17 (1997)
24. Lent, B., Swami, A., Widom, J.: Clustering association rules. Paper presented at the Proceedings of the 13th International Conference on Data Engineering, Birmingham, UK, April 7-11 (1997)
25. Mannila, H., Toivonen, H., Verkamo, A.I.: Efficient algorithms for discovering association rules. Paper presented at the Proceedings of AAAI 1994 Workshop on Knowledge Discovery in Databases Seattle, WA, USA, July 31- August 4 (1994)
26. Mannila, H., Toivonen, H., Verkamo, A.I.: Discovery of frequent episodes in event sequences. Data Mining and Knowledge Discovery 1(3), 259–289 (1997)
27. Morimoto, Y., Fukuda, T., Matsuzawa, H., Tokuyama, T., Yoda, K.: Algorithms for Mining Association Rules for Binary Segmentations of Huge Categorical Databases. Paper presented at the Proceedings of the 24th International Conference on Very Large Databases (VLDB), New York City, NY, USA, August 24-27 (1998)
28. Morishita, S.: On Classification and Regression. Paper presented at the Proceedings of the 1st International Conference on Discovery Science, Fukuoka, Japan, December 14-16 (1998)
29. Nakaya, A., Morishita, S.: Parallel Branch-and-Bound Graph Search for Correlated Association Rules. In: Zaki, M.J., Ho, C.-T. (eds.) KDD 1999. LNCS (LNAI), vol. 1759, pp. 127–144. Springer, Heidelberg (2000)
30. Piatetsky-Shapiro, G.: Discovery, analysis, and presentation of strong rules. In: Piatetsky-Shapiro, G., Frawley, W.J. (eds.) Knowledge Discovery in Databases, pp. 229–238. AAAI/MIT Press (1991)
31. Silverstein, C., Brin, S., Motwani, R., Ullman, J.: Scalable techniques for mining causal structures. Paper presented at the Proceedings of the 24th International Conference on Very Large Databases (VLDB), New York City, NY, USA, August 24-27 (1998)
32. Tan, H., Hadzic, F., Feng, L., Chang, E.: MB3-Miner: mining eMBedded subTREEs using tree model guided candidate generation. In: Proceedings of the 1st International Workshop on Mining Complex Data in Conjunction with ICDM 2005, Houston, Texas, USA, November 27-30, pp. 103–110 (2005)
33. Tan, H., Dillon, T.S., Hadzic, F., Chang, E.: SEQUEST: Mining Frequent Subsequences using DMA Strips. Paper presented at the Proceeding of the 7th International Conference on Data Mining and Information Engineering, Prague, Czech Republic, July 11-13 (2006)
34. Valentine, G.: Algorithms on Trees and Graphs. Springer, Berlin (2002)
35. Wang, J.T., Zhang, K., Jeong, K., Shasha, D.: A System for Approximate Tree Matching. IEEE Transactions on Knowledge and Data Engineering 6(4), 559–571 (1994)
36. Webb, G.I.: OPUS: An Efficient Admissible Algorithm for Unordered Search. Journal of Artificial Intelligence Research 3, 431–465 (1995)

37. Zaki, M.J.: Fast mining of sequential patterns in very large databases. University of Rochester Computer Science Department, New York (1997)
38. Zaki, M.J., Ogihara, M.: Theoretical Foundations of Association Rules. Paper presented at the Proceedings of the 3rd ACM SIGMOD Workshop on Research Issues in Data Mining and Knowledge Discovery, Seattle, Washington, USA, June 2-4 (1998)
39. Zaki, M.J.: Scalable Algorithms for Association Mining. IEEE Transactions on Knowledge and Data Engineering 12(3), 372–390 (2000)
40. Zaki, M.J., Lesh, N., Ogihara, M.: PlanMine: Predicting Plan Failures Using Sequence Mining. Artificial Intelligence Review 14(6), 421–446 (2000)
41. Zaki, M.J.: SPADE: An Efficient Algorithm for Mining Frequent Sequences. Machine Learning 42(1/2), 31–60 (2001)
42. Zaki, M.J., Hsiao, C.-J.: CHARM: An Efficient Algorithm for Closed Itemsets Mining. Paper presented at the Proceedings of the 2nd SIAM International Conference on Data Mining, Arlington, VA, USA, April 11-13 (2002)
43. Zaki, M.J.: Efficiently Mining Frequent Trees in a Forest: Algorithms and Applications. IEEE Transactions on Knowledge and Data Engineering 17(8), 1021–1035 (2005)

Chapter 3
Algorithm Development Issues

3.1 Introduction

In this chapter we discuss the main issues that arise when developing tree mining algorithms. The way that an algorithm is implemented often greatly determines its efficiency. The main aspects which affect the overall performance of the algorithm are the way that the document structure is represented at the algorithm level, the way that candidate subtrees are enumerated and counted and, in the case of unordered subtree mining, the canonical form ordering schemes.

Representation is one of the essential parts of any algorithm development, especially when it relates to data processing or manipulation. There are two representation problems: one which is related to how a tree is modeled and represented in memory or secondary storage, and a second which is related to how complex computations and data manipulations can be performed efficiently and effectively. Section 3.2 describes a way to uniquely represent subtrees in memory or secondary storage, while in Section 3.3 we address the second representation problem, by discussing the use of data structures. These data structures usually depend on the way in which the candidate generation is to be performed, and in the next chapter we describe some general characteristics of our tree model guided candidate generation, where we explain the data structures utilized within.

With the more complex relationships inherent in tree-structured data, enumeration of subtrees becomes more challenging than enumeration of itemsets from relational data. The relationships between items are no longer linear (1D) and there are many more candidates generated from a tree structure with N nodes than from relational data with N items (Tan et al. 2005). For association mining it is important that the enumeration technique should be complete so that association rules can be discovered. An optimal enumeration strategy should be non-redundant; that is, each candidate is enumerated only once, and complete in the sense that no candidates are missed (Tan et al. 2005; Zaki 2005a). The strategy used for enumerating all candidate subtrees in a unique and non-redundant manner is of critical importance for the efficiency of tree mining algorithms, and the different approaches will be explained in Section 3.4

F. Hadzic, H. Tan, and T.S. Dillon: Mining of Data with Complex Structures, SCI 333, pp. 41–65.
springerlink.com

Once the candidate subtrees have been enumerated, their frequency needs to be determined by counting the number of occurrences of these subtrees in the database. The infrequent subtrees need to be pruned throughout the process. Hence, as the number of candidates to be counted can be enormous, an efficient and rapid counting approach is extremely important. Furthermore, depending on whether the algorithm is required to handle the different support definitions, different information needs to be associated with a candidate subtree. For example, if one is using transaction-based support, then only the number of transactions in which the subtree occurs need to be counted, whereas for occurrence-match or weighted support, the different occurrences of the subtree within each transaction also need to be counted. The issue of frequency counting is discussed in Section 3.5.

To correctly enumerate and count all the unordered subtrees, the canonical form ordering process plays an important role. This is especially the case with long patterns as every candidate needs to be checked to ensure that it is in its canonical form and ordered if necessary. Canonical form ordering easily becomes one of the performance bottlenecks and the strategy adopted is usually dependent on the candidate enumeration used and the tree representation used. This aspect is discussed in Section 3.6. Having an overview of some of the main aspects that need to be taken into account during algorithm development, in Section 3.7 we provide an overview of the existing tree mining algorithms.

3.2 Tree Representation

A canonical representation for labeled trees is a unique way to represent a labeled tree and related to the way trees are represented in memory or secondary storage (Chi et al. 2005). A function f that maps labeled trees to its canonical representation should be *bijective*. The two well known representations for tree structure are the *adjacency-matrix* (Inokuchi et al. 2003) and the *adjacency-list* (Kuramochi & Karypis 2001). In the data mining community, a string-like representation is becoming very popular (Abe et al. 2002; Chi et al. 2005; Wang et al. 2004; Yang, Lee & Hsu 2003; Zaki 2005a). Each item in the string can be accessed in $O(1)$ time and the representation itself has been reported to be space efficient and easy to manipulate (Chi et al. 2005; Tan et al. 2005). In order to capture the hierarchical information of a tree, a string-like representation utilizes a notion of scope to denote the position of a node's right-most descendant node. Thus, the hierarchical structure embedded in tree data is semantically preserved and the original tree structure can be reconstructed from the string-like representation (Tan et al. 2005).

There are different forms of string encoding used for subtree mining. Zaki (2005) and Tan et al. (2005) utilize a depth-first string encoding. Luccio et al. (2001, 2004) have also independently defined this depth-first string encoding. Asai et al. (2003) and Nijssen & Kok (2003) independently defined a similar encoding for rooted ordered trees using depth sequences where the depth of a vertex is encoded in the string. They all encode the labels in the pre-order traversal manner. A depth-first string encoding, as described in (Zaki 2005a) can be generated by adding vertex

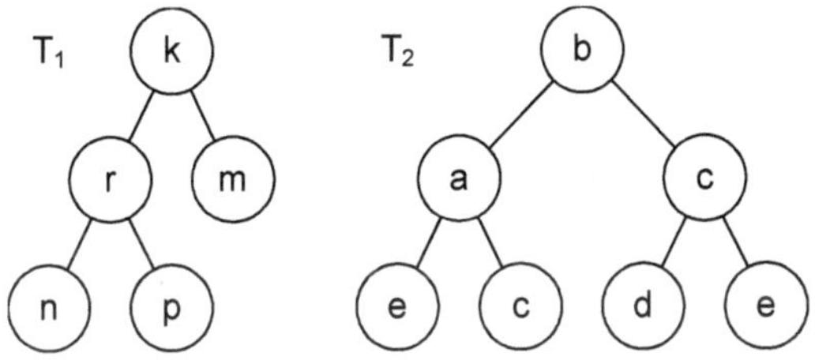

Fig. 3.1 Example trees T_1 and T_2

labels in a depth-first pre-order traversal of a tree $T(v_o,V,E,L)$, and appending a backtrack symbol (for example '/', and '/'$\notin L(V)$) whenever we backtrack from a child node to its parent node. In general, depth-first string encoding is good for describing vertical relationships (parent-child, ancestor-descendant). From Fig. 3.1, the depth-first string encoding for T_1 is 'k r n / p / / m /', and for T_2 is 'b a e / c / / c d / e / /'.

Others utilize breadth-first string encoding (Chi, Yang & Muntz 2004a). This encoding can be formed by traversing the tree in breadth-first order, level by level. Each sibling family is separated by the level coding (Nijssen & Kok 2003). In (Chi, Yang, & Muntz 2004b), special separating symbols '$' are used for separating each sibling family while the symbol '#' is used to indicate the end of the string encoding. The main difference is that with breadth-first, the encoding is constructed by iterating the tree in breadth-first order level by level where each level is encoded. For the example trees in Fig. 3.1, the breadth-first encoding as used in (Chi, Yang, & Muntz 2004b) is 'k \$ r m \$ n p #' for T_1, and 'b \$ a c \$ e c \$ d e #' for T_2. The breadth-first string encoding is good in capturing and grouping the sibling relationships between nodes in a tree, i.e. horizontal relationships. However, it is not so good when describing the parent-child and ancestor-descendant relationships (vertical relationships). It is more advantageous to utilize depth-first string encoding for techniques that exploit vertical relationships more than the horizontal relationships.

3.2.1 Efficient Representation for Processing XML Documents

Semi-structured documents tend to be large in size. The aim of this section is to discuss the way that the information in such documents can be represented so as to enable easier processing by the data mining algorithms. We will take XML documents as the case in point. Information encoded in XML documents is in the form of text format. Processing a large collection of text data can be very expensive. In the previous chapter, we explained the way in which XML document entities can be modeled, and the parallel between XML and tree structure. Even though the underlying structure of XML documents is a tree structure, processing the textual information describing XML document entities and their values can be very expensive.

To achieve efficient processing, the tree-structured data in an XML document can be transformed into a string-like representation.

```
<Book isbn="113-011-1911">
  <Author>
    <Name>Michael Foster</Name>
    <Company>IT Professional</Company>
  </Author>
  <Title>XML Bible</Title>
  <Price>49.90</Price>
</Book>
```

Fig. 3.2 A fraction of an XML tree

We mentioned earlier that depth-first string encoding is better in expressing vertical relationships. Since we would like to concentrate more on vertical relationships within a tree structure, we prefer to use the depth-first string encoding to represent the tree-structured items. In this case, the depth-first string encoding is constructed from the combination of XML elements, attributes and values by traversing the XML document tree model in depth-first (pre-order) traversal. Unless otherwise stated, all string encoding discussed in this book will refer to depth-first string encoding.

Rule 1: If structures and values are to be considered, XML can be transformed into a string such that the elements' names and their inner text (value) are appended as a single string label. For example, from Fig. 3.2, the element <Name> according to the said rule can be transformed into a single string 'Name["Michael Foster"]'. If only the structure is to be considered, then the element string (value/innertext) will be omitted and so we will use only the element name.

Rule 2: Similarly, if an element has one or more attributes, since we do not consider a node with multi labels, we will merge the element name-value with attributes name-value as a single string label. So from Fig. 3.2, for example, the element book according to the said rule can be transformed into a single string 'Book[isbn="113-011-1911"]'. If only the structure is to be considered, then the attribute name and value will be omitted.

However, the processing of long strings can still be very expensive. A common strategy used to expedite the processing of XML documents is to transform them into an integer-based form. With this approach, the textual context of each element node will be mapped into an integer number and the mapping is stored in an index table where the original string can be looked up on this index table at a later time for reporting purposes. This way the performance of the technique to discover frequent subtrees from XML documents can be improved greatly as reported in (Tan et al. 2005).

Fig. 3.3 is an example of an XML string index table computed from the XML tree in Fig. 3.2. One can use any hash function to map such strings into integer indexes. With this index table, the XML tree can be transformed into an integer-indexed tree as shown in Fig. 3.4.

By utilizing the same dataset format used by (Zaki 2005a), the above integer-indexed tree is then formatted as shown in Fig. 3.5. Please note that the second column (*cid*) could be used to refer to a specific entity which the record describes

string	*index*
Book[isbn="113-011-1911"]	100
Author	101
Name["Michael Foster"]	102
Company["IT Professional"]	103
Title["Xml Bible"]	104
Price["49.90"]	105

Fig. 3.3 XML string index table

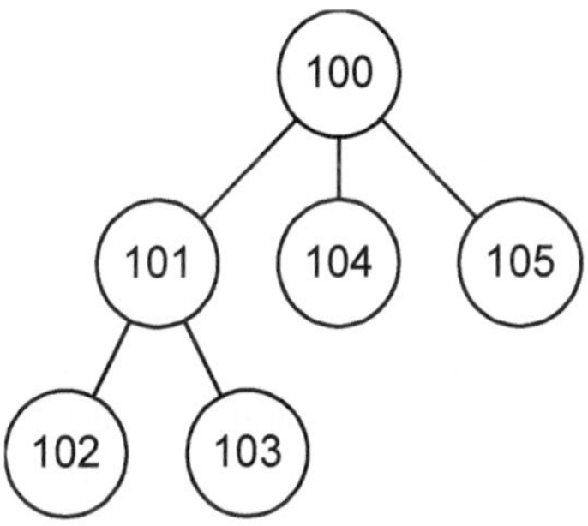

Fig. 3.4 Integer-indexed tree of XML tree in Fig. 3.1

(e.g. customer-id). However, in many domains such information is often unavailable, or it has been intentionally omitted or related through transaction-id. Hence, in most of the tree databases represented in this format, the *cid* column will simply be a repetition of the *tid* column.

tid	*cid*	*S*
0	0	11 100 101 102 -1 103 -1 -1 104 -1 105 -1
1	1	...
...	...	...
N	N	...

Fig. 3.5 An XML tree in Fig. 3.1 formatted as a string like representation as used in (Zaki 2005a). *tid*: transaction-id; *cid*: omitted (i.e. equal to *tid*); S: size of the string

A little extra post-processing will be required to look up the string table, but overall this technique is more scalable than processing string labels directly and improves frequency counting. This is the case when a hash table is used for candidate frequency counting. Hashing integer labels over string labels can have a significant impact on the overall candidates' counting performance as illustrated in the following example.

Suppose that a text '1234' is to be hashed as a string data, one has to compute the hash key from a string consisting of 4 characters '1','2','3','4'. On the other hand, suppose that a text '1234' is to be hashed as an integer data, one can simply treat '1234' as the hash key. This makes the latter cheaper to compute and this is advantageous for frequency counting in frequent pattern mining.

3.3 Data Structure Issues

An algorithm is a set of well-defined instructions for accomplishing a certain task. An algorithm works by modifying states of objects and transforming them into something tangible depending on the nature of the task it performs. We are concerned about two things here. An algorithm needs a way to store the state of the objects (*data container*) and to perform efficient operations on them (*method*). Both are the basis for what we commonly refer to as a data structure. The operations performed by an algorithm can be the insertion of new objects into the container, the deletion of objects from the container, the retrieval of objects value, the sorting of objects, etc. In this context, objects would refer to data.

A data structure is a data container with structured relationships that provides mechanisms for performing effective and efficient operations on the stored data while maintaining data integrity. A well-designed data structure allows a variety of critical operations to be performed economically, and is time, resource and space efficient. For example, an array allows a fast random read/write access; a binary tree allows efficient data retrieval, etc. Different kinds of data structures are suited to different kinds of applications, and some are highly specialized for certain tasks. For example, hash tables are good for the implementation of a dictionary, B-trees are particularly well-suited for storing data in secondary storage, and an array is good for simple random read/write access.

Algorithm performance and complexity, without precluding other factors, is heavily dependent on the *data structure* used. In the next chapter, we will provide a discussion of the data structures utilized in our tree model guided candidate generation framework. We will cover both static and dynamic aspects of these data structures. The static aspect is referred to as a *data representation*. The dynamic aspect is referred to as *data operation*. The latter will be discussed in subsequent chapters when we are discussing the mechanics of core algorithm. The choice of representation is normally affected by the goals. If the goal is to transfer information with emphasis on space efficiency, then we are interested in how we can represent data in an efficient compressed form that can be retrieved reliably and with less effort. However, compressed data will always require more processing and thus results in more time needed to consume the original data. If speed is a higher priority, then extended information might need to be encoded together with the main information. The extended information is often useful to speed up the processing as it remembers certain processing states thereby minimizing dynamic re-computation cost. However, superfluous extended information may also lead to overkill in performance. The preferred approach is to balance the amount of information so that it is not too much and not too little.

In this section, we have highlighted the importance of choosing appropriate data structures, and the next chapter will explain the core data structures used in our tree mining framework.

3.4 Enumeration Techniques

The two most commonly used enumeration strategies for tree-structured data are enumeration by extension and join (Chi et al. 2005). Another reported technique for mining frequent subtrees is to utilize the Pattern-Growth method (Wang et al. 2004), which is an extension of the previous work on relational data FP-growth (Han et al. 2004). Pattern-Growth does not perform level-by-level candidate enumeration as does the normal Apriori-based approach, and it is the only known method for obtaining frequent subtrees without a candidate generation process. Some other studies that utilized pattern-growth methods are COFI-Tree (El-Haji & Zaiane, 2003), TreeProjection (Agrawal, Aggarwal, & Prasad 2001) and H-mine (Pei et al. 2001). Other techniques utilize the structural model of the document for efficient candidate generation (Yang, Lee & Hsu 2003; Nijssen & Kok, 2003). In this section, we discuss each of these enumeration strategies in turn.

3.4.1 Enumeration by Join

The join approach was first utilized in the Apriori algorithm (Agrawal et al. 1993). In the Apriori-based technique, the 2-itemsets candidate generation, C_2, is done by joining L_1, frequent 1-itemsets, with themselves, $L_1 * L_1$. It means that if the order is important and $|L_1| = n$, apriori will generate n^2 candidates. For example, with $|L_1| = 3$, the Apriori-based technique will always generate $3^2 = 9$ candidates for 2-itemsets. Hence, the complexity of generating 2-itemsets in Apriori turns out to be $O(n^2)$. Secondly, the Apriori-based technique needs to scan the full database, compare and count the frequency of n^2 candidates generated with the database. In the worst case scenario, where the number of existing 2-itemsets patterns in the database is minimal, the Apriori-based technique generates the maximum combination of 2-itemsets. Therefore, there are n, where n = (maximum-minimum), number of 2-itemsets candidates wastefully generated which adds to I/O overhead. Let us take an example of $|L_1| = 10^4$, when order is important, 10^8 2-itemsets candidates will be generated. Assuming there are only 10^2 existing 2-itemsets in the database, there are $10^8 - 10^2 = 99{,}999{,}900$ wasteful candidates generated that add up to a huge I/O overhead as a result of the comparison cost. It is very costly to handle 2-itemsets candidate generation for a huge number of candidate sets in Apriori-like algorithms (Han & Kamber 2001).

Zaki (2005a) adapted the join enumeration strategy to the problem of mining frequent embedded rooted ordered subtrees. The join enumeration strategy for the enumeration of subtree/subgraph has become a common technique and is used in (Chi, Yang & Muntz 2004a; Inokuchi et al. 2000; Zaki 2005a, 2005b). However, employing the join enumeration technique for tree-structured data would result in the same computational cost problem for enumerating all frequent 2-itemsets (i.e. $O(n^2)$). Earlier Park, Chen, & Yu (1997) argue that accelerating the process of discovering frequent 2-itemsets can boost the overall performance of algorithm. This holds true at least for mining relational data where the number of 2-itemset candidates generated is normally larger than k-itemsets candidates for $k > 2$. For tree-structured data, however, this is not entirely true since the number of candidates generated for $k = 2$ might not be always larger than for $k > 2$ (Tan. et al. 2005). Whether or not the candidate generation process of 2-itemsets is the most expensive, accelerating this process will definitely contribute to the overall efficiency of the algorithm.

3.4.2 Enumeration by Extension

Another enumeration strategy is to use the extension technique. Essentially, an appropriate tree traversal method is used to expand candidate subtrees with all compatible nodes in the tree database. In the context of enumeration of a tree structure, the extension approach is more commonly known as a right-most-path extension, since the nodes are appended to the right-most-path of a candidate subtree (Abe et al. 2002; Nijssen & Kok 2003; Tan. et al. 2005). The right-most-path extension method is reported to be complete and all valid candidates are enumerated at most once (non-redundant) (Abe et al. 2002, Nijssen & Kok 2003; Tan et al. 2005).

3.4.3 Structure Guided Enumeration

The idea of utilizing a structural model for efficient enumeration appeared in (Yang, Lee & Hsu 2003; Nijssen & Kok, 2003; Tan et al. 2005; Tatikonda, Parthasarathy, & Kurc 2006). The approach uses the XML schema to guide the candidate generation so that all candidates generated are valid because they conform to the schema. Without utilizing schema-conscious (Tatikonda, Parthasarathy, & Kurc 2006) or structural information, a candidate generation technique could generate valid infrequent, valid frequent, invalid frequent, and invalid infrequent candidates. We do not want to generate invalid frequent and invalid infrequent candidates and this is where the join approach is lacking.

The concept of a schema-guided method can be generalized to a structure-guided enumeration strategy to improve the join approach for which enumerated candidates might not exist in the actual data. In this respect, a structure-guided approach will generate fewer candidates as opposed to the join approach. A candidate enumeration method known as *Tree Model Guided* (TMG) has been presented in (Tan et al. 2005, 2006a, 2008a). TMG can be applied to any data that has a model representation with clearly defined semantics that have tree-like structures. It ensures that only valid candidates which conform to the actual tree model structure of the data are generated. In the cases where the model representation of the tree structure is unavailable, the TMG approach will still perform the candidate generation according to the tree structure of the document, but may not be as efficient since the model needs to be obtained from the document itself. This indicates that a candidate subtree can be considered as valid in two ways: firstly, by conforming to an available model representation of the document tree structure (schema); and secondly, by conforming to the tree structure through which the information present in the examined document is represented.

By generating non-existent candidates with the join approach, the overall performance of the candidate generation process can be compromised because greater processing cost may be incurred by pruning those non-existent candidates. A study in (Tan et al. 2005, 2008a) reported that utilizing the structural information of tree-structured data can at least lower the complexity of generating candidate subtrees with length 2 (2-subtrees) to $O(C*n)$ from $O(n^2)$ for the join approach. (Tatikonda, Parthasarathy, & Kurc 2006) also developed an efficient left-most-path (LMP) based growth that takes the topology of the database trees into account and defines a candidate as *redundant* (invalid), when its support is zero or such a candidate does not exist in the actual database.

3.4.4 Horizontal vs. Vertical Enumeration

The enumeration approach can be further sub-divided into horizontal and vertical enumeration. Horizontal enumeration has the advantage of allowing level-by-level pruning to be performed completely. The disadvantage of this approach is that it requires more processing memory since at each enumeration the generated subtree would not yet have the final frequency count. This type of enumeration is also sometimes called breadth-first enumeration due to the fact that it computes the breadth before the depth. Note that the study in (Tan et al. 2005) shows that the level-by-level pruning which is commonly called full *k-1* pruning (Zaki 2005a) is important for tree-structured data to ensure that all generated *k*-subtrees are all frequent by making sure that all *(k-1)*-subtrees are also frequent. For relational data, this is more of an option for optimization purposes. For tree-structured data, whenever occurrence-match support is used and full pruning is not performed, some pseudo-frequent subtrees may be generated. The generation of too many pseudo-frequent subtrees could produce a performance issue as reported in (Tan et al. 2005, 2008a).

On the other hand, the vertical enumeration strategy, sometimes called the depth-first enumeration approach, enjoys good performance for processing long patterns (Bayardo 1998; Wang et al. 2004). This enumeration approach is also more space efficient (Chi et al. 2005; Zaki 2005a) because every enumeration will compute a frequency count of each generated subtree completely so that we do not have to retain the partial occurrence information until all subtrees with the same size have been counted. On the other hand, as mentioned previously, the problem with this enumeration approach is that full *(k-1)* pruning becomes a challenge because information of *(k-1)* subtrees is not guaranteed to be readily available. VTreeMiner for example, overcomes this issue by implementing opportunistic pruning (Zaki 2005a) which deals with over-performing full pruning. It has been shown that on some datasets, this technique can also have poor performance and can generate many pseudo-frequent subtrees (Tan et al. 2008a). In general, the performance of the technique which implements depth-first enumeration is well acknowledged.

3.5 Frequency Counting

The occurrences of candidate subtrees need to be counted in order to determine whether they are frequent, whilst the infrequent ones need to be pruned. As the number of candidates to be counted can be enormous, an efficient and rapid counting approach is extremely important. The efficiency of candidate counting is greatly determined by the data structure used. More conventional approaches use a direct checking approach. For each candidate generated, its frequency is increased by one if it exists in the transaction. A hash-tree (Agrawal & Srikant 1994; Chi et al. 2005) data structure can be used to accelerate direct checking. Another approach projects each candidate generated into a vertical representation (Chi et al. 2005; Zaki 2003, 2005a), which associates an occurrence list with each candidate subtree. One way to minimise the space problem of horizontal enumeration is to combine horizontal enumeration with vertical counting (Tan et al. 2006a). The advantage of retaining horizontal enumeration is that full (*k-1*) pruning can be performed completely. The trade-off is that you would enumerate the subtree whose support has been obtained more than once. However, this can be overcome by a caching technique, i.e. by avoiding the generation of the subtree whose support has been obtained.

One efficient way to expedite frequency counting performance is to store only the hyperlinks (Wang et al. 2004) of subtrees in the tree database instead of creating a local copy for each generated subtree. Subtree encoding is used to uniquely identify subtrees. To count the frequency of a subtree t is to count the number of subtrees that occur with the encoding t. The hashing of a string is more expensive than the hashing integer array (Tan et al. 2005). Thus, representing the string encoding as an integer array rather than as a string can have significant impact on the overall performance of the candidates counting process.

3.6 Canonical Form Ordering Schemes for Unordered Subtrees

In candidate generation, each subtree encoding should uniquely map to only one subtree. This enables the use of traditional hashing methods for efficient subtree counting. The group of possible trees obtained by permuting the sibling nodes in all possible ways is referred to as the automorphism group of a tree (Zaki 2005b). One of the trees from the automorphism group has to be chosen as the representative of the group. This is due to the fact that the subtrees as detected from a database of ordered labeled trees (e.g. XML) will by default have an ordering imposed on their sibling nodes. The generated subtrees need to be ordered according to the representative of the automorphism group. This ensures that the occurrences of one particular unordered subtree are correctly grouped so that the frequency can be easily determined. On the other hand, if this task is not performed, then the process will extract a set of ordered rather than unordered subtrees.

The main problem that needs to be considered is determining whether two trees are isomorphic to one another. A formal definition of the isomorphism problem for trees was given in Chapter 2. In general, two trees T_1 and T_2 are isomorphic, denoted as $T_1 \cong T_2$, if there is a bijective correspondence between their node sets which preserves and reflects the structure of the trees. In the field of mathematics, the term automorphism corresponds to an isomorphism of an object to itself. In the context of unordered subtree mining, an automorphism group of a tree T, denoted as $Auto(T)$, is a complete set of label preserving automorphisms of T, i.e. $Auto(T) = \{T_1, \ldots, T_n\}$ where $T_i \cong T_j$, for any $i = (1, \ldots, n)$ and $j = (1, \ldots, n)$.

During the pre-order traversal of a database, ordered subtrees are generated by default. In other words, the subtrees generated are in their canonical form for an ordered subtree. It is necessary to identify which of these ordered subtrees form an automorphism group of an unordered subtree. One tree needs to be selected to uniquely represent the unordered tree. This selected tree is known as the canonical form (CF) of an unordered tree. The problem of candidate generation is then to enumerate subtrees according to the canonical form so that the frequency of a candidate unordered subtree T, is correctly determined as the frequency of the members of $Auto(T)$.

With respect to the CF chosen, the aim should be that the enumerated candidate subtrees will require less sorting on average, since sorting the tree encodings usually becomes one of the performance bottlenecks. Different ways to define a canonical form for an unordered tree have been proposed. Chi, Yang, & Muntz (2004b) defined a canonical form based on depth-first traversal, depth-first canonical form (DFCF). They also defined a canonical form based on breadth-first traversal, breadth-first canonical form (BFCF), which was utilized in their HybridTreeMiner algorithm (Chi, Yang and Muntz 2004a). Asai et al. (2003) and Nijssen & Kok (2003) also defined equivalent canonical forms. These canonical forms essentially describe an ordering scheme that all the sibling nodes in the tree need to satisfy so that the tree can be considered as a canonical form representative of an unordered tree. As mentioned earlier, this is a necessity for unordered subtree mining as, whenever any instance of a tree is encountered, where the tree consists of the same node labels and the same parent-child, ancestor-descendant relationships, but different order of sibling nodes, it needs to be considered as the same candidate tree. In general, canonical forms of unordered subtrees can be classified into BFCF and DFCF categories, and we explain each of them in more detail next.

3.6.1 Depth-First Canonical Ordering Form

The DFCF ordering scheme as proposed in (Chi, Yang and Muntz 2004b), can be formally explained as follows:

Definition: Given two trees T_1 and T_2, with $root[T_2] = r_1$ and $root[T_2] = r_2$, let $descendants(r_1) : \{r_1d_1, \ldots, r_1d_m\}$ be the set of descendants of node r_1 and $descendants(r_2) : \{r_2d_1, \ldots, r_2d_n\}$ be the set of descendants of node r_2, ordered according to the pre-order traversal. Note that these sets can also contain the special backtrack symbol to indicate the backtracking during the traversal of the descendant nodes. Let $ST_x(r)$ denote the subtree of tree T_x with root node r, and $|descendants(r)|$ denote the number of descendants of node r. We define

$label(r_1d_i) < label(r_2d_j)$ if $label(r_1d_i)$ lexicographically sorts smaller than $label(r_2d_j)$, and so $T_1 < T_2$ *iff* :

a) $label(r_1) < label(r_2)$ or,
b) if $label(r_1) = label(r_2)$ and $|descendants(r_1)| = m$, $|descendants(r_2)| = n$, then either:

   i) $1 \leq i < j \leq min(m,n) \exists j$ such that $ST_1(r_1d_i) = ST_2(r_2d_i)$ and $ST_1(r_1d_j) < ST_2(r_2d_j)$
   ii) $m \leq n, \forall 1 \leq i \leq m$ and $m \leq n$, $ST_1(r_1d_i) = ST_2(r_2d_i)$

Within the described ordering scheme above, the smallest subtrees are placed to the left of the original tree. As an example, consider Fig. 3.6 where the steps are shown of sorting a tree T into its DFCF (T_{cf}) in a bottom-up manner. The bolded nodes indicate that the ordering scheme has just been applied to those nodes (i.e. all the nodes at that level). At the implementation level, the ordering may be performed differently depending on the way in which a tree is represented, and the choice that will lead to performing the task in the shortest time possible. The ordering process as it occurs at the implementation level within our general framework will be explained in Chapter 6.

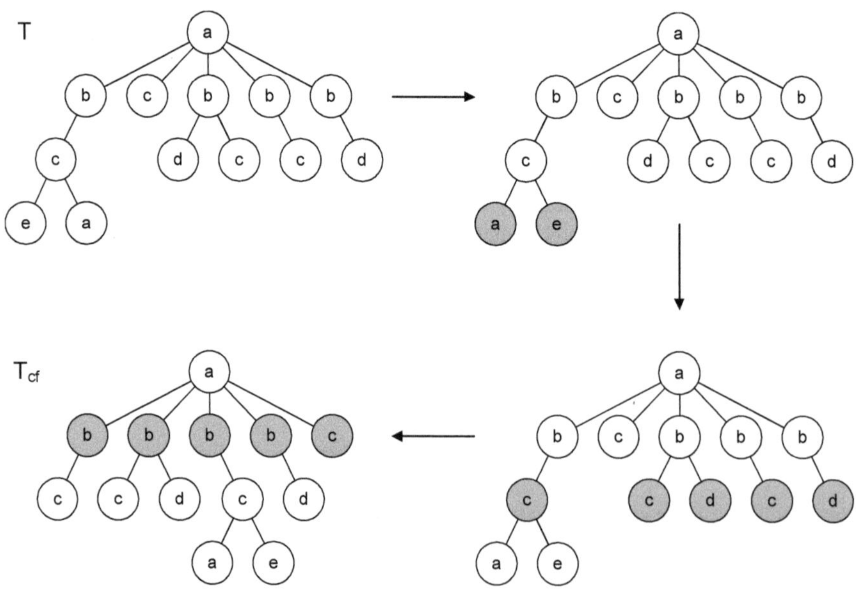

Fig. 3.6 Converting a tree T (top left) into its DFCF canonical form (T_{cf}) (bottom left)

3.6.2 *Breadth-First Canonical Form Ordering*

As described in (Chi, Yang, Muntz 2004a), and mentioned earlier in Section 3.2, this BFCF string encoding can be formed by traversing the tree in breadth-first order, level by level. Each sibling family is separated by the level coding (Nijssen & Kok 2003) or, as in (Chi, Yang, & Muntz 2004b), the special separating symbol '$' is used. We will use the encoding presented in (Chi, Yang, & Muntz 2004b) as a case in point to explain the BFCF ordering schema. For a rooted tree, the first character in the string is the root node label, and to indicate the end of the string, '#' symbol is used. Both '$' and '#' are not in the vertex labels set. In (Chi, Yang, & Muntz 2004b) the order is chosen so that symbol '#' sorts greater than symbol "$' and both sort greater than any other label from the vertex label set. Essentially, the same process of ordering nodes as performed for DFCF is done, which is ordering nodes level by level starting from the bottom. Nodes whose labels sort lexicographically smaller are placed to the left. The main difference is in the resolution of ordering when the labels are the same. While in DFCF we traverse the subtree rooted at the nodes being compared in a depth-first manner until we find a node label that is not the same, in BFCF we traverse the subtree in a breadth-first manner, thereby comparing the sibling nodes first rather than going down the descendant path of the tree. In Fig. 3.7, we provide an illustrative example to clarify this difference. When sorting the nodes at the second level, the subtrees rooted at these nodes would need to be analyzed since the labels are the same (i.e. 'a'). The BFCF ordering scheme would traverse the children of node 'a', i.e. its siblings and when it encounters nodes 'c' and 'd' during comparison it places the subtree containing node 'c' to the left (as 'c' < 'd'), as is shown on the left of Fig. 3.7. Hence, it traverses nodes in breadth first manner during comparison of labels. On the other hand, the DFCF ordering scheme would traverse the subtree in a depth-first manner and encounter nodes 'f' and 'e' first. Hence, the DFCF ordering scheme would place the subtree containing the node 'e' to the left (as 'e' < 'f'), as is shown on the right of Fig. 3.7.

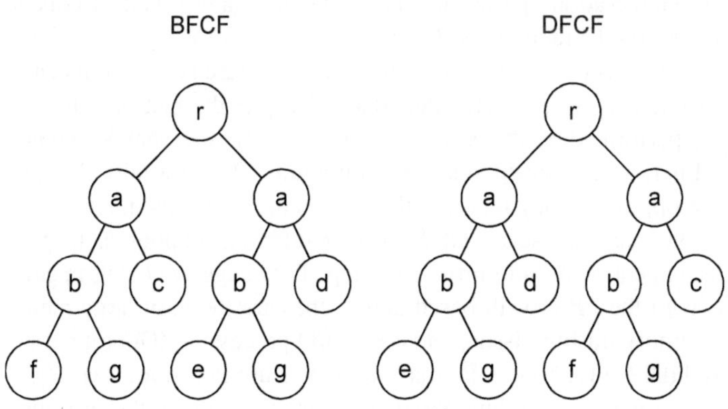

Fig. 3.7 Difference between BFCF and DFCF

3.7 Overview of Existing Tree Mining Algorithms

This section overviews some of the existing algorithms developed for the problem of frequent subtree mining. They are grouped according to the subtree type being mined, and the order in which the different groups are presented reflects the order of increasing complexity as one moves from one subtree type to another. At the end of the section, we will discuss in more detail one of the more popular approaches, while in the subsequent chapters, we will describe our general tree model guided framework (Tan et al. 2005, 2006a, Tan et al. 2008, Hadzic et al. 2010) using which all the different subtree types can be mined.

In regards to mining induced ordered subtrees, some of the available algorithms are FREQT (Asai et al. 2002), AMIOT (Hido & Kawano 2005), IMB3-Miner (Tan et al. 2006a) and PrefixTreeISpan (Zhou et al. 2006). FREQT (Asai et al. 2002) algorithm uses the rightmost extension enumeration technique and an optimized pruning technique. AMIOT (Hido & Kawano 2005) uses the 'right-and-left tree join' method to efficiently enumerate only those candidates that have a high probability of being frequent. IMB3-Miner (Tan et al. 2006a) is one of the algorithms developed within the TMG candidate enumeration framework and it utilizes the level of embedding constraint to mine induced subtrees. More recently, the algorithm PrefixTreeISpan (Zhou et al. 2006) extends the notations of prefix and post-fix from sequential mining and uses the idea of divide and conquer to find the complete set of frequent patterns. The main idea is to examine the prefix-tree subtrees and project their corresponding postfix-forests into the projected database.

Some of the existing algorithms capable of extracting frequent ordered embedded subtrees are discussed next. The TreeMiner (Zaki 2005a) algorithm uses a data structure called the vertical scope-list and utilizes the join approach for candidate generation. TreeMiner consists of two versions: one, which adopts a depth-first search approach and one which uses the breadth-first approach for candidate generation and counting. As expected, the depth-first search approach, VTreeMiner (Zaki 2005a), is more efficient than most of the horizontal approaches but, as previously mentioned, vertical enumeration approaches may generate many pseudo-frequent subtrees which can greatly degrade the performance. XSpanner (Wang et al. 2004) extends the FP-Growth concept from relational into tree-structured data and its enumeration model also generates only valid candidates. Despite the fact that the experimentation study performed by the XSpanner authors suggested that XSpanner outperforms TreeMiner, a very recent study by (Tatikonda, Parthasarathy & Kurc 2006) suggested the opposite. They reported that XSpanner performs much worse than TreeMiner for the many datasets used. XSpanner suffers from poor cache performance due to the expensive pseudo-projection step. They suggested that, in general, the problems with the FP-growth based approaches are the very large memory footprint, the memory trashing issue, and costly I/O processing (Ghoting et al. 2005). TRIPS and TIDES (Tatikonda, Parthasarathy & Kurc 2006) are two algorithms based on the sequential encoding based strategies to facilitate the fast generation of complete and non-redundant sets of candidate subtrees. The TRIPS algorithm is based on prüfer sequences, while the TIDES algorithm uses depth-first order sequences for candidate generation. MB3-Miner (Tan et al. 2005) employs the TMG candidate generation technique to only enumerate and count the candidate subtrees existing in the document. It introduces an efficient representation of a tree structure called *embedding list* so that the TMG enumeration method can be efficiently implemented. In (Tan et al. 2005), the TMG mathematical model for estimating the worst case complexity of enumerating all embedded subtrees has been presented. This motivated the strategy of tackling the complexity of mining embedded subtrees by introducing the Level of Embedding (Tan et al. 2006a) constraint. Thus, when it is too costly to mine all embedded subtrees, one can decrease the level of embedding constraint gradually down to 1, from which all the obtained subtrees are induced. The Razor algorithm (Tan et al. 2006b) is an extension to the IMB3-Miner algorithm and it mines distance-constrained embedded subtrees. Essentially, the distance-constraint allows the embedded subtrees to be distinguished based upon their node distance relative to the root node. Since many embedded subtrees can form one candidate, this adds more granularity since all of those subtrees are now grouped as different candidates when the distance among the nodes is different. Hence, mining distance constrained subtrees is more expensive in terms of space and time required (Tan et al. 2006b). From the application perspective, the work presented in (Hadzic et al. 2006) demonstrates the potential of the algorithms for mining ordered subtrees to provide interesting biological information when applied to tree-structured biological data.

To address the problem of extracting all frequent unordered induced subtrees, some of the existing algorithms are: Unot (Asai et al. 2003), RootedTreeMiner (Chi, Yang & Muntz 2004b), HybridTreeMiner (Chi, Yang & Muntz 2004a), the method presented by Nijssen & Kok (2003) and UNI3 (Hadzic, Tan & Dillon 2007). The uNot algorithm (Asai et al. 2003) mines induced unordered subtrees and uses a reverse search technique for incremental computation of unordered subtree occurrences. Breadth-first canonical form (BFCF) and depth-first canonical form (DFCF) for labeled rooted unordered trees have been presented in (Chi, Yirong & Muntz 2004b). In the same work, the authors proposed two algorithms: RootedTreeMiner which is the authors' re-implementation of uNot, a vertical mining algorithm based upon BFCF and FreeTreeMiner, based on extension of DFCF for discovering labeled free trees. As an extension to the work, efficient HybridTreeMiner (Chi, Yang & Muntz 2004a) algorithm was developed which systematically enumerates subtrees by traversing an enumeration tree which is defined based upon the BFCF for unordered subtrees. Nijssen & Kok (2003) present a bottom-up strategy for determining the frequency of unordered induced subtrees, and argue that the complexity of enumerating unordered trees as opposed to ordered is not much greater. All these approaches consider only the transaction-based support definition, while the UNI3 algorithm (Hadzic, Tan & Dillon 2007) which is an extension of the general TMG enumeration framework for the problem of unordered subtree mining, can use any of the three support definitions, discussed in the previous chapter.

There are not many algorithms available that mine unordered embedded subtrees. TreeFinder (Termier, Rousset & Sebag 2002) was the first attempt which uses inductive-logic programming for enumerating all candidate subtrees, but which in the process can miss many frequent subtrees. SLEUTH (Zaki 2005b) was the first complete approach proposed and it enumerates unordered subtrees by using unordered scope-list joins via the descendant and cousin tests. As an extension to the TMG framework for mining of unordered embedded subtrees, the U3 algorithm

(Hadzic et al. 2008) was proposed. The previously used *embedding list* (Tan et al. 2005) for representing hierarchical information was replaced by a compressed version called *recursive list* (Hadzic et al. 2008) that reduces the memory space consumption and has additional functionalities. U3 has the capability of restricting the level of embedding allowed in the extracted subtrees and can use any of the three support definitions. Unordered tree mining has been successfully applied in (Shasha, Wang & Zhang 2004) for the analysis of phylogenetic databases.

A final note from the reviewed approaches is that the general TMG approach for candidate enumeration has been extended for all the sub-problems of tree mining discussed in this book. Any of the subtree types can be mined and all the support definitions can be used. Furthermore, this adaptability to a variety of problems was never at the cost of a noticeable reduction in efficiency since the extended algorithms in most cases performed better than the state-of-the-art algorithms at that time. This was experimentally demonstrated in (Tan 2008; Hadzic 2008; Tan et al. 2005, 2006a, 2008a; Hadzic et al. 2007, 2008). For an overview of the current state-of-the-art techniques in the field of tree mining, please refer to (Tan et al. 2008b) as well as (Chi et al. 2005), where a number of popular algorithms are described in detail.

3.7.1 Algorithm Using the Join Candidate Enumeration Approach

This section discusses the TreeMiner algorithm (Zaki 2005a), which is the first algorithm developed for mining ordered embedded subtrees and it adopts the join approach for candidate enumeration. Despite the fact that using the join approach will unnecessarily generate many invalid subtree candidates that need to be pruned, the TreeMiner algorithm is still very efficient. We next describe its general working mechanism, focusing mainly on candidate generation and frequency counting.

3.7.1.1 TreeMiner: Candidate Generation

This section illustrates the candidate generation and frequency counting methods that are used by the TreeMiner.

The TreeMiner uses the join approach to generate candidate subtrees. To generate candidates systematically, it uses the concept of equivalent class. Two k-subtrees X, Y belong to the same prefix equivalence class if and only if they share a common prefix up to the $(k\text{-}1)^{th}$ node. An element of an equivalence class with label x and attached to position i^{th} is denoted as (x, i). Elements are kept sorted by node label as the primary key and position as the secondary key. An equivalence class with prefix P with elements $(x,i),(y,j),(z,k)$ is denoted as $[P]:(x,i),(y,j),(z,k)$.

The example provided in Fig. 3.8 illustrates the concept of *equivalence class* and *element*.

Consider the T_1 example in Fig. 3.8 which shows a class template for subtrees of size 4 with the same prefix subtree P of size 3, with string encoding P='*a c d*', denoted as [*a c d*], and two elements $(b,0)$ and $(e,1)$. The element $(b,0)$ refers to

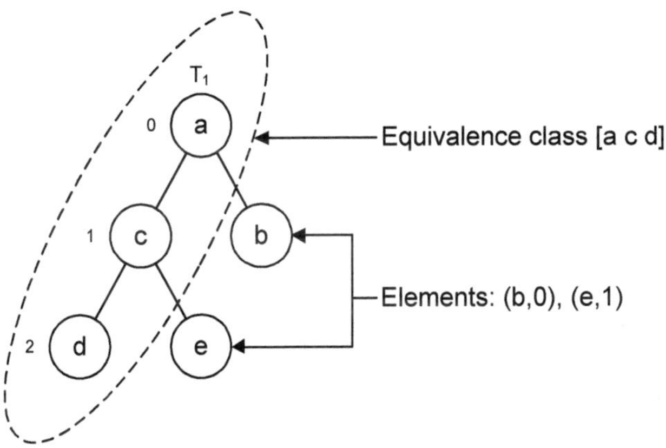

Fig. 3.8 Illustration of equivalence class [a c d] and element (b, 0)

a node with the label '*b*' and is connected to the tree at depth first search position 0 and equally $(e,1)$ refers to a node with label '*e*' that is connected to the tree at depth-first search position 1.

Given an equivalence class of k-subtrees, $(k+1)$-subtrees are obtained by applying the following rules:

Let P be a prefix class of a k-1 subtree with encoding P, and let (x,i) and (y, j) denote any two elements in P. Let the k-subtree $[P_x^i]$ be the extension of P with element (x,i). Let $[P_x^i]$ denote the class representing possible extensions of P_x^i. A join operator **x** on the two elements, denoted (x,i) **x** (y,j), is defined as follows:

Case 1 – $(i = j)$: (a) If $P \neq \emptyset$, add (y,j) and (y,n_i) to $[P_x^i]$ where n_i is the depth-first number for node (x,i) in tree P_x^i. (b) if $P = \emptyset$, add $(y,j+1)$ to $[P_x^i]$. The latter case (b), applies only when generating 2-subtrees candidates from 1-subtrees where $P = \emptyset$.

Case 2 – $(i > j)$: Add (y,j) to class $[P_x^i]$.

Case 3 – $(i < j)$: No new candidate is generated in this case.

By applying the join operator to each ordered pair of elements (x,i) and (y,j), all possible $(k+1)$-subtrees with the common prefix P will be enumerated.

In what follows we will show the step-by-step process of generating candidates with the join approach and exemplify the application of the above rules.

Consider the trees T_1 and T_2 from Fig. 3.9. Assume that the minimum support is 2 and *transaction support* is used. Initially, at $k = 1$, we start with equivalence class $P = \emptyset$ and elements $(a,\text{-}1)$, $(b,\text{-}1)$, $(c,\text{-}1)$, $(d,\text{-}1)$, $(f,\text{-}1)$. For now, we can omit the details of the way the frequency counting is performed. Filtering out elements with support count less than 2, $(c,\text{-}1)$ and $(f,\text{-}1)$ we obtain 1-subtrees with equivalence class $[P]$:$(a,\text{-}1)$,$(b,\text{-}1)$,$(d,\text{-}1)$ where $P = \emptyset$.

Since, $P = \emptyset$, we apply Case 1(b) to generate all 2-subtrees. Let E denote elements of equivalence class P, all candidates 2-subtrees are obtained by multiplying E x E. For example, we obtain an equivalence class $[a]$:$(a,0)$,$(b,0)$,$(d,0)$ after multiplying all elements of $[P]$: $(a,\text{-}1)$,$(b,\text{-}1)$,$(d,\text{-}1)$. Since inserting element $(a,0)$ will

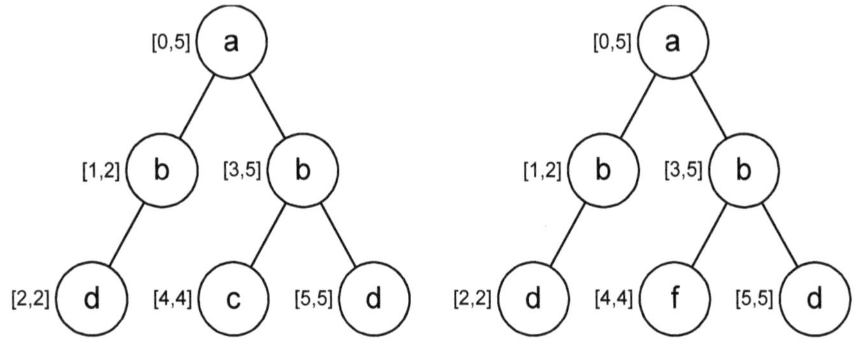

Fig. 3.9 Example trees, T_1 and T_2, for illustration of candidate generation by the join approach

result in an infrequent 2-subtree '*a a*', (*a*,0) is removed from the equivalence class [*a*]. Applying the same procedure, eliminating elements that will result in infrequent subtrees, to all generated equivalence classes leaves us with only two equivalence classes: [*a*]:(*b*,0),(*d*,0) and [*b*]:(*d*,0).

After generating 2-subtrees, then, all *k*-subtrees for $k = 3$ are generated by joining each element of each equivalence class [*a*] & [*b*]. This process is illustrated in Fig. 3.10, for equivalence class [*a*].

Starting from the box I (top-left of Fig. 3.10), we do a self-join of element (*b*,0) with itself, (*b*,0) **x** (*b*,0), and in this case since $i = j$, and $P \neq \emptyset$, Case I(a) applies: add (y, j) and (y, n_i) to $[P_x^i]$. In this case, $x = y = b$ and $i = 0$; $j = 0$; $n_i = 1$. Thus, $[P_x^i]$ = [*a b*] and $(y, j) = (b,0)$; $(y, n_i) = (b,1)$. From this join we obtain two 3-subtrees '*a b / b*' & '*a b b*'. From the T_1 and T_2 above, we know that only '*a b / b*' is frequent with support count = 2. Thus, after pruning the infrequent 3-subtrees, we obtain an equivalence class [*a b*]:(*b*,0).

Moving on to the box II (top-right of Fig. 3.10), again we perform a self-join of element (*d*,0) with itself, (*d*,0) **x** (*d*,0), and in this case since $i = j$, and $P \neq \emptyset$, Case I(a) applies: add (y, j) and (y, n_i) to $[P_x^i]$. In this case $x = y = d$ and $i = 0$; $j = 0$; $n_i = 1$. Thus, $[P_x^i]$ = [*a d*] and $(y, j) = (d,0)$; $(y, n_i) = (d,1)$. From this join we obtain two 3-subtrees 'a d / d' & 'a d d'. From the T_1 and T_2 above, we know that only 'a d / d' is frequent with support count = 2. Thus after pruning the infrequent 3-subtrees we obtain an equivalence class [*a d*]:(*d*,0).

Next, in the box III (bottom-left of Fig. 3.10), a join of (*b*,0) **x** (*d*,0) is performed. Applying the same procedure as above, an equivalence class [*a b*]:(*d*,0),(*d*,1) is obtained. None of the elements is discarded as none results in infrequent 3-subtrees.

Then, in the box IV (bottom-right of Fig. 3.10), a join of (*d*,0) **x** (*b*,0) is performed. Again, with the same procedure, an equivalence class [*a d*]:(*b*,0) is obtained. Element (*b*,1) is pruned because it results in infrequent 3-subtrees '*a d b*'.

After the above procedures have been completed, equivalence classes [*a b*]:(*b*,0), (*d*,0), (*d*,1) and [*a d*]:(*b*,0), (*d*,0) are obtained. Note that elements are sorted by the label as the primary key and the position as the secondary key.

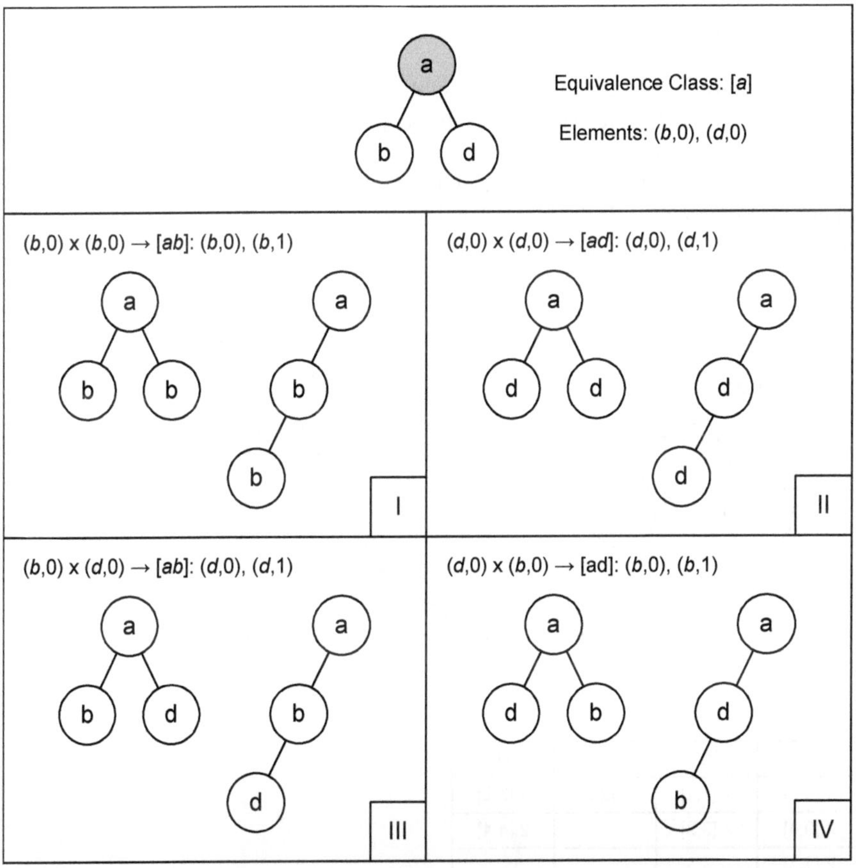

Fig. 3.10 Candidate generation of 3-subtrees from equivalence class [a]:(b,0),(d,0)

The process continues, with the generated equivalence classes at each step as the input for the subsequent step, generating *k*-subtrees for $k > 3$ until no more frequent *k*-subtrees are found. By applying the same procedure as above, we will eventually arrive at the generation of the *maximal* 5-subtree '*a b d* / / *b d*'. A *maximal* subtree is the subtree from which no more frequent super tree(s) can be generated. The *maximal* 5-subtree '*a b d* / / *b d*' is generated through the equivalence class [*a b*]:(*b*,0),(*d*,0),(*d*,1) which generates the equivalence class [*a b d*]:(*b*,0),(*d*,0),(*d*,1),(*d*,2) through joining (*d*,1) **x** (*d*,1), (*d*,1) **x** (*d*,0) and (d,1) **x** (b,0). The maximal equivalent class [*a b d* / / *b*]:(d,0),(d,1) is then obtained from [*a b d*]:(*b*,0),(*d*,0),(*d*,1),(*d*,2) by joining (*b*,0) x (*d*,0) from which the maximal 5-subtrees '*a b d* / / *b d*' can be generated by attaching element (*d*,1) to the equivalent class [*a b d* / / *b*].

3.7.1.2 TreeMiner: Frequency Counting

To obtain a frequency count of (k+1)-subtrees from two subtrees T_1 & T_2 in an equivalence class [P], TreeMiner joins the scope-list of T_1 and T_2, denoted as $\lambda(T_1)$ and $\lambda(T_2)$ respectively.

A scope-list of a tree T_k (a tree T with $|T|= k$) is defined as a 3-tuple (t, m, s) where t is the *tree id*, m is the match label of the (k-1) length of T_k and s is the scope of the *right most node* of T.

Fig. 3.11 shows examples of scope-lists of T_1 & T_2 subtrees.

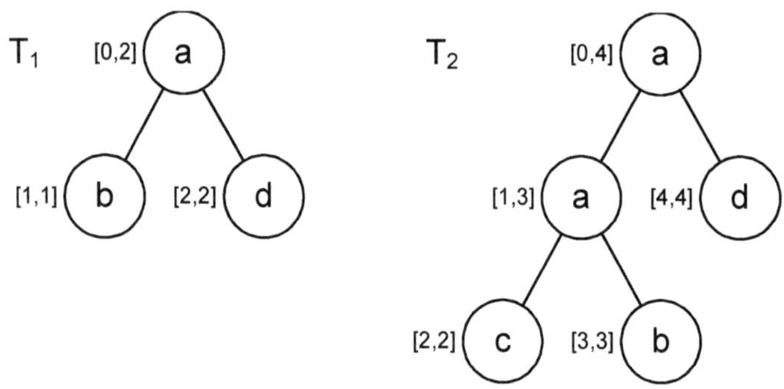

'a'	'b'	'c'	'd'
1,[0,2]	1,[1,1]	2,[2,2]	1,[2,2]
2,[0,4]	2,[3,3]		2,[4,4]
2,[1,3]			

'a b'	'a a c / b'	'a a b / / d'	'a b / d'
1,0,[0,2]	2,012,[3,3]	2,013,[4,4]	2,03,[4,4]
2,0,[3,3]			
2,1,[3,3]			

Fig. 3.11 Examples of scope-lists of T_1 & T_2 subtrees

Consider one example from Fig. 3.11: the scope-list of subtree with encoding '*a a b / / d*' of T_2 is given by (t, m, s) where t is the tree-id of T_2, that is 2 and the match label of the equivalence class of [a a b] is given by nodes at position 0, 1, and 3. Thus m = 013. Moreover, the *scope* of the last node 'd' is given by [4,4]. Hence, (2, 013, [4,4]) produces the scope-list of '*a a b / / d*'.

There are basically two types of scope-list join: (a) in-scope (b) out-scope. The in-scope join is used for *ancestor-descendant* or *parent-child* node extension whereas the out-scope is used for *sibling extension*. Let T_k be a tree T with $|T|= k$ and T_{k+1} is produced by extending node x. The former type of extension is when the relationship between the *right-most-node* of the k-subtree and the extending node is *ancestor-descendant* or *parent-child*. The latter is when the relationship between the *right-most-node* of the k-subtree and the extending node is not *ancestor-descendant or parent-child.*

General rules for scope-list joins of $\lambda(T_x)$ and $\lambda(T_y)$ are as follows:

1. The tree-id of $\lambda(T_x)$ and $\lambda(T_y)$ must be the same. Given (t_y, m_y, s_y) and $(t_x, m_x, s_x), t_y = t_x$.
2. The match label of $\lambda(T_x)$ and $\lambda(T_y)$ must be the same. Given (t_y, m_y, s_y) and $(t_x, m_x, s_x), m_y = m_x$.
3. a. ***in-scope*** join: $s_y \subset s_x$ (s_x contains s_y), that is given a scope of T_x denoted as $[l_x, u_x]$, and T_y denoted as $[l_y, u_y]$, $l_x \leq l_y$ and $u_y \leq u_x$.
   b. ***out-scope*** join: $s_x < s_y$ (s_x is strictly less than s_y), that is given a scope of T_x denoted as $[l_x, u_x]$, and T_y denoted as $[l_y, u_y]$, $u_x \leq l_y$. Note that there is no overlapping between the scope of T_x and T_y.

3.7.1.3 Computing Frequency of '*a b*' by Scope-List Join λ('*a*') x λ('*b*')

The computing frequency of '*a b*' can be obtained by joining the scope-list of 'a' and 'b', denoted as λ('*a*') x λ('*b*'). From the scope-list table in Fig. 3.12, multiplication is performed between each element of λ('*a*') and each element of λ('*b*'). λ('*a b*') can be obtained by following the scope-list join rules above.

λ('*a*')	λ('*b*')
1,[0,2]	1,[1,1]
2,[0,4]	2,[3,3]
2,[1,3]	

λ('*a*') x λ('*b*')
1,[0,2] x 1,[1,1]
2,[0,4] x 2,[3,3]
2,[1,3] x 2,[3,3]

λ('*a b*')
1,0,[1,1]
2,0,[3,3]
2,1,[3,3]

Fig. 3.12 λ('a') x λ('b') = λ('ab')

Following the rules above, the scope-list join between λ('*a*') and λ('*b*') is performed as follows.

First, we join the first element of λ('*a*'), (1,[0,2]), with any element of λ('*b*') that satisfies rules 1, 2 and 3. The only element of λ('*b*') that satisfies rules 1, 2 and 3 is (1,[1,1]), i.e. both tree-id = 1, (k-1) match prefix = Ø, and it satisfies the ***in-scope*** constraint. Note that we use ***in-scope*** join because the relationship between 'a' and 'b' in this case is of a *parent-child* relationship.

Next, move to the next element of λ('*a*'), (2,[0,4]). The only element of λ('*b*') with tree-id 2 is (2,[3,3]). The join can be performed between (2,[0,4]) and (2,[3,3]) because it satisfies rules 1, 2 and 3. Both elements have tree-id 2 and (k-1) match prefix = Ø and it satisfies the **in-scope** constraint.

The last join to be performed is between (2,[1,3]) and (2,[3,3]). This join also satisfies rules 1, 2 and 3.

Once all the possible joins have been performed, the frequency count of '*a b*' is equal to the size of λ('*a b*'), denoted as | λ('*a b*')|, that is 3.

3.7.1.4 Computing Frequency of '*a b / d*' by Scope-List Join λ('*a b*') x λ('*a d*')

In this example (Fig. 3.13), an **out-scope** join of λ('*a b*') and λ('*a d*') is illustrated.

λ('*a b*')	λ('*a d*')
1,0,[1,1]	1,0,[2,2]
2,0,[3,3]	2,0,[4,4]
2,1,[3,3]	

λ('*a b*') x λ('*a d*')
1,0,[1,1] x 1,0,[2,2]
2,0,[3,3] x 2,0,[4,4]

λ('*a b / d*')
1,01,[2,2]
2,03,[4,4]

Fig. 3.13 λ('*a b*') x λ('*a d*') = λ('*a b / d*')

Following rules 1, 2, and 3 above, the scope-list join between λ('*a b*') and λ('*a d*') is performed in the following way.

First, we join the first element of λ('*a b*'), (1,0,[1,1]), with any element of λ('*a d*') that satisfies rules 1,2 and 3. The only element of λ('*a d*') that satisfies rules 1,2 and 3 is (1,01,[2,2]), i.e. both tree-id = 1, (k-1) match prefix = 'a', and it satisfies the ***out-scope*** constraint. Note that we use ***out-scope*** join because the relationship between node '*b*' as the right-most-node and 'd' as the extending node in this case is not a descendant-ancestor or parent-child relationship.

Next, let us move on to the next element of λ('*a b*'), that is (2,0,[3,3]). The only element of λ('*a d*') with tree-id 2 is (2,0,[4,4]). The join can be performed between (2,0,[3,3]) and (2,0,[4,4]) because it again satisfies rules 1, 2 and 3. Both elements have tree-id 2 and (k-1) match prefix = '*a*' and it satisfies the **out-scope** constraint.

By now all joins have been performed and the frequency count of '*a b / d*' is equal to | λ('*a b / d*')|, that is 2.

3.8 Conclusion

This chapter has presented the current state-of-the-art of the existing work in the area of frequent subtree mining. The main issues that need to be addressed by each tree mining algorithm were explained and some common ways of approaching the problem reviewed. We then provided an overview of some of the main algorithms developed in the area of frequent subtree mining, and described one of the earliest approaches using the join candidate enumeration approach in more detail. In the following chapter, the focus will be on our work in the area characterized by a tree model guided candidate generation approach. It is a general framework that can be used for mining all subtree types under different constraints and support conditions, and as such is the focus of this book.

References

1. Abe, K., Kawasoe, S., Asai, T., Arimura, H., Arikawa, S.: Optimized substructure discovery for semistructured data. Paper presented at the Proceedings of the 6th European Conference on Principles of Data Mining and Knowledge Discovery, Helsinki, Finland, August 19-23 (2002)
2. Agrawal, R., Imieliski, T., Swami, A.: Mining Association Rules between Sets of Items in Large Databases. In: Proceedings of the ACM SIGMOD International Conference on Management of Data, Washington D.C., USA, May 26-28, pp. 207–216. ACM, New York (1993)
3. Agrawal, R., Srikant, R.: Fast Algorithms for Mining Association Rules. In: Proceedings of the 20th International Conference on Very Large Data Bases (VLDB), Santiago de Chile, Chile, Septemebr 12-15, pp. 487–499 (1994)
4. Agrawal, R.C., Aggarwal, C.C., Prasad, V.V.V.: A Tree Projection Algorithm for Generation of Frequent Item Sets. Journal of Parallel and Distributed Computing 61(3), 350–371 (2001)
5. Asai, T., Abe, K., Kawasoe, S., Arimura, H., Sakamato, H., Arikawa, S.: Efficient substructure discovery from large semi-structured data. Paper presented at the Proceedings of the 2nd SIAM International Conference on Data Mining (SIAM 2002), Arlington, VA, USA, April 11-13 (2002)
6. Asai, T., Arimura, H., Uno, T., Nakano, S.-i.: Discovering Frequent Substructures in Large Unordered Trees. In: Grieser, G., Tanaka, Y., Yamamoto, A. (eds.) DS 2003. LNCS (LNAI), vol. 2843, pp. 47–61. Springer, Heidelberg (2003)
7. Bayardo, R.J.: Efficiently mining long patterns from databases. Paper presented at the Proceedings of the ACM SIGMOD Conference on Management of Data, Seattle, USA, June2-4 (1998)
8. Chi, Y., Yang, Y., Muntz, R.R.: HybridTreeMiner: An Efficient Algorithm for Mining Frequent Rooted Trees and Free Trees Using Canonical Forms. Paper presented at the Proceedings of the 16th International Conference on Scientific and Statistical Database Management (SSDBM 2004), Santorini Island, Greece, June 21-23 (2004a)
9. Chi, Y., Yang, Y., Muntz, R.R.: Canonical forms for labeled trees and their applications in frequent subtree mining. Knowledge and Information Systems 8(2), 203–234 (2004b)
10. Chi, Y., Muntz, R.R., Nijssen, S., Kok, J.N.: Frequent Subtree Mining - An Overview. Fundamenta Informaticae, Special Issue on Graph and Tree Mining 66(1-2), 161–198 (2005)
11. El-Haji, M., Zaiane, O.R.: COFI-tree Mining: A New Approach to Pattern Growth with Reduced Candidacy Generation. Paper presented at the Workshop on Frequent Itemset Mining Implementations (FIMI 2003) in conjunction with IEEE-ICDM, Melbourne, Florida, USA, November 19-22 (2003)
12. Ghoting, A., Buehrer, G., Parthasarathy, S., Kim, D., Nguyen, A., Chen, Y.-K., Dubey, P.: Cache-conscious frequent pattern mining on a modern processor. Paper presented at the Proceedings of the 31st International Conference on Very Large Data Bases (VLDB), Trondheim, Norway, August 30 - September 2 (2005)
13. Hadzic, F., Dillon, T.S., Sidhu, A.S., Chang, E., Tan, H.: Mining Substructures in Protein Data. Paper presented at the IEEE Workshop on Data Mining in Bioinformatics (DMB 2006), in conjunction with IEEE ICDM 2006, Hong Kong, December 18-22 (2006)
14. Hadzic, F., Tan, H., Dillon, T.S.: UNI3 - Efficient Algorithm for Mining Unordered Induced Subtrees using TMG Candidate Generation. In: Proceedings of IEEE Symposium on Computational Intelligence and Data Mining (CIDM), Honolulu, Hawaii, USA, April 1-5, pp. 568–575. IEEE, Los Alamitos (2007)

15. Hadzic, F, Advances in knowledge learning methodologies and their applications. Curtin University of Technology, Perth (2008)
16. Hadzic, F., Tan, H., Dillon, T.S.: U3 – mining unordered embedded subtrees using TMG candidate generation. In: Proceedings of the IEEE / WIC / ACM International Conference on Web Intelligence, Sydney, Australia, December 9-12, pp. 285–292 (2008)
17. Hadzic, F., Tan, H., Dillon, T.S.: Tree Model Guided Algorithm for Mining Unordered Embedded Subtrees. Web Intelligence and Agent Systems: An International Journal (WIAS) 8(4) (2010)
18. Han, J., Pei, J., Yin, Y., Mao, R.: Mining Frequent Patterns without Candidate Generation: A Frequent-Pattern Tree Approach. Data Mining and Knowledge Discovery 8(1), 53–87 (2004)
19. Han, J., Kamber, M.: Data Mining: Concepts and Techniques, 2nd edn. Elsevier, Morgan Kaufmann Publishers, San Francisco, CA, USA (2006)
20. Hido, S., Kawano, H.: AMIOT: Induced Ordered Tree Mining in Tree-Structured Databases. In: Proceedings of the 5th IEEE International Conference on Data Mining (ICDM), Houston, Texas, USA, November 27-30, pp. 170-177 (2005)
21. Inokuchi, A., Washio, T., Motoda, H.: Complete Mining of Frequent Patterns from Graphs: Mining Graph Data. Machine Learning 50(3), 321–354 (2003)
22. Kuramochi, M., Karypic, G.: Frequent Subgraph Discovery. Paper presented at the Proceedings of the IEEE International Conference on Data Mining (ICDM 2001), San Jose, California, USA, November 29 - December 2 (2001)
23. Luccio, F., Enriquez, A.M., Rieumont, P.O., Pagli, L.: Exact Rooted Subtree Matching in Sublinear Time. Universita Di Pisa, Pisa (2001)
24. Luccio, F., Enriquez, A.M., Rieumont, P.O., Pagli, L.: Bottom-up Subtree Isomorphism for Unordered labeled Trees. Universita Di Pisa, Pisa (2004)
25. Nijssen, S., Kok, J.N.: Efficient discovery of frequent unordered trees. In: Proceedings of the 1st International Workshop on Mining Graphs, Trees, and Sequences, Dubrovnik, Croatia (2003)
26. Park, J.S., Chen, M.-S., Yu, P.S.: Using a Hash-Based Method with Transaction Trimming for Mining Association Rules. IEEE Transactions on Knowledge and Data Engineering 9(5), 813–825 (1997)
27. Pei, J., Han, J., Lu, H., Nishio, S., Tang, S., Yang, D.: Hmine: Hyper-structure mining of frequent patterns in large databases. Paper presented at the Proceedings of the 1st International Conference on Data Mining, San Jose, California, USA, November 29-December 2 (2001)
28. Shasha, D., Wang, J.T.L., Zhang, S.: Unordered Tree Mining with Applications to Phylogeny. Paper presented at the Proceedings of the 20th International Conference on Data Engineering, Boston, MA, USA, March 30-April 2 (2004)
29. Tan, H., Hadzic, F., Feng, L., Chang, E.: MB3-Miner: mining eMBedded subTREEs using tree model guided candidate generation. In: Proceedings of the 1st International Workshop on Mining Complex Data in conjunction with ICDM 2005, Houston, Texas, USA, November 27-30, pp. 103–110 (2005)
30. Tan, H., Dillon, T.S., Hadzic, F., Chang, E., Feng, L.: IMB3-Miner: Mining Induced/Embedded Subtrees by Constraining the Level of Embedding. In: Ng, W.-K., Kitsuregawa, M., Li, J., Chang, K. (eds.) PAKDD 2006. LNCS (LNAI), vol. 3918, pp. 450–461. Springer, Heidelberg (2006a)
31. Tan, H., Dillon, T.S., Hadzic, F., Chang, E.: Razor: mining distance-constrained embedded subtrees. In: Proceedings of the Workshop on Ontology Mining and Knowledge Discovery from Semistructured documents (MSD) in conjunction with ICDM 2006, Hong Kong (2006a)

32. Tan, H.: Tree Model Guided (TMG) enumeration as the basis for mining frequent patterns from XML documents. University of Technology Sydney, Sydney (2008)
33. Tan, H., Hadzic, F., Dillon, T.S., Feng, L., Chang, E.: Tree Model Guided Candidate Generation for Mining Frequent Subtrees from XML. ACM Transactions on Knowledge Discovery from Data 2(2) (2008a)
34. Tan, H., Hadzic, F., Dillon, T.S., Chang, E.: State of the art of data mining of tree structured information. International Journal of Computer Systems Science and Engineering 23(2), 255–270 (2008b)
35. Tatikonda, S., Parthasarathy, S., Kurc, T.: TRIPS and TIDES: new algorithms for tree mining. Paper presented at the Proceedings of the 15th ACM International Conference on Information and Knowledge Management (CIKM), Arlington, Virginia, USA, November 6-11 (2006)
36. Termier, A., Rousset, M.-C., Sebag, M.: TreeFinder: a First Step towards XML Data Mining. In: Proceedings of the 2nd IEEE International Conference on Data Mining (ICDM), Maebashi City, Japan, December 9-12, pp. 450–458 (2002)
37. Wang, C., Hong, M., Pei, J., Zhou, H., Wang, W., Shi, B.: Efficient pattern-growth methods for frequent tree pattern mining. In: Dai, H., Srikant, R., Zhang, C. (eds.) PAKDD 2004. LNCS (LNAI), vol. 3056, pp. 441–451. Springer, Heidelberg (2004)
38. Yang, L.H., Lee, M.L., Hsu, W.: Efficient Mining of XML Query Patterns for Caching. In: Proceedings of the 29th International Conference on Very Large Data Bases (VLDB), Berlin, Germany, September 9-12, pp. 69–80 (2003)
39. Zaki, M.J., Gouda, K.: Fast vertical mining using diffsets. Paper presented at the Proceedings of the 9th ACM SIGKDD International Conference on Knowledge Discovery and Data Mining, Washington D.C., USA, August 24-27 (2003)
40. Zaki, M.J.: Efficiently Mining Frequent Trees in a Forest: Algorithms and Applications. IEEE Transactions on Knowledge and Data Engineering 17(8), 1021–1035 (2005a)
41. Zaki, M.J.: Efficiently Mining Frequent Embedded Unordered Trees. Fundamenta Informaticae 66(1), 33–52 (2005b)
42. Zhou, L., Lu, Y., Zhang, H., Hu, R., Zhou, C.: Mining Frequent Induced Subtrees by Prefix-Tree-Projected Pattern Growth. In: Proceedings of the 7th International Conference on Web-Age Information Management (WAIM) Workshops, Hong Kong, June 17-19 (2006)

Chapter 4
Tree Model Guided Framework

4.1 Introduction

In this chapter, we describe the main characteristics of the Tree Model Guided (TMG) Framework for frequent subtree mining. This framework has good extendibility to all of the current problems for frequent subtree mining (Hadzic 2008; Tan 2008). An algorithm is considered as extendible in the sense that minimal effort is required to adjust the general framework so that different but related problems can be solved. Furthermore, the results presented in works such as (Tan et al. 2005; 2006a, 2008, Hadzic et al. 2007, 2010) indicate that it currently exhibits the best or comparable performance among the current state-of-the-art methods. The TMG framework is also conceptually simple to understand, especially with respect to the small adjustments required to address different sub-problems within the tree mining field. The remainder of the algorithm development issues are addressed in such a way as to accommodate the most efficient execution of the TMG candidate generation. Hence, as mentioned in the previous chapter, the important aspects that need to be taken into account in addition to the candidate enumeration strategy are: tree representation, representative data structures and their operational use, and the frequency counting of generated candidate subtrees. As mentioned in Chapter 3, in the tree mining field a string-like representation is the most popular representation because each item in the string can be accessed in O(1) time, it is space efficient and easy to manipulate. In our framework, we utilize the depth-first or pre-order string encoding as described in Chapter 3. The problem of candidate subtree enumeration is to efficiently extract a complete and non-redundant set of subtrees from a given document tree. We explain the TMG approach to candidate subtree enumeration in Section 4.2. As the name implies, the enumeration phase is guided by the tree model of the document in order to generate only valid candidate subtrees. This tree model corresponds to the underlying structure of the document and a subtree is considered valid by conforming to it.

As mentioned in the previous chapter, choosing appropriate data structures for representing the database information is essential for solving the problem in an efficient manner. In the context of tree mining, as a first requirement, the nodes'

F. Hadzic, H. Tan, and T.S. Dillon: Mining of Data with Complex Structures, SCI 333, pp. 67–86.
springerlink.com

related information should be represented globally, so that the tree database does not need to be scanned multiple times. The information should be easy to access and the cost associated with it should be minimal. For this purpose, a global pre-order sequence in memory is created which is referred to as *dictionary*. The *dictionary* stores each node in the tree database following the pre-order traversal indexing. The nodes' related information can be directly accessed. To employ the TMG candidate generation, a suitable representation of structural aspects of the tree database was required for performing the task in an efficient and effective manner. The *embedding list* (*EL*) representation was first developed for this purpose and through *EL*, all the structural relationships between the nodes in the original tree are preserved. Despite the usefulness of the *EL* structure, it occupies extra memory storage to store the nodes coordinates or hyperlinks of the generated candidate subtrees. This can be an issue when processing a very large tree database. To improve the space efficiency property of *EL*, it was modified into a *recursive list* (*RL*) representation, which is a fusion of the *dictionary* and the *EL* structure. It is a more compact representation of the *EL* that reduces the memory space consumption with additional functionalities that previously belonged only to the *dictionary*. The *RL* serves as a global look-up list at the same time as it encodes the embedding relationships of the subtrees to be mined. Please note that, even though the final version of the algorithm uses the latest, more space efficient *RL* representation, both representations will be explained in Section 4.3 for clarity.

Frequency counting is an important task in mining frequent patterns that can become a serious issue if not performed efficiently. After candidate subtrees have been enumerated, the next task is to determine whether those subtrees are frequent. This is done by computing the support of each subtree and involves counting the occurrences of that subtree in the database of trees. The frequent subtrees are those whose support is greater than or equal to the specified minimum support σ. In a database of labeled trees, many instances of subtrees can occur with the same encoding. Depending on which support definition is used, the support of each subtree can be computed differently. However, regardless of which support definition is used, the frequency of subtrees that share common encoding φ must be determined. The problem of candidate subtree counting corresponds to the means by which the frequency of a subtree is determined. The way that this handled within the TMG framework is explained in Section 4.4, where we describe the data structure used for this purpose.

Mining large and complex data can be very expensive and intractable. *Constraints* play an important role in performing complexity control (reduced search space) in the context of frequent itemsets, associations, correlations, sequential patterns, and many other interesting patterns in large databases (Pei, Han & Laksmanan 2001; Tan et al. 2006). The basic idea of the constraint is to narrow down the search space such that one can focus on a more desirable result and filter out those that are not of interest. Mining without imposing user-specified constraints may generate numerous patterns and often becomes intractable. Moreover, many of the generated patterns may not be interesting to the user or useful for a particular application. In the tree mining field, one can mine many different types of subtrees, all of which imply certain structural relationships between the data objects in the association found. Depending on the way the information is organized in the document and the nature of the data and application, different structural relationships are tolerated in different applications. Hence, imposing constraints is at times motivated by a property or information that a pattern needs to possess for a particular application. In Section 4.5, we elaborate further on this issue, and look at the types of constraints used in the TMG framework, and their purpose. Their implications for knowledge analysis related tasks will be further discussed in Chapter 9. We will give an overview of some of the existing work on constrained-based tree mining in Chapter 12 where we discuss some recent trends and open issues.

The purpose of this chapter is to describe some general properties of the TMG approach and introduce its main ideas, focusing on the handling of the important aspects discussed in earlier chapters. In the subsequent chapters, we will describe how the TMG approach is used for the mining of subtrees of increasing complexity (i.e. with extensions to the general framework).

4.2 Tree Model Guided Candidate Subtree Enumeration

The purpose of candidate subtree enumeration is to extract a complete and non-redundant set of subtrees from a given document tree. For unordered subtrees, it becomes a problem to enumerate subtrees according to the chosen canonical form so that each candidate is assigned to a group (automorphism group) and thereby the frequency correctly determined. The complexity introduced by the fact that items are organized in a tree structure calls for an efficient strategy that, throughout the process, will generate only those candidate subtrees that actually exist in the document trees. The two commonly used enumeration strategies are enumeration by extension and join (Chi et al. 2005). While the join approach works well for relational data (Agrawal & Srikant 1994), it may encounter difficulties when applied to tree-structured data. Using the join approach, many invalid candidates can be generated throughout the process which can slow down the execution time, or become unsolvable in finite time. The approach described in (Yang, Lee, & Hsu 2003) uses the XML schema to guide the candidate generation so that all candidates generated are valid because they conform to the schema.

As mentioned earlier, TMG is an enumeration technique that uses the underlying structure of the tree database to generate only valid candidates. Hence, the concept of schema-guided is generalized into tree model-guided (TMG) candidate generation. TMG can be applied to any data that has a model representation with clearly defined semantics that have tree-like structures. It ensures that only those valid candidates which conform to the actual tree model structure of the data are generated. In the cases where the model representation of the tree structure is unavailable, the TMG approach will still perform the candidate generation according to the tree structure of the document. This indicates that a candidate subtree can be considered valid in two ways. Firstly, by conforming to an available model representation of the document tree structure (schema), and secondly, by conforming to the tree structure through which the information in the examined document is represented. To illustrate this, in Fig. 4.1 we show a simple graphical representation of

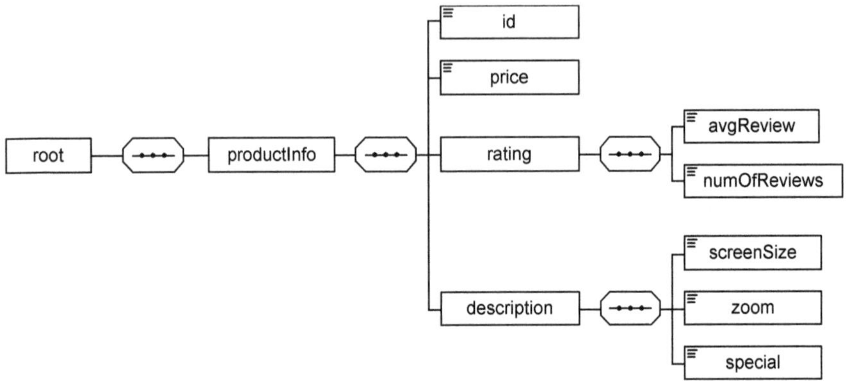

Fig. 4.1 Simple XML schema tree describing product information

an XML schema describing some product information. Starting from the 'root' the hierarchical properties can be easily seen.

Fig. 4.2 depicts two valid and invalid subtrees, of the XML schema tree from Fig. 4.1. The top two trees are valid in the sense that they both conform to the schema (i.e. information is organized in the same way) and also represent valid substructures of the underlying tree structure of the schema. The first invalid subtree displayed is invalid because it does not conform to the schema, in the sense that the information is not represented in the way as specified by the schema. The attribute 'price' is meant to be a child node of 'productInfo', and not of node 'description', as is the case in the subtree. However, this subtree is not structurally invalid, since the structure of a node having three child nodes is valid in the underlying structure of the schema. The second invalid subtree at the bottom of Fig. 4.2, is invalid in both senses. Firstly, it violates the schema specifications of how the information should be organized. Secondly, its structural properties are not present in the underlying structure of the schema. The tree rooted at the 'root' node in Fig. 4.1, is only two levels deep, while the subtree is four levels deep, and hence the structural aspects are violated.

A tree database has, by default, an ordering imposed on its sibling nodes. Therefore, the TMG candidate generation approach will result in a set of ordered subtrees that preserves the original sibling order in the database. If unordered subtrees are to be mined, these candidates need to be ordered into their canonical form so that all candidates are grouped correctly into their automorphism group (see Section 3.6 from Chapter 3) of an unordered subtree. Enumerating unordered subtrees directly would be a more complex process, especially if the underlying structure of the document (which is ordered) is to be used to guide the candidate generation.

The TMG candidate generation is a non-redundant and systematic approach. The candidates are enumerated using a bottom-up approach so that all the smallest candidate subtrees are enumerated first. Hence, at each step, a set of k-subtrees is enumerated, the starting set consisting of 1-subtrees (i.e. subtrees consisting of one node). To obtain the $(k+1)$-subtrees, the subtrees in the k-subtree set are used and

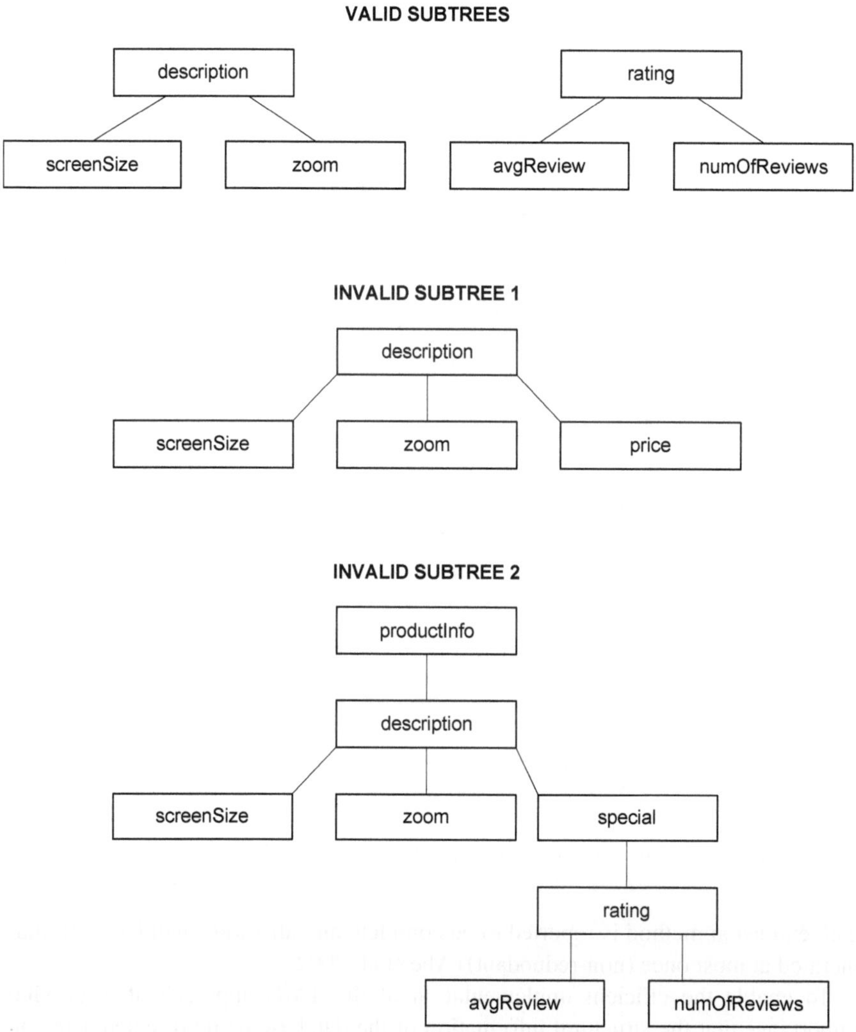

Fig. 4.2 Examples of valid and invalid subtrees of XML schema tree from Fig. 4.1

a new (k+1)-subtree is formed for each of the possible extensions of the k-subtree with a node in the database. Hence, TMG adapts a horizontal (breadth-first) rather than vertical (depth-first) enumeration strategy (Chapter 3, Section 3.4.4).

In Fig. 4.3, we show all valid k-subtrees of a tree T, for varying k.

The underlying structure of the document is used as a guide so that all the extensions performed result in a valid k-subtree, i.e. it exists in the document structure. The enumeration strategy used by TMG is a specialization of the right-most path extension approach (Abe et al. 2002, Zaki 2005). However, it is different from the one that is proposed in FREQT (Abe et al. 2002) in the sense that TMG enumerates embedded subtrees and FREQT enumerates only induced subtrees. The right-most

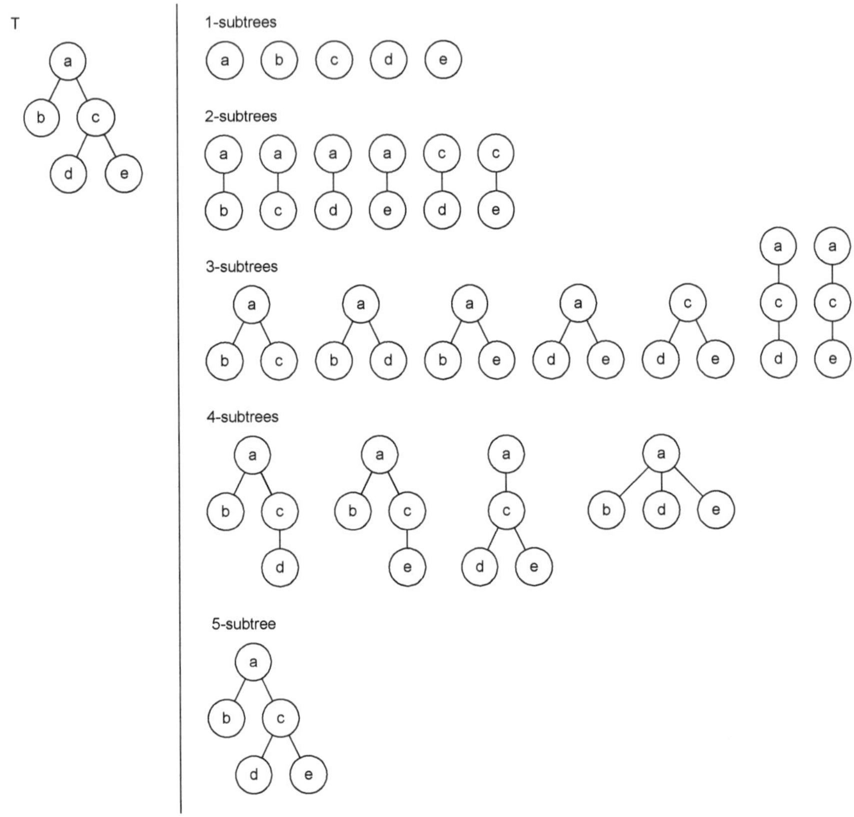

Fig. 4.3 All valid embedded *k*-subtrees of tree T

path extension method is reported to be complete and all valid candidates are enumerated at most once (non-redundant) (Abe et al. 2002).

To enable the efficient implementation of the TMG approach, it is of vital importance that the structural information of the database be represented using an appropriate structure. Furthermore, once a set of candidate *k*-subtrees has been enumerated, their frequency needs to be determined and their occurrence information stored for the generation of (*k*+1)-subtrees. To satisfy these requirements, a number of representational structures are used within the implemented framework and these are discussed in the next section.

4.3 Efficient Representations and Data Structure for Trees

In general, processing of tree structures is computationally complex and expensive. The standard ways to represent a graph is through a collection of the *adjacency list* and the *adjacency matrix*. A tree structure is an acyclic graph. The adjacency list, in general, uses less space than the *adjacency matrix*. The *adjacency list* is preferable for representing sparse graphs. The *adjacency matrix*, on the other hand, is preferable for representations when the graph is dense because it gives faster data access. Ideally, a better representation would have the desirable characteristics of both structures, i.e. it is space efficient with fast access time.

The efficiency of an apriori-based frequent subtree mining technique is measured by how well it performs candidate generation and frequency counting tasks. Due to the large number of candidate subtrees that can be generated, it is important that the candidate generation and frequency counting process be performed in a very efficient way in terms of time and space. Thus, the choice of data structures becomes a very important factor.

Several data structures are presented in this section, whose purpose is to enable the efficient candidate enumeration and frequency counting process. A *dictionary* structure is used as a global lookup structure that allows fast data access and reduces multiplication of the same information that would otherwise be stored in each generated subtree locally. A novel *embedding list* is proposed to provide an efficient candidate enumeration process. The *embedding list* and the *dictionary* are the basic constructs for the candidate generation process. A vertical structure, *vertical occurrence list*, is proposed for fast frequency counting. To take the space efficiency of the *embedding list* further, a new structure called a *recursive list* which fuses the *embedding list* and the *dictionary* is proposed. Furthermore, the efficiency of the candidate enumeration and frequency counting process is enhanced by storing only the right most path coordinate of each enumerated subtree. We refer to the structure that holds only the right most path occurrence coordinate(s) as the *RMP coordinate list*. Each data structure is discussed in detail below and subsequent chapters will describe them in the context of the working mechanism of the tree mining algorithms.

4.3.1 Dictionary

A *dictionary* is an intermediate structure that is used to capture hierarchical relationships inherent in the tree-structure. During the initial transformation step, the external data is transformed into this structure. The *dictionary* is a horizontal representation of the external data. When generating candidate subtrees, it will be too costly if each instance of a subtree stored in memory is stored and duplicated locally. Instead, it is more efficient to generate a global structure and store only the hyperlinks (Tan et al. 2005; Wang et al. 2004) to such a global structure. This structure allows direct access to the global shared information and thereby avoids extra space cost, which would be caused if duplicate information were to be copied (stored) locally.

It is unrealistic to generate a *dictionary* from an extremely large database. To deal with such a situation, one could consider constructing a *dictionary* in secondary storage with a performance trade-off. An alternative solution is to consider a distributed system where the database is partitioned into several segments from which a corresponding *dictionary* structure will be generated, so that the *dictionary* is split over a number of machines. However, this is considered as possible future work.

This *dictionary* structure is useful when it is utilized for frequency counting and pruning purposes. We represent this global structure as an array object so that random access to this structure can be performed in $O(1)$ time. Each cell of the *dictionary* stores information such as *label, level, scope* and a *link* to a parent node pre-order position. We refer to this link information as *direct parent pointer* (*dpp*). The *dpp* of the root node is equal to -1. The index of each cell refers to the position of each node in the original tree. Thus, each cell in the *dictionary* will contain a tuple of {*label, level, scope, dpp*}. The *level* information of each node stored in the *dictionary* is used only when we mine induced subtrees. This will become clearer when a strategy to mine induced subtrees using the *level of embedding* constraint is discussed in the next chapter. The example from Fig. 4.4 illustrates a tree *T* and a *dictionary D* that is generated from tree *T*.

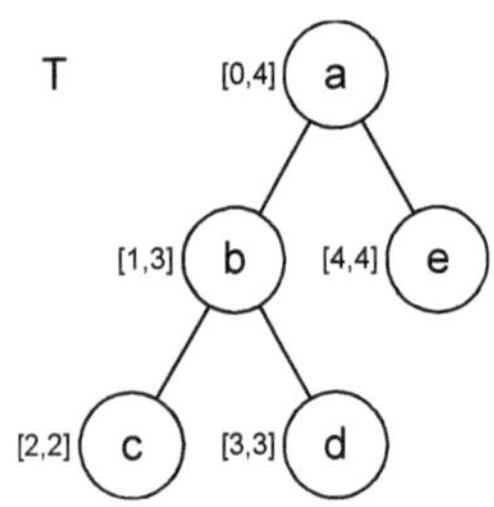

[0]	[1]	[2]	[3]	[4]
a,0,4,-1	b,1,3,0	c,2,2,1	d,2,3,1	e,1,4,0

Fig. 4.4 Illustration of generating a dictionary *D* from tree *T* where each cell in *D* has {*label*, level, *scope, dpp*} tuple

The *scope* of a node refers to the position of its rightmost leaf node with the given node as a root or its own position if it is a leaf node itself (Tan et al. 2005; Zaki 2005). This *scope* information is useful for a number of things. One can determine whether a node is a leaf node if its position is equal to its scope. The scope can be also utilized to determine whether a node is a descendant node or sibling node by doing a scope check (Zaki 2005). Given node *A* has position *i*, and its scope is $i+n$, node *B* with a position *j* such that $j > i$ and $j \leq i+n$, we know that node *B* is a descendant of node *A*, otherwise if $j > i+n$ and the *dpp* of both nodes are equal, then nodes *A* and *B* are siblings. In addition, *scope* information encodes the hierarchical notion of tree structures. In Fig. 4.4, the *scope* of each node is shown together with each node's *position*. For example, the root node of tree *T* at position 0 has a scope value of 4, denoted as 0,4.

An item in the *dictionary D* at position *i* is referred to as *D[i]*. The notion of position of an item refers to its index position in the *dictionary*. For example, *D[3]* stores the structural information {*label, level, scope, dpp*}={*d*,2,3,1} of node with label *d*, and so on.

Later, when mining frequent subsequences, this global structure is modified into a two-dimensional array so that each row will correspond to a *dictionary* of sequences of objects local to each transaction to which those objects belong. This removes the need to keep a global index. This structure is referred to as a *DMA-list* (Tan et al. 2006b), and it will be discussed in detail in Chapter 10. This approach has the advantage of a better locality with the price of more clock cycles to access a higher dimension array.

4.3.2 Embedding List

An embedded subtree is more complex in comparison to an induced subtree (Tan et al. 2006a). The framework proposed in this study is built with a capacity to tackle the complexity of mining embedded subtrees. A novel and unique *embedding list* (*EL*) representation suitable for describing embedded subtrees was proposed in (Tan et al. 2005). This representation is adapted from the conventional *adjacency list* format by relaxing the adjacency notion into an embedding notion in order to capture the embedding relationships between nodes in trees. The list not only contains adjacent nodes (children), but also takes all its descendants and forms an *embedding list* in pre-order ordering. Although we call this structure an *embedding list*, it can consist of multiple *embedding lists* just as the *adjacent list* is a collection of adjacent lists.

For speed considerations, each embedded list is represented as an array of integers so that each item in the list can be accessed in *O(1)* time. This representation is completely different from the *string-like* (Zaki 2005), *adjacency matrix* (Inokuchi et al. 2003) or *adjacency list* (Kuramochi et al. 2001) representation utilized previously for trees. This representation is also different from the *embedding lists* structure that has been very recently discussed in (Tatikonda, Parthasarathy, & Kurc 2006). The commonality between their structure and ours (Tan et al. 2005) is that it is an array-based structure.

The integer values in the *EL* correspond to a node's position in the corresponding trees that the *EL* represents. Thus, the *EL* captures only the structural aspect of tree structures. One can obtain information about nodes from the *dictionary.* The interaction between the *dictionary* and the *EL* was an essential part of the candidate generation strategy in work such as (Tan et al. 2005, 2006a, 2008). In the candidate generation process, the *dictionary* functions as a lookup structure so that each occurrence of a subtree can be represented as an integer occurrence coordinate rather than being used to store every node's information locally. As the same information can be copied locally multiple times, the latter approach is more expensive.

Fig. 4.5 illustrates a tree *T* with all of its 2-subtree candidates and the *EL* of tree *T*. The main idea of an *EL* is that we create a list of nodes that share the same root node of a tree *T*. We use notation *EL[i]* to refer to an *embedding list* with a node at position *i* as its root key. An *embedding list* of the root node is called the main *embedding list.* It is the longest generated list in an *EL*. A slot refers to an array position in the *EL*. Thus, *EL[i][n]* refers to a value stored at slot *n* in *EL[i]*, whereas |*EL[i]*| refers to the size of the *embedding list* of the node at position *i*. So, for example from Fig. 4.5, *EL[0]* refers to the *embedding list* with root-key 0,

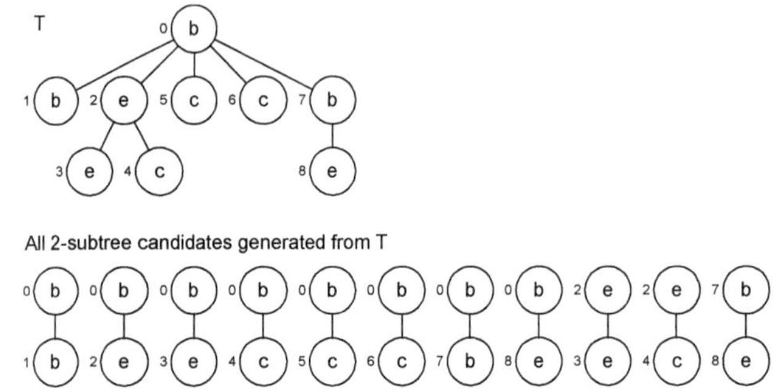

0:	1	2	3	4	5	6	7	8
2:	3	4						
7:	8							

Fig. 4.5 A tree *T* with all of its 2-subtree candidates and the *EL* of tree *T*

i.e. 0:[1,2,3,4,5,6,7,8]. *EL[0][0]* refers to a value stored in *EL[0]* at slot 0. From Fig. 4.5 *EL[0][0]* = 1, *EL[0][6]* = 7, *EL[7][0]* = 8, and so on.

A *k*-subtree is a subtree that consists of *k* nodes. An *EL* of tree *T* can be constructed by joining all 2-subtree candidates of a tree *T* that are rooted on the same node and order the list in a way that it forms a sorted list that follows the tree pre-order traversal sequence. Please note that the ordering is important for mining ordered subtrees. Each list will have its own key, which is the root node of the joined 2-subtrees. Thus, an *EL* of tree *T* is actually a compact representation of 2-subtree candidates that can be generated from a tree *T*. On the other hand, one can easily construct the original tree by simply connecting the root node and each node in the main *embedding list* with an edge.

EL transforms the hierarchical nature of a tree structure into multiple list structures. Such a conceptualization simplifies the enumeration process as it reduces the dimension of generating candidates from the hierarchical structure into a flat structure. The hierarchical structure is virtually a 2-dimensional structure, whereas the flat structure is a 1-dimensional structure. The enumeration of 1-dimensional structures can be done much more simply by iterating through it in a non-recursive manner. However, despite its usefulness, *EL* requires extra memory storage to store the nodes' coordinates or hyperlinks of the generated candidate subtrees. This can be an issue when processing a very large tree database. This is what motivated the development of a more compressed structure that is discussed next.

4.3.3 Recursive List

To help in data compression, a fusion of the *dictionary* structure and *EL* is introduced. This resulting structure is called a *recursive list* (*RL*). In a way, it is a more compact representation of the *EL* with additional functionality that previously belongs only to the dictionary. *RL* serves as a global lookup list at the same time it encodes the embedding relationships of the subtrees to be mined. Utilizing *RL* can result in space saving and slightly better performance as reported in our study (Hadzic et al. 2008, 2010). This optimization will be subsequently referred to as the *RL optimization*.

RL collapses multiple *embedding lists* of a tree *T*, *EL*[*i*],*EL*[*j*],..,*EL*[*z*] into *RL*[*i*] (i.e. *RL* starts at *i*) where $i < j < z$ and $EL[j] \subseteq EL[i]$ and $EL[z] \subseteq EL[i]$. *RL* utilizes the notion of scope (as explained in Section 4.3.1) to segment each list within a flat structure. It allows us to simulate the hierarchical nature captured by the *EL*. In *RL*, there are lists within a list, and hence the name 'recursive list'.

In Fig. 4.6, the *scope* of each node is shown together with each node *position* for the example tree. For example, the root node of example tree is at position 0 with scope 6, denoted as 0,6. A comparison of Fig. 4.6 and Fig. 4.7 illustrates the difference between the *RL* and *EL*. Please note that in Fig. 4.6 we do not show all the arrows reflecting the start and the end of an *EL* within *RL*. When a particular node has no descendants (i.e. scope is equal to position, *EL[2]*, *EL[4]*, *EL[5]* and *EL[6]*) there will be no extensions taking place. However, the arrows for *EL[2]* and *EL[4]* are shown in the figure for illustrative purposes.

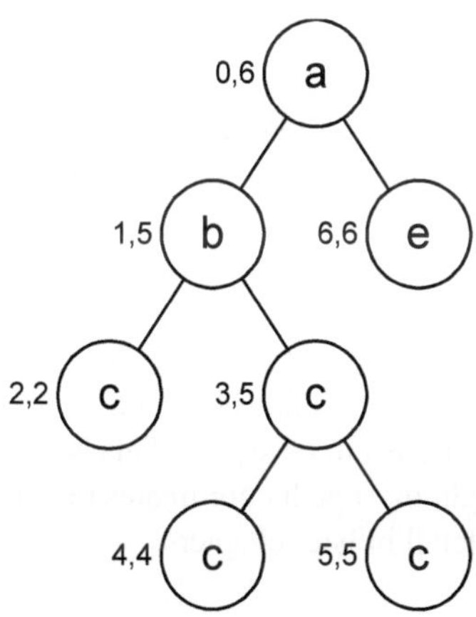

0	1	2	3	4	5	6	
1	2	3	4	5			
2	ϕ						
3	4	5					
4	ϕ						
5	ϕ						
6	ϕ						

Fig. 4.6 *EL* generated from example tree

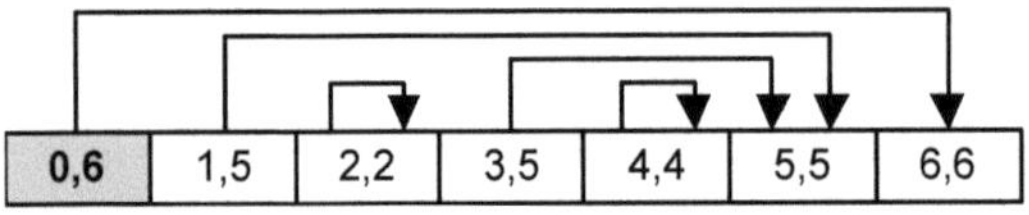

Fig. 4.7 *RL* for example tree from Fig. 4.6 with line and arrow indicating the start and end of each list. Each cell denotes (pos, scope) of nodes of tree.

Due to its compact format, utilizing the *RL* can result in space saving and offers slightly better performance than utilizing an *EL*. Hence, in subsequent chapters when the details of the TMG framework are provided, we will discuss the use of *RL* only within the whole mechanism as that is the latest, most optimal version of the implementation of the TMG framework.

4.4 Frequency Counting

A particular subtree can occur at many places in a tree database, and for each subtree these occurrences need to be stored in an appropriate structure. For this purpose, the *vertical occurrence list* (*VOL*) structure was used, and is discussed first. The frequency of the subtree is determined through the size of the *VOL* for occurrence match support or by the number of unique transaction identifiers when the transaction-based support definition is used. Each subtree is presented using the unique depth-first (pre-order) string encoding so that conventional hashing methods can be applied for efficient counting. The occurrences of a candidate subtree, are what is necessary for expanding a frequent k-subtree (subtree consisting of k nodes) to the next (k+1)-subtree. As in the systematic enumeration approach of TMG, candidate subtrees are generated by extending the nodes in the right-most path of the subtree, it is only the right-most path coordinates that need to be stored. This is explained in Section 4.4.2. The *VOL* actually stores right most path coordinates rather than full coordinates, as will be explained in more detail in later chapters.

4.4.1 Vertical Occurrence List

For efficient frequency counting, a vertical structure, called a *vertical occurrence list* (*VOL*), is used to store an *occurrence coordinate(s)* (denoted as *oc*) of each generated subtree. An *occurrence coordinate* is basically an integer array of varying size whose values are integer hyperlinks to coordinates of nodes following the preorder traversal ordering that are indexed in the *dictionary*. Unless otherwise stated, we will assume this definition of *occurrence coordinate*.

The main characteristics of *VOL* are that it can efficiently construct a *vertical list* of *oc* and compute its size vertically with almost no cost. With this vertical structure, the frequency count of each subtree is equal to the size of the structure vertically. The advantage of using this *VOL* is that the frequency count does not need to be updated separately in addition to inserting the *oc*, whereas an *oc* is needed for the candidate generation process.

Each occurrence of a subtree is stored as an *oc* as previously described. The *vertical occurrence list* of a subtree groups the *oc* of the subtree by its encoding. Computing the frequency of a subtree can be easily determined from the size of the *VOL*. We use the notation *VOL(L)* to refer to the *VOL* of a subtree with encoding *L*. Consequently, the frequency of a subtree with encoding *L* is denoted as $|VOL(L)|$. We can use *VOL* to count the *occurrence-match* support and *transaction-based* support. For *occurrence-match* support, we suppress the notion of the transaction id (*tid*) that is associated with each *oc*. For transaction-based and hybrid support, the notion of *tid* of each *oc* is accounted for when determining the support count. Hence, for occurrence match support, the frequency count is the size of the *VOL*; for transaction-based support the frequency count is equal to the number of unique transactional identifiers (*tid*) in the *VOL*, and for hybrid support the frequency count is determined based on the number of *oc* associated with each unique *tid*.

As an example, consider an ordered embedded subtree from tree *T* in Fig. 4.5 with encoding φ:'b c / e'. Its *oc* are {0,6,8},{0,5,8}, and {0,4,8}. This information is represented by *VOL*, as is shown in Fig. 4.8. When *occurrence-match* support is used, the frequency of this subtree is computed by computing the size of the *VOL*, $|$*VOL('b c / e')*$|$, i.e. 3. When *transaction-based* support is used (Fig. 4.9), the frequency of the example subtree is equal to 1 because from the *transaction-based* support definition given in Chapter 2, the support of a subtree *t* is equal to the numbers of transactions that support subtree *t*. In the example from Fig. 4.5, there is only a single tree *T*, and hence all *oc* come from the same tree that corresponds to a single transaction with id 0 (first column of *VOL* from Fig. 4.9).

0	6	8
0	5	8
0	4	8
b c / e		

Fig. 4.8 *VOL*('b c / e') of a subtree φ:'b c / e' when occurrence-match support is used

0	0	6	8
0	0	5	8
0	0	4	8
b c / e			

Fig. 4.9 *VOL*('b c / e') of a subtree φ:'b c / e' when transaction-based support is used

With numerous occurrences of subtrees, the list can increase to very large proportions and for long subtrees, the storage of full *oc* can be quite expensive, and this is not necessary as is discussed next.

4.4.2 RMP Coordinate List

When mining sequences, it is sufficient that only the last node coordinate be stored for the purpose of generating candidate sequences (Tan et al. 2006b). However, storing only the last node coordinate is not sufficient for subtree mining because the way the string encoding is generated for a subtree is different from that for sequences. In the case of sequences, there is no concept of backtracking (Zaki 2005). In a conventional approach, when mining tree structures, full occurrence coordinates must be stored so that the parent-child or the ancestor-descendant relationship between nodes can be preserved. This becomes one of the performance degrading factors as this can become very expensive for long subtrees. Thus, a good strategy to overcome the problem is considered important.

As defined in Chapter 3, the right-most-path of a tree is the (shortest) path connecting the right-most leaf with the root node. The usage of the right-most-path for candidate subtree generation has been discussed in (Abe et al. 2002; Tan et al. 2005). The right-most-path is also the basis for right-most-path expansion candidate generation family (Abe et al. 2002). While the utilization and applications of the right-most-path have been numerous, to our knowledge it has never been sufficiently discussed or exploited for tackling the space optimization problem.

We utilized the right-most-path concept for generating candidate subtrees. With such a candidate generation technique, the right-most-path information is required to generate candidate $(k+1)$-subtrees from a k-subtree. However, the right-most-path information is not automatically obvious since, in a conventional approach, full occurrence coordinates are stored and so no mechanism is in place which allows for an efficient retrieval of the right-most-path coordinates for each subtree. Instead, the *direct parent pointer* (*dpp*) was utilized for each node, which enables the upward traversal of the right-most-path during the candidate generation process. The drawback of this approach is that the process does not assume that every node belongs to the right-most-path and hence extra checking was required during the traversal. An optimization can be attained only if the right-most-path coordinates are stored so that further checking is not required and *dpp* would no longer be needed as a work around.

By storing only the right-most-path coordinates (denoted as *RMP-oc*), significant reduction in the coordinate storage requirement is achieved. Especially for subtrees with high fan-out, this can be a performance boost since the length of the right-most-path of a high fan-out subtree is normally much less than its length. The processing cost of a subtree with a shorter coordinate list is also substantially reduced. This structure is basically a *VOL* that contains right-most-path coordinates.

Fig. 4.10 illustrates how the right-most-path can help reduce the storage needed to store the (occurrence) coordinates of a subtree. Note that the right-most-path of T_1, T_2 and T_3 is the path between node P (right-most-node) and node Q, node R and S, and T and U respectively. Assume that we give a pre-order number to each node in the trees.

As evident from the subtrees in Fig. 4.10, if we have to store full (occurrence) coordinates of tree T_1 it will require us to store 12 coordinates, $\{0,1,2,3,4,5,6,7,8,9,10,11\}$. In contrast, storing the right-most-path coordinates requires only 2 coordinates, $\{0,11\}$. However, it is interesting to note that the space gain can be achieved only if the subtrees have a high fan-out like in T_1. For T_2 which basically is just a sequence, storing the right-most-path does not provide a space advantage as there

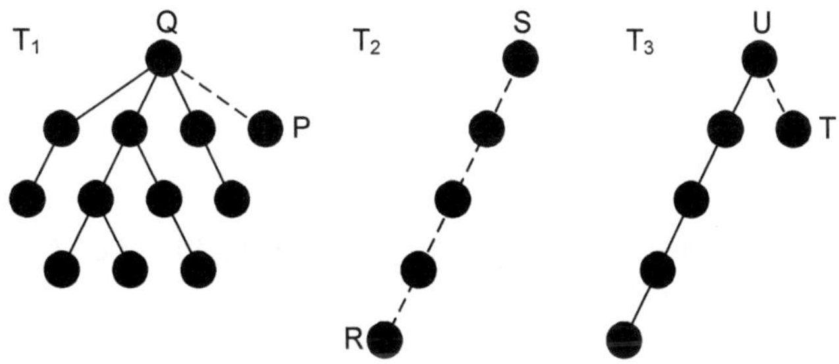

	T_1	T_2	T_3
right most path coordinate	{0,11}	{0,1,2,3,4}	{0,5}
conventional coordinate	{0,1,2,3,4,5,6,7,8,9,10,11}	{0,1,2,3,4}	{0,1,2,3,4,5}

Fig. 4.10 Illustration of the right-most-path coordinate (*RMP-oc*) of tree T_1, T_2 and T_3

are the same numbers of coordinates to be stored as if the right-most-path was not utilized at all. For T_3 although it is structurally quite similar to T_2, we can see a large space advantage by storing its right-most-path, i.e. {0,5} instead of {0,1,2,3,4,5}. This advantage seems to benefit both wide trees and deep trees. In general, we can say that the advantage is greater when extending a *k*-subtree that has fewer nodes on the right side of its root node, especially when extending a node to a root node for which the length of the right-most-path will become 2. With the right-most-path coordinate optimization, regardless of how many nodes do not belong to the right-most-path of such a subtree, the number of coordinates to be stored is always equal to the length of the right-most-path. Thus, utilizing the right most path concept for storing subtrees' coordinates provides a significant advantage over space processing and speed over our previous implementation since less checking is required, and the use of *dpp* during the candidate generation process is unnecessary.

4.5 Constraints

Applying constraints is useful to help us limit excessive exposure to something that would otherwise be uncontrollable or difficult to handle (Tan et al. 2006a). In addition, constraints can be also used to split an entity into many entities with a more granular property so that further distinction of such entities can be obtained. In this section, the focus is narrowed on the actual constraints where the general problem of frequent subtree mining is not modified, but the traditional definition of an embedded subtree has been changed according to the added constraint. The TMG framework uses two types of such constraints: *level of embedding* and *distance constraint*. The level of embedding constraint helps us devise a strategy reflecting a unified view of tackling the problem of mining frequent induced/embedded subtree. The distance constraint is used to split embedded subtrees into more granular distance-constrained embedded subtrees. This might be useful for certain applications where the distance between the nodes in a hierarchical structure could be considered important, and two embedded subtrees with different distance relationships among the nodes need to be considered as separate entities.

4.5.1 Feasible Computation through Level of Embedding Constraint

There are different tree-structured patterns. Patterns can appear as sub-patterns of the other patterns; therefore, we usually refer to patterns within a database as sub-patterns. Mining different patterns means different complexity when processing them. The two major types of subtrees are induced and embedded subtrees. Given a set of embedded subtrees E and induced subtrees I, which are subtrees of tree T, then $I \subseteq E$. Mining embedded subtrees is naturally more complex than mining induced subtrees. The TMG framework is designed to tackle the problems of mining both induced and embedded subtrees. The other motivation for mining embedded subtrees is that it represents the most complex type within the context of mining frequent subtrees. In general, by being able to tackle the more complex and difficult problem, one can resolve the less complex and easier problem.

An induced subtree by definition is a subset of an embedded subtree. The formal definition of an induced subtree was given in Chapter 2. Most of the other studies adopt a similar formal definition. However in this study, we develop a definition that shows a closer relationship between the induced and embedded subtrees, rather than considering one as just a subset of the other. The TMG framework is unique in using this concept to tackle the mining of both induced and embedded subtrees. Utilizing the notion of level of embedding, we propose a framework that can be flexibly switched between mining the two major types of subtrees. Most of the known approaches will normally adopt a specialized approach for tackling each subtree variant. Motivated by such a proposition, we utilize the concept of level of embedding that connects induced and embedded subtrees in a different light than they were traditionally defined.

The level of embedding is defined as the length of a path between two nodes that form an ancestor-descendant relationship. Intuitively, when the level of embedding inherent in the database of trees is high, numerous numbers of embedded subtrees exist. Thus, when it is too costly to mine all frequent embedded subtrees, one can decrease the level of embedding constraint gradually down to 1, from which all the obtained frequent subtrees are induced subtrees. With such a definition, we would see the induced subtree as an embedded subtree where the maximum level of embedding that is allowed to occur is constrained to 1. Hence, mining frequent induced subtrees can be formulated in the same way as mining embedded subtrees where the level of embedding is constrained to 1 (Tan. et al. 2006a). We next provide a formal definition of the level of embedding and the constraint.

If $S = (vs_0, V_S, L_S, E_S)$ is an embedded subtree of $T = (vt_0, V_T, L_T, E_T)$ (see Chapter 2), and two vertices $p \in V_S$ and $q \in V_S$ form an ancestor-descendant relationship, the ***level of embedding*** (Tan et al. 2006a), between p and q, denoted by $\Delta(p,q)$, is defined as the length of the path between p and q. With this observation, a ***maximum level of embedding constraint*** δ can be imposed on the subtrees extracted from T, such that any two ancestor-descendant nodes present in an embedded subtree of T, will be connected in T by a path that has the maximum length of δ. In this regard, we could define an induced subtree S_I as an embedded subtree where the *maximum level of embedding* that can occur in T is equal to 1, since the level of embedding of two nodes that form a parent-child relationship is equal to 1.

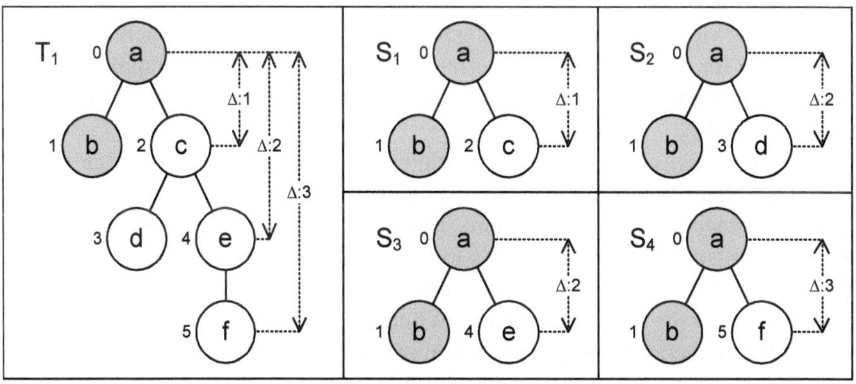

Fig. 4.11 Illustration of restricting the level of embedding

If we denote a node at position x in a tree as n_x, then from Fig. 4.11 the level of embedding, Δ, between node at position 0, n_0, and node at position 5, n_5, in tree T is 3, whereas $\Delta(n_0,n_3) = \Delta(n_0,n_4) = 2$. According to our previous definition of induced and embedded subtrees, subtree S_1 is the only example of an induced subtree, whereas the rest are examples of embedded subtrees.

4.5.2 Splitting Embedded Subtree through Distance Constraint

The embedded subtrees extracted using the traditional definition are incapable of being further distinguished based upon the node distances within that subtree. For certain applications the distance between the nodes in a hierarchical structure could be considered important and two embedded subtrees with different distance relationships among the nodes need to be considered as separate entities. The distances of nodes relative to the root (node depth) of a particular subtree will need to be stored and used as an additional equality criterion for grouping the enumerated candidate subtrees. The updated definition of an embedded subtree can be formally expressed as follows for unordered and ordered embedded subtrees:

Given a tree $S = (vs_0, V_S, L_S, E_S)$ and tree $T = (vt_0, V_T, L_T, E_T)$, S is an ***unordered distance-constrained embedded subtree*** of T *iff* (1) $V_S \subseteq V_T$; (2) $L_S \subseteq L_T$, and $L_S(v) = L_T(v)$; (3) if $(v_1, v_2) \in E_S$ then $parent(v_2) = v_1$ in S and v_1 is ancestor of

v_2 in T; and (4) $\forall v \in V_S$ there is an integer stored indicating the level of embedding between v and the root node of S in the original tree T.

Given a tree $S = (vs_0, V_S, L_S, E_S)$ and tree $T = (vt_0, V_T, L_T, E_T)$, S is an ***ordered distance-constrained embedded subtree*** of T *iff* (1) $V_S \subseteq V_T$; (2) $L_S \subseteq L_T$, and $L_S(v) = L_T(v)$; (3) if $(v_1, v_2) \in E_S$ then $parent(v_2) = v_1$ in S and v_1 is ancestor of v_2 in T; (4) $\forall v \in V_S$ there is an integer stored indicating the level of embedding between v and the root node of S in the original tree T; and (5) the left-to-right ordering among the sibling nodes in T is preserved in S.

This notion of distance-constrained embedded tree mining will have important applications in biological sequences, web information systems, conceptual model analysis, knowledge matching and for general knowledge management related tasks by allowing for more specialized queries. Examples of a number of application areas and implications for knowledge analysis will be discussed and illustrated in Chapters 7 and 9.

The major extension requirement to implement such a constraint is for an appropriate candidate encoding scheme to distinguish subtrees based upon structure and the node distances within the structure. The structural aspects need to be preserved and extra distance information needs to be stored. We present an encoding strategy to efficiently enumerate candidate subtrees taking into account the distances of nodes relative to the root of the subtree. While the number of unique candidate subtrees to be enumerated will grow because of the added granularity, this constraint can be used to filter out certain embedded subtrees based on the distance criterion. In addition, this constraint can be used to further distinguish into its distance-coded variants what would otherwise be just an embedded subtree.

We slightly modify the string encoding discussed earlier in order to implement the distance constraint. The distance to the root is worked out from the node depths stored in the *dictionary/recursive list*, where the root of the subtree is assigned the depth of 0 and all other nodes are assigned the difference between their depth and the original depth of the new subtree root. Hence, the additional information corresponds to the depths of nodes within the newly encoded subtree. Further modification of the encoding consists in storing a number next to each backtrack '/' symbol indicating the number of backtracks in the subtree, as opposed to storing each of those backtracks as a separate symbol. This representation allows for easier string manipulation due to uniform block size.

4.6 Conclusion

In this chapter, the focus is on giving formal definitions and contexts of representations, data structures and constraints for efficiently mining frequent subtrees as the basis for discussion in later chapters.

The first representation problem that has been addressed in this chapter relates to ways of efficiently representing subtrees in memory or secondary storage. We have given the rationale of using depth-first string encoding to represent subtrees, and have discussed a strategy to expedite the processing of XML documents using the indexing technique.

The second representation problem that the framework tries to address is related more to the way that complex computations and data manipulations can be performed efficiently and effectively. The choice of data structures used is critically important for efficient candidate generation and frequency counting. Several data structures were proposed to address this issue. A number of constraints that are used in the TMG framework were also discussed.

References

1. Abe, K., Kawasoe, S., Asai, T., Arimura, H., Arikawa, S.: Optimized substructure discovery for semistructured data. Paper presented at the Proceedings of the 6th European Conference on Principles of Data Mining and Knowledge Discovery, Helsinki, Finland, August 19-23 (2002)
2. Agrawal, R., Srikant, R.: Fast Algorithms for Mining Association Rules. In: Proceedings of the 20th International Conference on Very Large Data Bases (VLDB), Santiago de Chile, Chile, Septemebr 12-15, pp. 487-499 (1994)
3. Chi, Y., Muntz, R.R., Nijssen, S., Kok, J.N.: Frequent Subtree Mining - An Overview. Fundamenta Informaticae, Special Issue on Graph and Tree Mining 66(1-2), 161–198 (2005)
4. Hadzic, F., Tan, H., Dillon, T.S.: UNI3 - Efficient Algorithm for Mining Unordered Induced Subtrees using TMG Candidate Generation. In: Proceedings of IEEE Symposium on Computational Intelligence and Data Mining (CIDM), Honolulu, Hawaii, USA, April 1-5, pp. 568–575. IEEE, Los Alamitos (2007)
5. Hadzic, F., Tan, H., Dillon, T.S.: U3 – mining unordered embedded subtrees using TMG candidate generation. In: Proceedings of the IEEE / WIC / ACM International Conference on Web Intelligence, Sydney, Australia, December 9-12, pp. 285–292 (2008)
6. Hadzic, F.: Advances in knowledge learning methodologies and their applications. Curtin University of Technology, Perth (2008)
7. Hadzic, F., Tan, H., Dillon, T.S.: Tree Model Guided Algorithm for Mining Unordered Embedded Subtrees. Web Intelligence and Agent Systems: An International Journal (WIAS) 8(4) (2010)
8. Inokuchi, A., Washio, T., Motoda, H.: Complete Mining of Frequent Patterns from Graphs: Mining Graph Data. Machine Learning 50(3), 321–354 (2003)
9. Kuramochi, M., Karypic, G.: Frequent Subgraph Discovery. Paper Presented at the Proceedings of the IEEE International Conference on Data Mining (ICDM 2001), San Jose, California, USA, November 29 - December 2 (2001)
10. Pei, J., Han, J., and Lakshmanan, L.V.S, Mining frequent itemsets with convertible constraints. Paper presented at the Proceedings of the 17th International Conference on Data Engineering, Heidelberg, Germany, April 2-6 (2001)
11. Tan, H., Hadzic, F., Feng, L., Chang, E.: MB3-Miner: mining eMBedded subTREEs using tree model guided candidate generation. In: Proceedings of the 1st International Workshop on Mining Complex Data in conjunction with ICDM 2005, Houston, Texas, USA, November 27-30, pp. 103–110 (2005)
12. Tan, H., Dillon, T.S., Hadzic, F., Chang, E., Feng, L.: IMB3-Miner: Mining Induced/Embedded Subtrees by Constraining the Level of Embedding. In: Ng, W.-K., Kitsuregawa, M., Li, J., Chang, K. (eds.) PAKDD 2006. LNCS (LNAI), vol. 3918, pp. 450–461. Springer, Heidelberg (2006a)

13. Tan, H., Dillon, T.S., Hadzic, F., Chang, E.: SEQUEST: Mining Frequent Subsequences using DMA Strips. Paper presented at the Proceeding of the 7th International Conference on Data Mining and Information Engineering, Prague, Czech Republic, July 11-13 (2006b)
14. Tan, H.: Tree Model Guided (TMG) enumeration as the basis for mining frequent patterns from XML documents. University of Technology Sydney, Sydney (2008)
15. Tan, H., Hadzic, F., Dillon, T.S., Feng, L., Chang, E.: Tree Model Guided Candidate Generation for Mining Frequent Subtrees from XML. ACM Transactions on Knowledge Discovery from Data 2(2) (2008)
16. Tatikonda, S., Parthasarathy, S., Kurc, T.: TRIPS and TIDES: new algorithms for tree mining. Paper presented at the Proceedings of the 15th ACM International Conference on Information and Knowledge Management (CIKM), Arlington, Virginia, USA, November 6-11 (2006)
17. Wang, C., Hong, M., Pei, J., Zhou, H., Wang, W., Shi, B.: Efficient Pattern-Growth Methods for Frequent Tree Pattern Mining. In: Dai, H., Srikant, R., Zhang, C. (eds.) PAKDD 2004. LNCS (LNAI), vol. 3056, pp. 441–451. Springer, Heidelberg (2004)
18. Yang, L.H., Lee, M.L., Hsu, W.: Efficient Mining of XML Query Patterns for Caching. In: Proceedings of the 29th International Conference on Very Large Data Bases (VLDB), Berlin, Germany, September 9-12, pp. 69–80 (2003)
19. Zaki, M.J.: Efficiently Mining Frequent Trees in a Forest: Algorithms and Applications. IEEE Transactions on Knowledge and Data Engineering 17(8), 1021–1035 (2005)

Chapter 5
TMG Framework for Mining Ordered Subtrees

5.1 Introduction

In this chapter, we will elaborate on the overall TMG framework for mining ordered subtrees as described in Chapter 4 (Tan 2008). In an ordered tree, for each internal node, the order of its children is fixed. Such trees have found many useful applications in areas such as vision, natural language processing, molecular biology, programming compilation etc. (Wang, Zhang, Jeong & Shasha 1994). In the research on automatic natural language processing, the *dictionary* definitions are represented syntactically as trees. Computational linguists extract semantic information about these definitions and in the process construct semantic taxonomies (Chodorow & Klavans, 1990; Neff, Roy & Omneya 1998). In the molecular biology field, large amounts of analyzed RNA structures are collected and stored in the form of ordered labeled trees. When the researchers want to acquire information about a new RNA structure, it is compared against those already in the database in order to detect structural similarities and thereby relate different RNA structures (Shapiro & Zhang 1990). Since the researchers will maintain the RNA-related information in the same order, a comparison of ordered subtrees is sufficient. A general observation about applications where ordered subtree mining is suitable is that the left-to-right order among sibling nodes is commonly fixed and known beforehand. Ordered subtree mining is useful for querying a single database where the order restriction can be placed on the query subtree because it is known beforehand.

In the previous chapter, we described some of the general characteristics of the TMG tree mining framework. The characteristics of the various representative structures were explained as were the optimizations that took place within our framework. This chapter describes how the framework is used for the mining of ordered subtrees, assuming that the more optimal implementation using the *recursive list* is used. The corresponding algorithm has been referred to as IMB3-Miner (Tan et al. 2006). As such, the previously explained *dictionary* and *embedding list* (*EL*) structures will not be discussed in the implementation details of this chapter, as both of their functionalities are handled by the *recursive list* (*RL*) structure

F. Hadzic, H. Tan, and T.S. Dillon: Mining of Data with Complex Structures, SCI 333, pp. 87–138.
springerlink.com

(see Chapter 4). We will, however, provide some experiments that indicate the difference in the implementation choices, and show the memory saving through *RL* optimization. However, as the *EL* is conceptually simpler than *RL*, it has enabled us to realize the mathematical representation of the enumeration process and carry out worst case performance analysis. For this reason, *EL* will still be mentioned later in the chapter since it makes the mathematical analysis easier to understand.

The algorithm described focuses on discovering all ordered induced and embedded subtrees, from a database of rooted ordered labeled trees. Formal definitions of both types of subtrees have been given in Chapter 2. Fig. 5.1 illustrates the difference between induced and embedded subtrees. The label of each node is shown as a single-quoted symbol inside the circle, whereas its pre-order position is shown as an index at the left side of the circle. We make use of the term ‘occurrence coordinate(s)’ (*oc*) to denote the occurrence(s) of a particular node or a subtree in the tree database. The *oc* of each node from *T* is shown on the left of the node. For a subtree, *oc* is a sequence of *oc* from nodes that belong to that particular subtree. Hence, one can easily see the difference between ordered induced and embedded subtrees from the tree database *T* in Fig. 5.1. For example, T_3 with *oc*:’0 3 4’ is an embedded subtree because node ‘b’ with *oc* 0, is an ancestor of node ’e’ with *oc*:3, and node ‘c’ with *oc*:4 in the original tree *T*, while it is a parent in the subtree. All the subtrees maintain the same order of sibling nodes as in *T*; this is what makes them ordered subtrees.

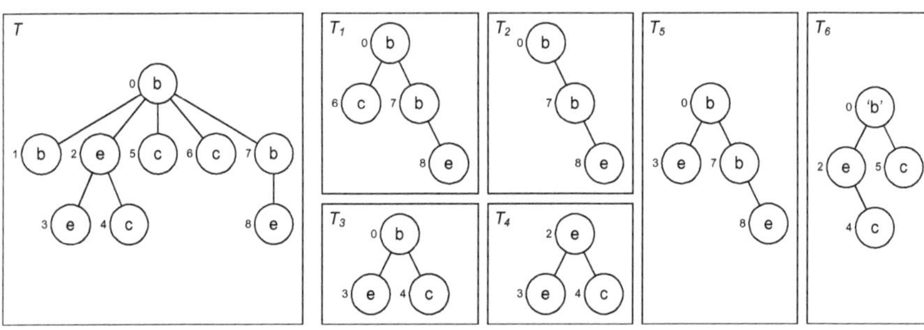

Fig. 5.1 Example of ordered induced subtrees (T_1, T_2, T_4, T_6) and ordered embedded subtrees (T_3, T_5) of tree *T* (note that induced subtrees are also embedded)

The chapter starts with a high level description of the TMG framework for the mining of ordered induced/embedded subtrees (Section 5.2). Each of the steps involved is described in detail in Section 5.3. Section 5.4 provides a mathematical analysis of the worst case performance of the TMG candidate enumeration technique, for both induced and embedded ordered subtrees. A number of experiments using both synthetic and real-world data are described in Section 5.5. These serve the purpose of evaluating the performance of the TMG framework with respect to the current state-of-the-art algorithms in the field as well as highlighting the strengths and weaknesses of the different implementation choices. The important issues are highlighted and a discussion is provided explaining each of the findings. Section 5.6 summarizes the chapter.

5.2 Overview of the Framework for Mining Ordered Induced/Embedded Subtrees

A high level description of the general steps taken by the proposed approach is provided, and in the sections which follow, each step is explained in elaborate detail taking implementation considerations into account. We are concerned here with the optimized version of the TMG framework. The basic steps taken by the IMB3-Miner algorithm are presented in the flowchart of Fig. 5.2. The tree database is first transformed into a database of rooted integer-labeled trees. The *recursive list* (*RL*) is constructed which is a global sequence of encountered nodes in the pre-order traversal together with the necessary node information (label, level, scope) During this process, the node labels are hashed to obtain the set of frequent 1-subtrees (F_1) and then the set of frequent 2-subtrees is also obtained. TMG candidate generation using the *RL* structure takes place. The string encodings of candidate subtrees are hashed to the C_k (candidate k-subtrees) hash table, and the right-most path (*RMP*) occurrence

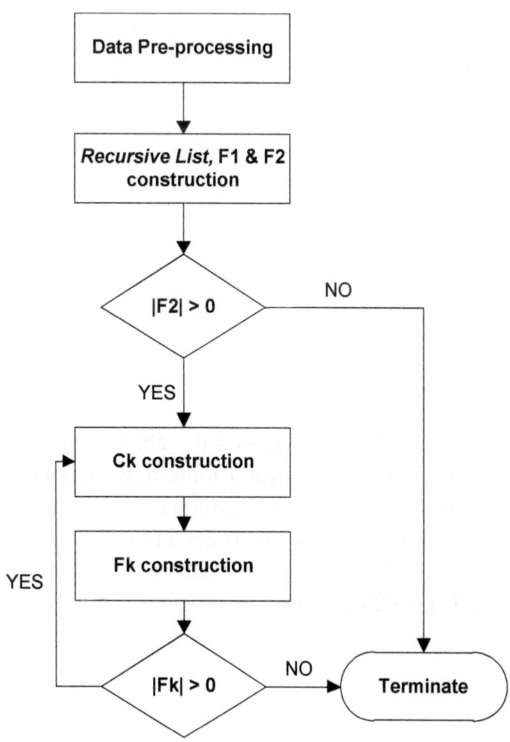

Fig. 5.2 General steps of the IMB3-Miner algorithm

coordinates are stored. During C_k construction, each frequent (k-1)-subtree is extended one node at a time, starting from the last node of its *RMP* (right-most node), up to its root. The whole process is repeated until all frequent k-subtrees have been enumerated.

Whilst we have explained these steps using the *RL* structure, implementations using the *dictionary* and *embedding list* structures (see Chapter 4) are given in (Tan 2008; Tan et al. 2006, 2008), and the reader is referred to these for details.

5.3 Detailed Description of the Framework

This section describes in more detail each of the steps taken by the IMB3-Miner algorithm. We discussed various canonical forms for trees in the last chapter. Before the whole process of the IMB3-Miner algorithm can be clearly explained, we first explain the way that a subtree is represented at the implementation level in the TMG framework.

5.3.1 Tree Representation

The pre-order string encoding (φ) as used in (Zaki 2005) is utilized. It is a sequential representation of the nodes of a tree as encountered during the pre-order traversal of that tree. In addition, the backtrack symbol ('/') is used when moving up a node in the tree during the pre-order traversal. The encoding of a subtree T is denoted as $\varphi(T)$, and as an example consider Fig. 5.1 again. The string encoding of T is, $\varphi(T)$:'b b / e e / c / / c / c / b e / / '. The backtrack symbols can be omitted after the last node. The string encoding of the subtrees T_1 to T_6 from Fig. 5.1 are:

$\varphi(T_1)$:*'b c / b e'*
$\varphi(T_2)$:*'b b e'*
$\varphi(T_3)$:*'b e / c'*
$\varphi(T_4)$:*'e e / c'*
$\varphi(T_5)$:*'b e / b e'*
$\varphi(T_6)$:*'b e c / / c'*

A k-subtree. is a subtree consisting of k nodes. Throughout the chapter, the '+' operator is used to denote the operation of appending two or more tree encodings. However, this operator should be contrasted with the conventional string append operator, as in the encoding used, the backtrack ('/') symbols need to be computed accordingly. One subtree can occur at many places in the tree database. For example, the subtree T_6 occurs at 2 positions in T, i.e. 0245, 0246.

5.3.2 Data Pre-processing

When determining subtree frequency using a hash table, the processing of integer-labeled trees has a computational advantage over string-labeled trees, especially when the labels are long (Tan et al. 2005a). To expedite the frequency counting, the database of XML documents (or any other tree-structured documents) can be transformed into a database of rooted integer-labeled ordered trees. One format to represent the database of rooted integer-labeled ordered trees is proposed in (Zaki 2005). Each tag in an XML document can be encoded as an integer. Each integer will identify each tag uniquely. To encode a particular tree, these integers are used in the same way that string labels were used in the string encoding explained in the previous chapter. The only difference is that the backtrack ('/') symbol is replaced by a negative one ('-1'). For example, if labels *b*, *c* and *e* are mapped to integers 1, 2, 3, respectively, then the encodings of tree *T* and subtrees T_1 to T_6 from Fig. 5.1, become:

$\varphi(T)$:*'1 1 -1 3 3 -1 2 -1 -1 2 -1 2 -1 1 3'*
$\varphi(T_1)$:*'1 2 -1 1 3'*
$\varphi(T_2)$:*'1 1 3'*
$\varphi(T_3)$:*'1 3 -1 2'*
$\varphi(T_4)$:*'3 3 -1 2'*
$\varphi(T_5)$:*'1 3 -1 1 3'*
$\varphi(T_6)$:*'1 3 2 -1 -1 2'*.

For XML documents, we should expect the string labels to be long strings, so mapping string labels to unique integers will optimize the frequency counting process and avoid additional hash key computation. For each XML tag, we consider tag-name, attribute(s) and value(s), as was explained in Section 3.2.1 of Chapter 3. Hence, each unique system-generated integer corresponds to each unique tag combination. To mine the structure of XML documents only, one can modify this easily by omitting the presence of attribute(s) and value(s) for each tag.

5.3.3 Generating the Recursive List (RL), F_1 and F_2

The main purpose of this step is to store the pre-order positions of the nodes in the tree database in such a manner that candidate subtrees can be efficiently enumerated. The tree database, T_{db}, is scanned once to create the global sequence *RL* in memory, through which nodes' related information can be directly accessed. Each node is stored following the pre-order traversal of the T_{db}. Position, label, scope, and level information is stored for each node. The scope of a node refers to the position of its right-most leaf node or its own position if it is a leaf node itself (Zaki 2005, Tan et al. 2006). The level refers to the level in the T_{db} tree where this node occurs. An item in *RL* at position *i* is referred to as *RL*[*i*]. Every time a node is inserted into the *RL*, we generate a candidate 1-subtree. Based on its label, we increment its support count in the C_1 (candidate 1-subtrees) hash table. If its support count is $\geq \sigma$ (user-specified minimum support count), we insert the candidate 1-subtree to the frequent 1-subtree set, F_1. An example *RL* structure representing a tree (T_1) is displayed in Fig. 5.3. The pre-order position of a node in the tree database is equal to the index of the *RL* at which that node is stored, and the label, scope and level are shown in that order underneath the entry. All this information is necessary to enumerate only valid

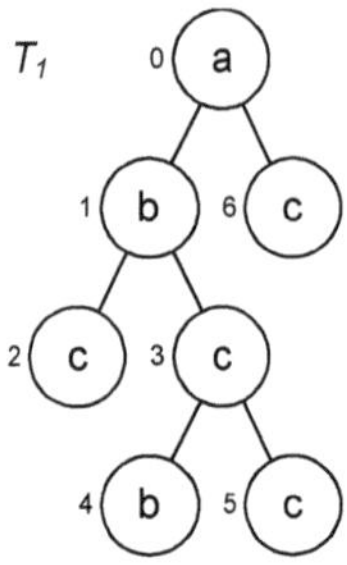

[0]	[1]	[2]	[3]	[4]	[5]	[6]
a,6,0	b,5,1	c,2,2	c,5,2	b,4,3	c,5,3	c,6,1

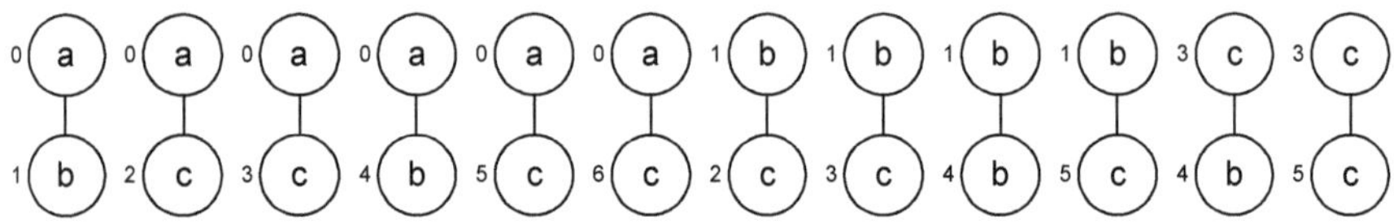

Fig. 5.3 *RL* structure for tree T_1 and generated candidate embedded 2-subtrees from T_1 at $\sigma = 1$

subtree candidates and is accessed during the TMG candidate enumeration process explained next.

During the process of enumerating a $(k+1)$-subtree from a k-subtree, the nodes that can form part of the extension are those that are contained within the scope of the node being extended (extension point). This process is explained in detail in the next section. During the construction of the *RL* structure, the set of candidate 2-subtrees are also generated. It was mentioned in Chapter 4 that, for every candidate k-subtree generated, its occurrence in the database is stored in the *vertical occurrence list* (*VOL*) structure. Hence, every item from F_1 will have a label and a set of occurrences associated with it. The set of candidate 2-subtrees is obtained as follows. For every item in F_1, the set of its occurrences is obtained which corresponds to the entries in the *RL* structure. For each entry in *RL*, a number of candidate 2-subtrees are generated by appending the label of the node contained in the *RL* entry to each of the labels of the nodes (*RL* entries) that are within the scope of the node being processed. For example, let us say that 1-subtree with label 'a' has an occurrence at position p in *RL*, where the scope is equal to $p+n$. The set of candidate 2-subtrees that can be obtained by extending this particular node, is formed by appending the label 'a' to each of the node labels from all the positions x in *RL*, where $p < x \leq p + n$. The resulting encoding is then hashed to the C_2 hash table and the occurrences for each candidate are stored in the *VOL* structure. The set of all candidate 2-subtrees from tree T_1 obtained in this way, are displayed at the bottom of Fig. 5.3. Going back to the example above, the node with label 'a' occurs at position 0 in *RL* and it is extended with nodes from position 1 to position 6 in *RL* since its scope is equal to 6. As a result, 6 candidate 2-subtrees are formed by extending the node with label 'a', which are the first 6 subtrees displayed at the bottom of Fig. 5.3. The occurrences are displayed on the left of the nodes. As noted earlier, at the implementation level it is the actual string encoding (see Section 5.3.1) which is used to represent these subtrees and the set of node occurrences are stored in the *VOL*. Please note that the example illustrates the set of embedded 2-subtrees. If induced subtrees were mined, then the level information of a node would be used for ensuring that only those candidate 2-subtrees are generated where the level of embedding between the two nodes is equal to 1.

Please note that all the node-related information is always obtained from the *RL*. The node-related information is obtained based upon the stored occurrence in the *VOL* which corresponds to an entry in *RL*. The scope information is always utilized, so that the complete set of valid candidate subtrees is enumerated. For simplicity, in the following section the fact that all this information is extracted from the *RL* will not always be repeated. Rather, the discussion will just indicate that certain node information has been obtained, and it is known that this is obtained from the *RL* structure.

5.3.4 Enumerating Ordered Induced/Embedded k-Subtrees Using the TMG Enumeration Technique

In what follows, the process of generating all candidate k-subtrees is described. This is the core process of the described framework and, as such, a number of aspects need to be explained. The section starts with an explanation of how the different occurrences of a subtree are stored and then proceeds to explain the general TMG candidate generation framework for enumerating all the frequent k-subtrees. At the end of the section, we explain the process of pruning which ensures that no pseudo-frequent subtrees are generated.

5.3.4.1 Right-Most Path Occurrence Coordinates

A subtree can occur at different positions in a database and the occurrence coordinates of a subtree need to be stored in order to distinguish it from other subtrees having the same string encoding. We store the right-most path occurrence coordinate (*RMP-oc*) since by definition, the right-most path is the shortest path from the right-most node to the root node. Thus, the storage of *RMP-oc* is always guaranteed to be maximal, and it constitutes all the required information for enumerating k+1-subtrees from a k-subtree. The worst case of storing the *RMP-oc* would be equal to storing every coordinate of a node in a subtree; i.e. when the subtree becomes a sequence (each node has degree 1). The best case of storing *RMP-oc* for k-subtrees where $k > 1$ is in storing only 2 coordinates; i.e. whenever the length of the *RMP* is equal to 1. Given a k-subtree T with occurrence coordinates o_c $[e_0,e_1,\ldots,e_{k-1}]$, the *RMP-oc* of T, denoted by $\Psi(T)$, is defined by $[e_0,e_1,\ldots,e_j]$ such that $\Psi(T) \subseteq oc(T)$; e_j=e_{k-1}; and $j \leq k$-1 and the path from e_j to e_0 is the right-most path of tree T.

5.3.4.2 TMG Enumeration

In this discussion, the right-most path of subtrees is referred to as *RMP*, RMP occurrence coordinate(s) of subtrees as *RMP-oc* and the occurrence coordinate(s) of a subtree or a singular node as *oc*. To enumerate all *k*-subtrees from a (*k*-1)-subtree, the TMG enumeration approach extends one node at a time for each node in the *RMP* of (*k*-1)-subtree. Suppose that nodes in the *RMP* of a subtree are defined as extension points and the level of embedding (see Chapter 4) between two nodes at position *n* and *t* is denoted by $\Delta(n,t)$. The TMG can be formulated as follows. Let $\Psi(T_{k-1})$:$[e_0,e_1,\ldots e_j]$ denote the *RMP-oc* of a frequent (*k*-1)-subtree T_{k-1}, Φ the scope of the root node e_0 and δ the maximum level of embedding constraint; the choice of the δ parameters allows us to mine both induced and embedded subtrees using the same framework. When no constraint is imposed on δ, we are mining embedded subtrees and induced subtrees can be mined when δ is set to 1 (i.e. only parent-child relationships are allowed). TMG generates *k*-subtrees by extending each extension point $n \in \Psi(T_{k-1})$ with a node with *oc t* for which the following conditions are satisfied: (1) $n < t = \Phi$, (2) $\Delta(n,t) \leq \delta$. Suppose that the encoding of T_{k-1} is denoted by $\varphi(T_{k-1})$ and $l(e_j,t)$ be a labelling function for extending extension point *n* with a node at position *t*. $\varphi(T_k)$ would be defined as $\varphi(T_{k-1})+l(e_j,t)$, where $l(e_j,t)$ determines the number of backtrack symbols '/' to be appended before the label of the new node is added to $\varphi(T_k)$. The number of backtrack symbols is calculated as the shortest path distance between the extension point *n* and the right-most node *r*, (notation $pl(n,r)$).

To generate *RMP* at each step of candidate generation, the computed number of backtrack symbols *b* that need to be appended before the new node with *oc t*, is added to the encoding. Given that the $\Psi(T_{k-1})$ is $[e_0,e_1, \ldots, e_j]$, the *RMP* of the *k*-subtree ($\Psi(T_k)$) is generated by appending *t* at position $(j+1) - b$ of the ($\Psi(T_{k-1})$) and removing any *RMP-oc* that occur after *t*, thereby making *t* the right-most node of T_k. This will ensure that at each extension of (*k*-1)-subtree, *RMP-oc* of *k*-subtree are appropriately stored.

To provide an illustrative example, let us say that we are extending the T_{k-1} subtree from Fig. 5.4, where $\Psi(T_{k-1})$:[0,4,5], $\varphi(T_{k-1})$:'a b / b c', and right-most node 'c' (*oc*:5). If we are extending T_{k-1} from extension point node 'b' (*oc*:4) with node 'e' (*oc*:8) then $l(5,8)$ will append one backtrack symbol ($pl(4,5) = 1$) and the label 'e' to $\varphi(T_{k-1})$. The new encoding $\varphi(T_k)$ becomes 'a b / b c / e', and $\Psi(T_k)$:[0,4,8] (i.e. inserting 8 at position $(j+1) - b = (3+1) - 1 = 3$ of $\Psi(T_{k-1})$). Similarly, if we were extending the same T_{k-1} subtree from extension point node 'a' (*oc*:0) with node 'c' (*oc*:9) then $l(5,9)$ will append two backtrack symbols ($pl(0,5) = 2$) and the label 'c' to $\varphi(T_{k-1})$. The new encoding $\varphi(T_k)$ would become 'a b / b c / / c', and $\Psi(T_k)$:[0,9] (i.e. inserting 9 at position $(j+1) - b = (3+1) - 2 = 2$ of $\Psi(T_{k-1})$).

5.3.4.3 Pruning

To ensure that all generated subtrees do not contain infrequent subtrees, full (*k*-1) pruning must be performed. This implies that at most (*k*-1) numbers of

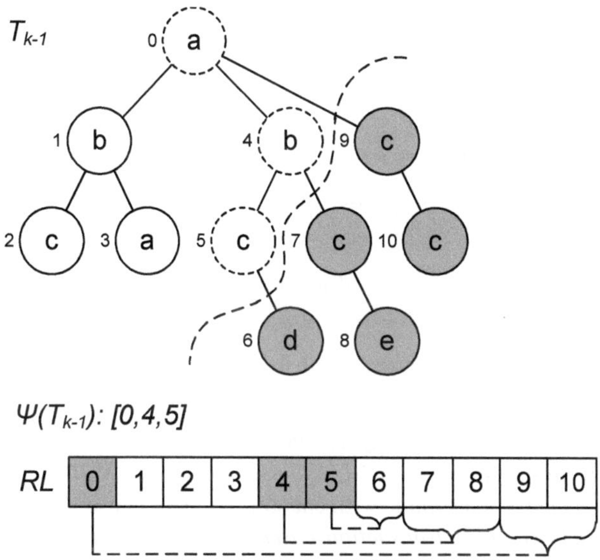

Fig. 5.4 TMG enumeration: extending (k-1)-subtree T_{k-1} where $\varphi(t_{k-1})$:*'a b / b c'* occurs at position *(0,1,4,5)* with nodes at positions 6, 7, 8, 9, and 10

(k-1)-subtrees need to be generated from the currently expanding k-subtree. An exception is made whenever the δ constraint is set to 1, i.e. mining induced subtrees, as we need to generate only n numbers of (k-1)-subtrees where $n < (k\text{-}1)$ and is equal to the number of leaf nodes in the k-subtree. When the removal of the root node of k-subtree does not generate a disconnected tree or a forest (Zaki 2005), then an additional (k-1)-subtree is generated by removing the root node from the expanding k-subtree. Each of these generated k-1-subtrees is checked for existence in the F_{k-1} set. The expanding k-subtree is pruned when at least one of its (k-1)-subtrees cannot be found in the F_{k-1} set (i.e. it is infrequent). Otherwise, it is added to the F_k set. This ensures that the downward-closure lemma (Agrawal & Srikant 1994) is satisfied and in the case of occurrence-match support, the method will not generate any pseudo-frequent subtrees (see Section 2.6.1 from Chapter 2). Performing full (k-1) pruning is quite time consuming and expensive. To accelerate full pruning, a caching technique is used by checking whether a candidate is already in the frequent k-subtree set (F_k). In this way, if a k-subtree candidate is already in the frequent k-subtree set, it is known that all its (k-1)-subtrees are frequent, and hence only one comparison is made.

5.3.5 *Frequency Counting*

Each candidate subtree is uniquely represented by its pre-order string encoding (see Section 5.1), with which many *RMP-oc* can be associated. *VOL* is used to store the *RMP-oc* of a particular subtree and to determine the support of a subtree using any of the current support definitions (see Sections 4.4.1 and 4.4.2). *VOL* of a subtree

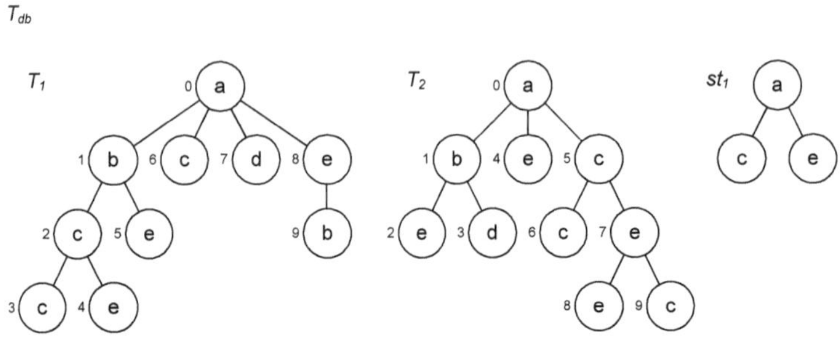

1	0	5
1	0	8
1	0	4
1	0	5
1	0	8
1	0	8
2	0	7
2	0	8
'a c \ e'		

Fig. 5.5 Example tree database ($T_d b$) and its subtree st_1 with its *VOL*

groups the *RMP-oc* of the subtree by its encoding. Hence, the frequency of a subtree can be easily determined from the size of the *VOL*. We use the notation *VOL(L)* to refer to the *vertical occurrence list* of a subtree with encoding *L*. We can use *VOL* to count the *occurrence-match (weighted), transaction-based* and *hybrid* support. For *occurrence-match* support we ignore the transaction id (*tid*) that is associated with each occurrence coordinate. For *transaction-based* and *hybrid* support the notion of *tid* of each occurrence coordinate is accounted for when determining the support.

To provide an illustrative example, in Fig. 5.5 we show an example tree database T_{db} consisting of two transactions T_1 and T_2, of which an example subtree st_1 is shown on the right together with its *VOL*. In this example, the *VOL* generated is for the case when ordered embedded subtrees are mined. The first column of the *VOL* corresponds to the transaction identifier (*tid*) of where the specific instance of the subtree 'a c / e' has occurred. The other two columns contain the *RMP-oc* of that specific instance. We can see that the *RMP-oc* '0 5' and '0 8' repeats, the reason being that they come from a different instance of the subtree in T_1. For example, the instances of st_1 with *oc*:'0 2 5', and *oc*:'0 3 5', have the same *RMP* but the *oc* of node 'c' was different. We do not store this *oc* of node 'c' because 'c' is not in *RMP* of st_1, and hence, is not an extension point from which this particular subtree will be extended to form the 4-subtree candidate. Please note that the shown *VOL* is for the cases when transaction-based or hybrid support definitions are used. In the case of occurrence-match support definition, the first column containing the transaction identifiers would not be needed and is omitted as we are counting the total number of occurrences in the database and the information about the transaction in which an instance of the subtree was found is not necessary.

When the occurrence-match support is used, the frequency of ordered embedded subtree st_1, is equal to the size of the *VOL*, i.e. 8 (Fig. 5.5). When transaction-based support is used, the support of 'st_1' is 2 since there are two transactions (*tid*:1, *tid*:2) that support subtree 'st_1'. When hybrid support definition is considered, two values are taken into account so that, given a hybrid support of $x|y$, a subtree will be considered frequent if it occurs at least y times in x transactions. To enforce this support constraint, a transaction will be considered to support a subtree if there are at least y number of identifiers of that transaction in *VOL*. In other words, a subtree

```
Inputs: Tdb(tree database), σ(min.support), δ(max. level of embedding)
Outputs: Fk(frequent subtrees), RL(recursive List)
{RL, F1, F2}: DatabaseScanningRLConstruction(Tdb, δ)
k = 3

while(|Fk| ≥ 0) {
  Fk = GenerateCandidateSubtrees(Fk-1, δ)
  k = k + 1
}

GenerateCandidateSubtrees(Fk-1, δ):
for each frequent k-subtree tk-1 ∈ Fk-1 {
Lk-1 = GetEncoding(tk-1)
VOL-tk-1 = GetVOL(tk-1)
for each RMP occurrence coordinate RMP-ock-1 (r:[m,…n]) ∈ VOL-tk-1 {
  for (j = n + 1 to scope(r)) {
    {extpoint,slashcount} = CalcExtPointAndSlashcount(RMP-ock-1, j);
    Lk = Lk-1 + append('/', slashcount) + Label(j)

    if(EmbeddingLevel(extpoint, j) ≤ δ) {
      RMP-ock = TMG-extend(RMP-ock-1, j)
      if(Contains(Lk, Fk))
        Insert(h(Lk), RMP-ock, Fk )
    else
      if(AllSubpatternFrequent(Lk)) // all k-1 patterns frequent?
        Insert(h(Lk), RMP-ock, Fk )
      }
    }
  }
}
return Fk

TMG-extend(RMP-ock-1, j):
// right-most-path computation
|RMP-ock|= |RMP-ock-1| + 1 - slashcount; // compute size of RMP coordinate
α = |RMP-ock| - 1; // update the tail index α
RMP-ock[α] = j; // store the new node pos at tail index α
return RMP-ock
```

Fig. 5.6 IMB3-Miner pseudo code

will be considered frequent if there are at least x unique identifiers in *VOL* which repeat at least y times within the *VOL*. Hence, from Fig. 5.5, the hybrid support of the subtree 'st_1' is equal to 2|2 since there are two unique transaction identifiers (*tid*) and they both repeat at least twice. In this example, we have assumed that ordered embedded subtrees were mined. If induced subtrees were mined by setting the maximum level of embedding constraint to 1, then the list would consist of only one occurrence of the subtree st_1 with *oc*:'0 6 8' in T_1 from Fig. 5.5.

The cost of the frequency counting process comes from at least two main areas. First, it comes from the *VOL* construction itself. With numerous occurrences of subtrees, the list can grow to an enormous size. Secondly, for each candidate generated, its encoding needs to be computed. Constructing an encoding from a long tree pattern can be very expensive. An efficient and fast encoding construction can be employed by a step-wise encoding construction so that at each step the computed value is remembered and used in the next step. In this way, a constant processing cost that is independent of the length of the encoding is achieved. Thus, fast candidate counting can occur.

5.3.6 *Pseudo Code*

The pseudo-code of the IMB3-Miner algorithm is given in Fig. 5.6. For mining an embedded subtree, the maximum *level of embedding* is set to $\geq$ the maximum level of embedding in the database or we can set it to a very large number. For mining induced subtrees the maximum of *level of embedding* is set to 1.

5.4 Mathematical Analysis

In this section, we will discuss the mathematical model of the *TMG* enumeration technique and complexity analysis of enumeration of induced and embedded subtrees. Deriving the mathematical model of an approach is always useful for a number of reasons. It can be used in complexity analysis and for enhancing the predictability of the approach. Also, it can be used as a means of validating the correctness of the approach. Furthermore, it functions as an implementation-independent model of the approach. On the other hand, the complexity analysis of enumerating induced and embedded subtrees is provided as a theoretical basis to underline the importance of the *level of embedding* constraint to help us constrain the complexity of mining embedded subtrees to a manageable level. Since an induced subtree is a subset of an embedded subtree, mining an induced subtree is less complex if not much less complex than mining an embedded subtree. The mathematical model of the *TMG* enumeration technique will be presented in Section 5.4.1, followed by the complexity analysis of enumerating induced and embedded subtrees in Section 5.4.2.

5.4.1 *Mathematical Model of TMG*

The TMG enumeration approach belongs to the family of horizontal enumeration approaches. It is an optimal enumeration strategy as it always generates unique subtrees (non-redundant) and it exhausts the search space completely. Due to the fact that the TMG enumeration approach is optimal and all candidate subtrees enumerated by the TMG approach are valid, it is most likely that any other horizontal enumeration approach would need to enumerate at least as many candidates if not more.

There is no simple way to parameterize a tree structure unless it is specified as a *uniform tree*. A *uniform tree T(n,r)* is a tree with height equal to *n* where all its internal nodes have degree *r*. The *closed form* of an arbitrary tree *T* is defined as a

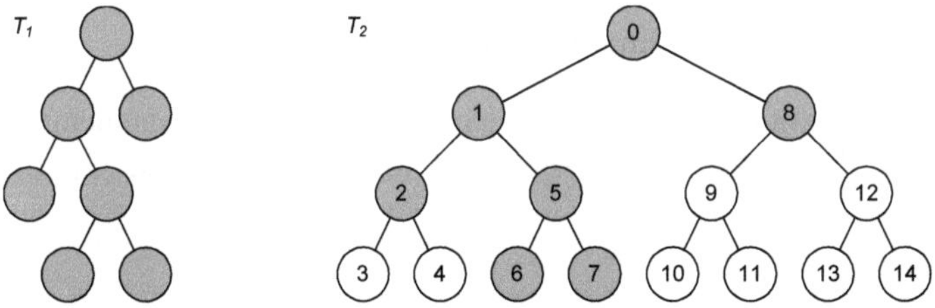

Fig. 5.7 Example of an arbitrary tree T_1 and its closed form T_2 (3,2)

uniform tree with height equal to the height of T and degree equal to the maximum degree of all internal nodes in T. Thus, the worst case complexity of enumerating embedded subtrees from any arbitrary trees is given by their *closed form* (Fig. 5.7). The size of a uniform tree *T(n,r)* can be computed by counting the number of nodes at each depth. For a uniform tree with degree r at depth d there will be r^d numbers of nodes. Hence, there are $r^0 + r^1 + r^2 + \cdots + r^d$ number of nodes in a uniform tree *T(n,r)*. This can be computed using the geometric series formula $(1-r^{n+1})/(1-r)$. When the root node is omitted, the following formula is used, $r(r^n-1)/(r-1)$. If r = 1, the size of the uniform tree is equal to its height n and it becomes a sequence.

Given that that the TMG enumeration approach uses the structural aspect of tree structures to guide the enumeration of subtrees, the enumeration complexity of the TMG enumeration approach is bounded by the actual tree structures rather than by its labelling set.

We define the cost of enumeration as the number of candidate instances enumerated throughout the candidate generation process as opposed to the number of candidate subtrees generated. The mathematical model of the TMG enumeration is formulated as follows: given a *uniform tree T(n,r)*, the worst case complexity of candidate generation of T is expressed mathematically in terms of its height n and degree r.

Throughout this chapter, we assume that all candidate subtrees generated are embedded subtrees. Therefore, unless otherwise stated, candidate subtrees are of embedded subtree type.

The task of frequent subtree mining is to discover all candidate subtrees whose support is equal to or greater than the user-specified minimum support σ. Since we are considering labeled trees, in order to discover such candidate subtrees, we have to count their support by counting the number of occurrences of the candidate subtrees that have the same string encoding. This means that for one candidate subtree with an encoding φ, there can be many instances of this subtree with the same encoding. Henceforth, we refer to an instance of a candidate subtree as a *candidate (subtree) instance*. So, logically the enumeration complexity of enumerating embedded subtrees using the TMG enumeration approach would be expressed as the number of candidate instances rather than the candidate subtrees.

5.4.1.1 Complexity of 1-Subtree Enumeration

Since there are $|T|$number of candidate 1-subtree instances that can be enumerated from a uniform tree T, the complexity of 1-subtree enumeration, denoted as $\|T_1\|$, is equal to the size of the tree $|T|$.

5.4.1.2 Complexity of 2-Subtree Enumeration

In Chapter 4, we mentioned that the *EL* can be constructed by joining all the 2-subtree candidates that have a common root node position and inserting each leaf node in the list with the same *root key*. In other words, all 2-subtrees with root key n are enumerated by constructing the *EL* with root key n and the size of the $EL[n]$ equates to the number of 2-subtrees with the root key n. Therefore, the total sum of the lists' sizes in the *EL* reflects the total number of 2-subtree candidates. Please note that we use notation *EL[n]* to refer to an *embedding list* with a node at position n as its root key, which in this case is the actual root node of the tree.

Let s be a set with n objects. The combinations of k objects from this set s ($_sC_k$) are subsets of s having k elements each (where the order of listing the elements does not distinguish two subsets). The combination $_sC_k$ formula is given by $s!/(s\text{-}k)!k!$. Thus, for 2-subtree enumeration, the following relation exists. Let an $EL[r]$ consist of l number of items; each item is denoted by j. The number of all generated valid 2-subtree candidates (r:[j]) rooted at r is equal to the number of combinations of l nodes from $EL[r]$ having 1 element each. As a corollary, the complexity of 2-subtree enumeration, denoted as $\|T_2\|$, of tree T with size $|T|$is equal to the sum of all generated 2-subtree candidates given by expression 5.1 below.

$$\sum_{r=1}^{|T|} {}_{|EL[r]|}C_1 \tag{5.1}$$

5.4.1.3 Complexity of *k*-Subtree Enumeration

The generalization of 2-subtrees enumeration complexity can be formulated as follows. Let an $EL[r]$ consists of l number of items; each item is denoted by j. The number of all generated valid k-subtree candidates ($[r, e_1, \ldots, e_{k-1}]$) rooted at r is equal to the number of combinations of l nodes from $EL[r]$ having k-1 elements each, $|EL[r]C_{k-1}|$. As defined in Chapter 4, a valid occurrence coordinate of valid candidates has the property that $e_1 < e_2 < \ldots < e_{k-1}$. Thus, the enumeration complexity of k-subtree enumeration can be computed as the sum of all of the generated k-subtree candidates and this is given by the expression 5.2:

$$\sum_{r=1}^{|T|} {}_{|EL[r]|}C_{k-1} \tag{5.2}$$

In the expressions 5.1 and 5.2, the size of each *EL* (*EL*[*r*]) is unknown. If we consider *T* as a uniform tree *T(n,r)*, a relationship between the height *n* and the degree *r* of a uniform tree *T* with the size of each *EL* can be derived.

5.4.1.4 Determining ${}_r\delta^{n-d}$ of the Uniform Tree

A uniform tree *T*(*n*,*r*) can be represented as an *embedding list EL*[*r*]. The size of a uniform tree *T*(*n*,*r*) is governed by the geometric series formula $r(r^n-1)/(r-1)$. Consequently, the corresponding *embedding list EL*[*r*] that represents the uniform tree *T*(*n*,*r*), is the same size as the uniform tree *T*(*n*,*r*). We refer to the *EL*[*r*] as an *embedding list* with the *root key r*.

Let ${}_r\delta^{n-d}$ denote the size of an *embedding list* rooted at a node with depth/level *d* of a uniform tree *T*(*n*,*r*). Since, in a uniform tree *T*(*n*,*r*), there are r^d number of nodes at each level *d* and for each node with pre-order position *i* an *embedding list EL*[*i*] is created, for each level in *T(n,r)* there are r^d number of *embedding lists* that have the same size ${}_r\delta^{n-d}$, as given by expression 5.3.

$$r^d\,{}_r\delta^{n-d} \tag{5.3}$$

Using the fact that for each level in *T(n,r)* there are r^d number of *embedding lists* that have the same size ${}_r\delta^{n-d}$ and there are *n* levels, therefore the total number of candidate *k*-subtree instances that can be generated from a uniform tree *T*(*n*,*r*) can be expressed as shown in expression 5.4.

$$r^0 \cdot {}_{{}_r\delta^n}C_{k-1} + r^1 \cdot {}_{{}_r\delta^{n-1}}C_{k-1} + \ldots + r^n \cdot {}_{{}_r\delta^0}C_{k-1} \tag{5.4}$$

Further, expression 5.4 can be rewritten as expression 5.5:

$$\sum_{i=0}^{n-1} r^i \cdot {}_{{}_r\delta^{n-i}}C_{k-1}, \text{ for } {}_r\delta^{n-i} \geq (k-1) \tag{5.5}$$

Substituting ${}_r\delta^{n-d}$ with $r(r^{(n-d)}-1)/(r-1)$ in expression 5.5 gives us expression 5.6.

$$\sum_{i=0}^{n-1} r^i \cdot {}_{\frac{r(r^{n-i}-1)}{r-1}}C_{k-1}, \text{ for } \frac{r(r^{n-i}-1)}{r-1} \geq (k-1) \tag{5.6}$$

Please note that if the $|EL| < (k-1)$, no candidate subtrees would be generated; thus the constraint ${}_r\delta^{n-i} \geq (k-1)$ takes care of this condition. Hence, using the expressions developed, the complexity of total *k*-subtree candidates from a uniform tree *T(n,r)* for *k*=1,...,|*T(n,r)*|is given by equation 5.7.

$$\sum_{k=1}^{|T(n,r)|} \|T(n,r)\|_k = \|T(n,r)\|_1 + \sum_{k=2}^{|T(n,r)|} \|T(n,r)\|_k \tag{5.7}$$

From expression 5.6, the second term of equation 5.7 can be further expanded as follows to give expression 5.8.

$$\sum_{k=2}^{|T(n,r)|} \|T(n,r)\|_k = r^0 \cdot {}_{\frac{r(r^n-1)}{r-1}}C_{k-1} + \ldots + r^{n-1} \cdot {}_r C_{k-1}, \text{for } \frac{\mathrm{r(r^{n-i}-1)}}{\mathrm{(r-1)}} \geq \mathrm{k-1}$$

$$\sum_{k=2}^{|T(n,r)|} \|T(n,r)\|_k = \sum_{i=0}^{n-1} r^i \left\{ \sum_{k=2}^{\frac{1-r^{n-i+1}}{1-r}} {}_{\frac{r(r^{n-i}-1)}{r-1}}C_{k-1} \right\}$$

$$\sum_{k=2}^{|T(n,r)|} \|T(n,r)\|_k = \sum_{i=0}^{n-1} r^i \left\{ \sum_{k=2}^{A} {}_B C_{k-1} \right\}, A = \frac{1-r^{n-i+1}}{1-r}; B = \frac{r(r^{n-i}-1)}{r-1} \quad (5.8)$$

Finally, equation 5.7 can be restated as equation 5.9:

$$\sum_{k=1}^{|T(n,r)|} \|T(n,r)\|_k = \frac{(1-r^{n+1})}{(1-r)} + \sum_{i=0}^{n-1} r^i \left\{ \sum_{k=2}^{\frac{1-r^{n-i+1}}{1-r}} {}_{\frac{r(r^{n-i}-1)}{r-1}}C_{k-1} \right\} \quad (5.9)$$

Thus, given an arbitrary tree *T* and its closed form *T'(n,r)*, the worst case complexity of enumerating embedded subtrees using the TMG approach can be computed using Equation 5.9 where *n* is the height of *T'* and *r* is the degree of *T'*.

Given a complete tree with degree 2 and height 3 denoted by *T(3,2)*, using Equation 5.9, we could compute that the enumeration cost for generating all possible subtrees is 16,536, i.e. there are 16,536 subtrees enumerated. When the height of the tree is increased by 1, *T(4,2)*, the enumeration cost for generating all possible subtrees is 1,073,774,896. Further, if we increase the degree by 1, *T(3,3)*, the number of subtrees generated blows out to 549,755,826,275.

Although the TMG enumeration approach is optimal, the formula clearly demonstrates that the complexity of generating embedded subtrees from a complete tree structure can be intractable in certain cases. It also suggests that the worst case complexity of enumerating all possible candidates from data in a tree structure form is mainly determined by the structure of the tree (*height* and *degree)*.

Fig. 5.8 shows the enumeration cost graph of a uniform tree *T(3,2)*. The produced curve is not exactly symmetric, i.e. the left hand side of the curve (from the beginning to the peak of the curve) has a slightly higher enumeration cost than does the right hand side of the curve (from the peak to the end of the curve).

It is interesting to analyze the curve produced by equation 5.9 and shown in Fig. 5.8 above for *T(3,2)*. To help us understand the curve, we will use the *embedding list* representation of *T(3,2)* as shown in Fig. 5.9. Please see Fig. 5.7 for the topological structure of the uniform tree *T(3,2)*.

To illustrate how the enumeration cost of *T(3,2)* can be computed, using expression 5.6, we show the computation for k = 2, 3 and 4 and omit the computation for $k > 4$ as this can be easily obtained using the same procedure.

$$\|T(3,2)\|_2 = 2^0 .{}_{14}C_1 + 2^1 .{}_6C_1 + 2^2 .{}_2C_1 = 34$$
$$\|T(3,2)\|_3 = 2^0 .{}_{14}C_2 + 2^1 .{}_6C_2 + 2^2 .{}_2C_2 = 125$$
$$\|T(3,2)\|_4 = 2^0 .{}_{14}C_3 + 2^1 .{}_6C_3 = 404$$

The enumeration cost is governed by the size of each embedding list. Intuitively one can see that for *k*=4, *EL*[2], *EL*[5], *EL*[9], *EL*[12] no longer contribute to the enumeration cost of *T(3,2)* since their size is less than *(k-1)* as explained above (expression 5.5 & 5.6). For the same reason, *EL*[1] and *EL*[8] would cease to contribute to the enumeration cost of *T(3,2)* for $k > 7$. Then, for $7 < k \leq 15$ the enumeration cost simply comes from *EL*[0]. If the lists in the *EL* (Fig. 5.9) are grouped by their size, 3 different categories are obtained. The graphs in Fig. 5.10 are obtained by plotting the enumeration cost for those 3 different categories separately.

Expression 5.1 suggests that the complexity of enumerating 2-subtrees from a tree *T* using the TMG enumeration approach is equal to the sum of each list size in the *EL* of *T*. In other words, the complexity of enumerating 2-subtrees from a tree *T* using the TMG enumeration approach is bounded by $O(C*n)$ where *C* is a constant $< |T|$. Using the join approach (Zaki 2005), the complexity of enumerating

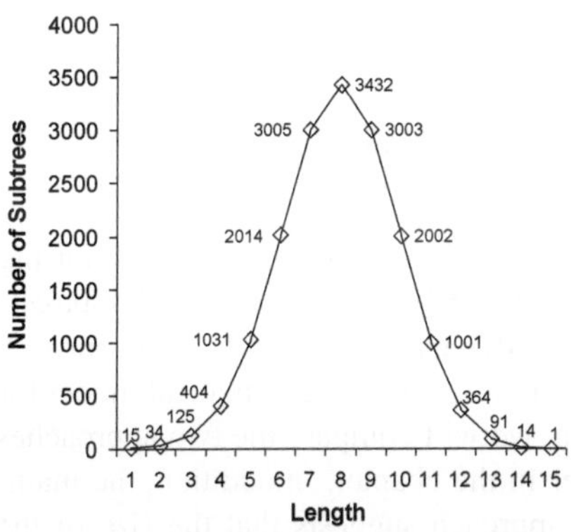

Fig. 5.8 Enumeration cost graph of uniform tree *T(3,2)*

0:	1	2	3	4	5	6	7	8	9	10	11	12	13	14
1:	2	3	4	5	6	7								
2:	3	4												
5:	6	7												
8:	9	10	11	12	13	14								
9:	10	11												
12:	13	14												

Fig. 5.9 Embedding List of a uniform tree *T(3,2)*

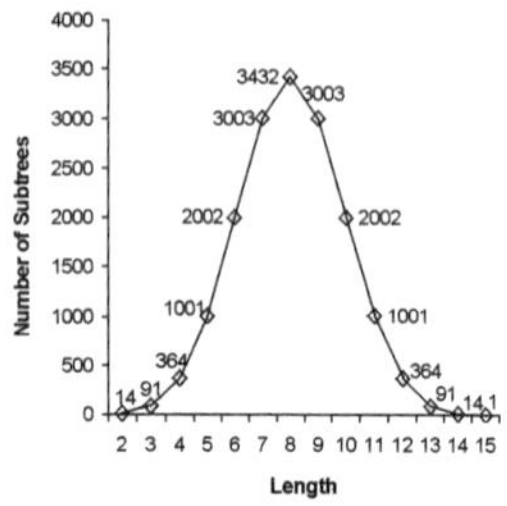

(a) size:14

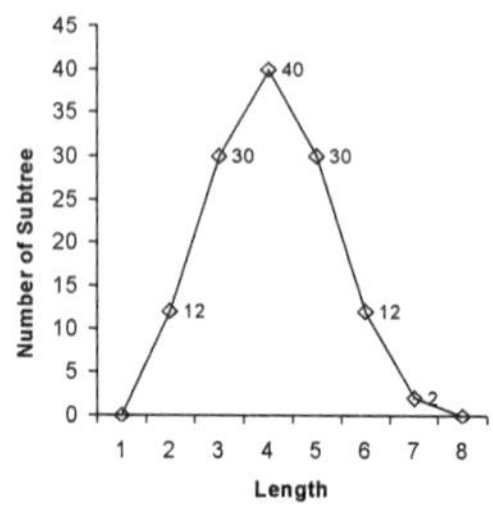

(b) size:6

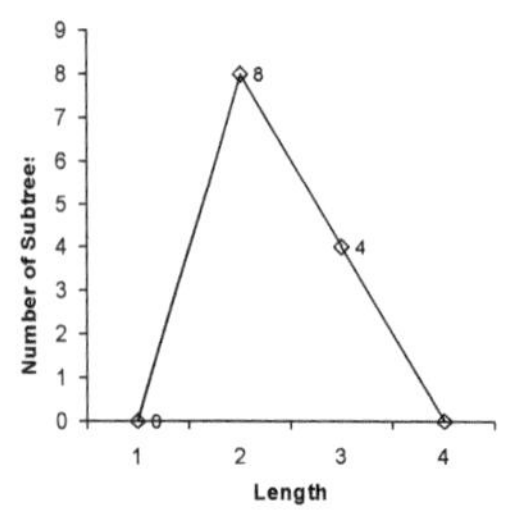

(c) size:2

Fig. 5.10 Enumeration Cost of *T(3,2)* from embedding list in Fig. 5.8

2-subtrees is bounded by O($|T|*|T|$). Thus, the TMG enumeration approach using the tree model guided concept has a lower complexity for enumerating 2-subtrees. It is, however, rather difficult to compare the TMG approach with the join approach for enumerating k-subtrees for $k > 2$, as we do not have a mathematical model for the join approach. For this particular reason, we will compare the two approaches directly through experiments, provided later in the chapter. In addition, the mathematical model of the TMG enumeration approach suggests that the size of the database of trees is not the main factor determining the complexity of enumerating embedded subtrees. The structural property of a tree structure plays a more important role here.

Moreover, equation 5.9 helps us understand that it is far more difficult to mine frequent embedded subtrees from a single transaction of a uniform tree with a large height n and degree r than thousands of transactions of a uniform tree with a small height n and degree r. Let us consider an example. The enumeration cost of *T(3,2)* is 16,536 and the enumeration cost of *T(3,3)* is 549,755,826,275. In other words, the enumeration cost of a database of trees that contain 1 transaction of *T(3,3)* is comparable to the enumeration cost of a database of trees with 33,245,998 transactions of *T(3,2)*. This indicates that mining frequent embedded subtrees is a more difficult problem to solve than mining frequent sequences or induced subtrees.

To conclude this section, it is worth mentioning that even though we have shown here that the complexity of the task can sometimes make it unfeasible, in reality the developed algorithms are still well scalable even for quite large databases. To be able to obtain a mathematical formula for worst case analysis of the *TMG* enumeration approach, we had to assume a uniform tree with support set to one. In real world cases, the tree databases would have varying depths and degrees, and support thresholds would be set higher. The number of candidate subtrees to be enumerated would decrease since many infrequent candidates would be pruned throughout the process. As the frequency distribution is generally not known beforehand, the support threshold cannot be integrated into the TMG enumeration approach mathematical formula. At this stage, we aimed at obtaining some insight into the worst-case complexity of the task and in the future we will strive to obtain an analysis on an average case basis. Hence, despite the large complexity indicated by the formula, the developed tree mining algorithms are still well scalable for large databases of varying depth and degree, as is demonstrated in experiments provided at the end of the chapter.

5.4.2 Complexity Analysis of Candidate Generation of an Embedded/Induced Subtree

This section provides a mathematical analysis for the task of mining induced and embedded subtrees. It starts by deriving a formula for counting the number of induced subtrees from a uniform tree, from our previously obtained formula for counting the number of embedded subtrees. Throughout this process, the complexity relationship between the induced and embedded subtree mining tasks is revealed, and the effect of introducing the level of embedding constraint is discussed. The time and space complexity for the problem of mining frequent subtrees mostly comes from the candidate enumeration and counting phase. Hence, we have used the term complexity here to refer to the total number of candidate subtrees that need to be generated. Because all candidate subtrees enumerated by the TMG enumeration approach are valid, it is most likely that any other method would need to enumerate at least as many candidates. Hence, the mathematical formulas presented here give a good indication of the complexity involved in the task of mining induced/embedded subtrees.

5.4.2.1 Counting Embedded k-Subtrees

The number of different ways of picking p items from n items where the order is not important and each item can be chosen only once, is expressed through the formula given in equation 5.10:

$$_nC_p = \frac{n!}{p!(n-p)!} \tag{5.10}$$

The number of candidate embedded k-subtrees instances that can be generated from a uniform tree T with a known *degree r* and *depth d*, denoted as *T(r,d)* can be computed using equation 5.9.

$$|C_k| = r^0 \cdot_{\frac{r(r^d-1)}{r-1}} C_{k-1} + r^1 \cdot_{\frac{r(r^{d-1}-1)}{r-1}} C_{k-1} + r^{d-1} \cdot_r C_{k-1} \tag{5.11}$$

As already discussed in Chapter 2, an induced subtree preserves the parent-child relationships of each node in the original tree but does not have the concept of ancestor-descendant. Therefore, the visibility between two adjacent nodes (from different levels) is limited by only 1 level. In Chapter 4 (Section 4.5.1), we redefined an induced subtree as a specialization of an embedded subtree where the maximum *level of embedding* (δ) between any two adjacent nodes from different levels is equal to 1. Unless otherwise specified, we will use δ to denote the maximum *level of embedding* between two adjacent nodes from different levels. On the other hand, an embedded subtree additionally preserves the ancestor-descendant relationships that could span over several levels, i.e. $\delta > 1$. In this section, we will derive an equation to compute the total number of candidate induced subtrees from a uniform tree *T(r,d)*. To achieve our objective, we will first express equation 5.11 in a different form. Deriving the induced subtree equation from the embedded subtree equation will allow for a clearer comparison of the complexity of each.

Since any differences between the induced and embedded k-subtree generation start only from $k \geq 2$, we will first formulate the common equation to compute the total number of 1-subtrees from a uniform tree *T(r,d)*. The other way to compute the number of embedded k-subtrees to be generated from a uniform tree *T(r,d)*, is to use a probabilistic approach, i.e. by counting all the possible unique combinations. For the generation of C_1, each node has no concept of children or descendants and there is a ${}_1C_1$ possibility for each node. Thus, the C_1 generation is formulated as given in equation 5.12:

$$|C_1| = r^0 \cdot_{r^0} C_1 + r^1 \cdot_{r^0} C_1 + \ldots + r^d \cdot_{r^0} C_1 = \left(\frac{1-r^{d+1}}{1-r}\right)_{r^0} C_1 = \left(\frac{1-r^{d+1}}{1-r}\right) \tag{5.12}$$

The formula can be abstracted into the form of $r^x \, ({}_nC_p)^+$. To simplify the discussion, we refer to the first term as r^x term and $({}_nC_p)^+$ term for the second. The '+' sign for the second term indicates that there can be one or more combinatorial terms. Equation 5.12 indicates the total number of 1-subtrees $|C_1|$that can be generated by using a geometric series formula. $|C_1|$can be also computed using equation 5.11 by substituting d with 0 and k with 1. The total number of 2-subtrees $|C_2|$is given by the equation 5.13:

$$\begin{aligned} |C_2| = {} & r^0 \cdot {}_1C_1 \cdot_{r^1} C_1 + r^1 \cdot {}_1C_1 \cdot_{r^1} C_1 + \ldots + r^{d-1} \cdot {}_1C_1 \cdot_{r^1} C_1 \\ & + r^0 \cdot {}_1C_1 \cdot_{r^2} C_1 + r^1 \cdot {}_1C_1 \cdot_{r^2} C_1 + \ldots + r^{d-2} \cdot {}_1C_1 \cdot_{r^2} C_1 \\ & + r^0 \cdot {}_1C_1 \cdot_{r^d} C_1 \end{aligned} \tag{5.13}$$

The above equation can be simplified into equation 5.14:

$$|C_2| = \left(\frac{1-r^{(d-\lambda+1)}}{1-r}\right) {}_1C_1 \cdot_{r^\lambda} C_1, \lambda = 1, \ldots, d \tag{5.14}$$

The equation above can be explained as follows. Let us consider a uniform tree with degree 2 and depth 2, *T(2,2)*. Starting with the root node, i.e. the only node at level 0, there are ${}_1C_1 . {}_2C_1$ ways of generating 2-subtrees. Continuing with the next level, level 1, there are r^1 nodes from which we can generate 2-subtrees. Hence, we multiply recursively r^1 again with ${}_1C_1 . {}_2C_1$ number of nodes at the next level, level 2. The recursive stopping condition is met here since no further 2-subtrees can be generated at level = d - 1 = 1. Up to now, we have computed the first two terms from equation 5.13. Since every node in an embedded subtree can have ancestor-descendant relationships, the combinatorial multiplication continues over multiple levels. The number of nodes at each level l can be determined by r^l. So if we substitute l with 2 it will give us 4 nodes at level 2. Hence, going back to level 0, we multiply the root node (${}_1C_1$) with the nodes in level 2 (${}_{2^2}C_1$). As there are no other possibilities, the recursive stopping condition is met and we obtain the total number of embedded 2-subtrees.

The recursive stopping condition in equation 5.14 is met when the r^x term meets the following conditions, $x = (d - \Delta)$ where d is the depth of the tree and Δ is the maximum *level of embedding* between the current root and leaf nodes. In order to determine the Δ, the previous equation needs to be re-factored so that each combinatorial term will have an association with the notion of level (l) such that a multiplication of two combinatorial terms has the following form ${}_{n_i}C_{p_i}^{\lambda_i} \cdot {}_{n_{i+1}}C_{p_{i+1}}^{\lambda_{i+1}}$. Please note that λ is not an absolute level. λ indicates a relative level in respect to the level of the root node of the subtree. The root node of the subtree can be any node from any level and is not necessarily the node from level 0 in the source tree. However, λ_0 will always be 0. Unless otherwise explicitly stated, any reference to the word *level* in our discussion refers to the relative level from the root node. Thus, Δ corresponds to the difference between the level of leaf nodes (last combinatorial term) and the level of the root node (first combinatorial term), which is simply just the level of the leaf nodes. For example, from equation 5.14 the recursive stopping condition for the last combinatorial expression $r^0 . {}_1C_1^0 \cdot {}_{r^2}C_{k-1}^2$ is 0, i.e. the x of the r^x term is equal to 0 as calculated from $d - \Delta = 2 - 2 = 0$.

For obtaining the number of k-subtrees for $k \geq 3$ we can extend equation 5.14. From equation 5.12 and 5.14, we can see that for 1-subtree there is only one ${}_nC_p$ combinatorial expression and likewise for 2-subtree there are two ${}_nC_p$ combinatorial expressions involved respectively. However, the relationship does not say there is 1 combinatorial expression for 1-subtrees and 2 for 2-subtrees and k for k-subtrees. The relationship hints that the terms p in the first, second and k combinatorial expressions of equation 5.14 should add up to 1, 2 and k respectively. Therefore, for k-subtrees there will be up to Ω combinatorial expressions involved such that the terms $p_0+p_1+\ldots+p_\Omega$ add up to k. Also, since each combinatorial expression corresponds to each level in the tree, Ω will be bounded to the depth of the tree, $\Omega \leq d$. With such properties and constraints, i.e. taking into consideration how the recursive stopping condition and Ω should be computed, as well as the fact that the multiplication between nodes for embedded subtrees can span over multiple levels, $\delta > 1$, a formula to compute all embedded 3-subtrees from a uniform tree *T(r,d)* can be developed as follows.

$$
\begin{aligned}
|C_3| &= r^0 \cdot {}_1C_1^0 \cdot {}_{r^1}C_2^1 + r^1 \cdot {}_1C_1^0 \cdot {}_{r^1}C_2^1 + \ldots + r^{d-1} \cdot {}_1C_1^0 \cdot {}_{r^1}C_2^1 \\
&+ r^0 \cdot {}_1C_1^0 \cdot {}_{r^2}C_2^2 + r^1 \cdot {}_1C_1^0 \cdot {}_{r^2}C_2^2 + \ldots + r^{d-2} \cdot {}_1C_1^0 \cdot {}_{r^2}C_2^2 \\
&\ldots \\
&+ r^0 \cdot {}_1C_1^0 \cdot {}_{r^d}C_2^d. \\
&+ r^0 \cdot {}_1C_1^0 \cdot {}_{r^1}C_1^1 \cdot {}_{r^2}C_1^2 + \ldots + r^{d-2} \cdot {}_1C_1^0 \cdot {}_{r^1}C_1^1 \cdot {}_{r^2}C_1^2 \\
&+ r^0 \cdot {}_1C_1^0 \cdot {}_{r^1}C_1^1 \cdot {}_{r^2}C_1^3 + \ldots + r^{d-3} \cdot {}_1C_1^0 \cdot {}_{r^1}C_1^1 \cdot {}_{r^3}C_1^3 \\
&\ldots \\
&+ r^0 \cdot {}_1C_1^0 \cdot {}_{r^1}C_1^1 \cdot {}_{r^d}C_1^d \\
&+ r^0 \cdot {}_1C_1^0 \cdot {}_{r^2}C_1^2 \cdot {}_{r^3}C_1^3 + \ldots + r^{d-3} \cdot {}_1C_1^0 \cdot {}_{r^2}C_1^2 \cdot {}_{r^3}C_1^3 \\
&\ldots \\
&+ r^0 \cdot {}_1C_1^0 \cdot {}_{r^2}C_1^2 \cdot {}_{r^d}C_1^d \\
&\ldots \\
&+ r^0 \cdot {}_1C_1^0 \cdot {}_{r^{d-1}}C_1^{d-1} \cdot {}_{r^d}C_1^d
\end{aligned}
$$

It can be seen from the above equation that each $r^x.({}_nC_p)^+$ term can occur recursively and as described earlier, the recursive stopping condition is met when the x in r^x term = d - Δ such that we can compute such recursive occurrence by using the geometric series $\left(\frac{1-r^{(d-\Delta)+1}}{1-r}\right)$. Hence, the above equation can be simplified to give equation 5.15:

$$
\begin{aligned}
|C_3| &= \left(\frac{1-r^{(d-\lambda)+1}}{1-r}\right) {}_1C_1^0 \cdot {}_{r^\lambda}C_2^\lambda + \left(\frac{1-r^{(d-\theta)+1}}{1-r}\right) {}_1C_1^0 \cdot {}_{r^\phi}C_1^\phi \cdot {}_{r^\theta}C_1^\theta; \\
&\phi < \theta; \phi = 1, \ldots, (\theta - 1); \theta = (\phi + 1), \ldots, d; \lambda = 1, \ldots, d
\end{aligned}
\tag{5.15}
$$

The above equation assumes that the depth of the subtrees is ≥ 3 as otherwise there will not be the multiplication between 3 different levels ${}_{n_1}C_{p_1}^0 \cdot {}_{n_2}C_{p_2}^\phi \cdot {}_{n_3}C_{p_3}^\theta$. If the depth of the tree is d then there can be a maximum d number of multiplications between combinatorial terms.

Now to further generalize for all k where k=1,... $|T(r,d)|$for $|T(r,d)|$ can be computed using the geometric series $\left(\frac{1-r^{d+1}}{1-r}\right)$, the combinatorial expression can be given by equation 5.16.

$$
\prod_{i=0}^{\Omega,\alpha} \left({}_{n_i}C_{p_i}^{\lambda_i}\right) = {}_{n_0}C_{p_0}^{\lambda_0} \cdot {}_{n_1}C_{p_1}^{\lambda_1} \cdots {}_{n_\Omega}C_{p_\Omega}^{\lambda_\Omega}
\tag{5.16}
$$

Where $n_i \geq p_i$ and Ω is a positive integer such that $\Omega \leq d$ and $p_0 + p_1 + \cdots + p_\Omega = k$. Additionally, α is an update function such that $n_i = n_{i-1} r^\lambda$ where $1 \leq \lambda \leq d$ and $\lambda_i < \lambda_{i+1}$.

Since a subtree can occur recursively $\left(\frac{1-r^{d-\Delta+1}}{1-r}\right)$ times within the original tree where the recursive stopping condition depends on the level of embedding between the root node and the leaf nodes (Δ), integrating Δ into equation 5.16 determines all possible embedded k-subtrees that can be generated from a uniform tree $T(r,d)$, and this is given in equation 5.17.

$$|C_k| = \left(\frac{1-r^{(d-\lambda_\Omega)+1}}{1-r}\right)\left\{\prod_{i=0}^{\Omega,\alpha}\left({}_{n^i}C_{p_i}^{\lambda_i}\right)\right\}, \Delta = \lambda_\Omega - \lambda_0 = \lambda_\Omega \qquad (5.17)$$

5.4.2.2 Counting Induced Subtrees by Enforcing Constraint

An induced subtree is an embedded subtree where the maximum level of embedding between any two adjacent nodes (from different levels) is equal to 1. To compute all induced k-subtrees that can be generated from a uniform tree $T(r,d)$, the maximum level of embedding constraint (δ = 1) between any two adjacent nodes needs to be enforced. In the following section, we will show how such a constraint can be integrated with the previous equations derived for embedded subtrees, starting with $k = 2$ and $k = 3$ and then generalizing for all k.

$$|C_2| = \left(\frac{1-r^{(d-\Delta)+1}}{1-r}\right){}_1C_1^{\lambda_0} \cdot {}_{r^1}C_1^{\lambda_0+1}, \Delta = (\lambda+1)-\lambda = 1 \qquad (5.18)$$

We will use the example similar to the one used for embedded subtrees above to illustrate how we can enforce the level of embedding and explain 5.18 above. Suppose that we are using a uniform tree $T(2,2)$ as an example and set δ to 1. Starting with the root node, i.e. the only node at level 0, there are ${}_1C_1{}_2C_1$ ways of generating 2-subtrees. Continuing with level 1, there are r^1 nodes from which we can generate 2-subtrees, and hence we multiply r^1 again with ${}_1C_1 \cdot {}_2C_1$ number of nodes at the next level, level 2. The recursive stopping condition is met here since no further 2-subtrees can be generated at level (d - 1) = 1. When δ is constrained to 1, by definition, the multiplication between two nodes can only happen between nodes from two consecutive levels. This reduces the scope of multiplication between nodes from different levels, which implies that the number of subtrees generated will be less when such a constraint is enforced. We will be using the example from Fig. 5.11, to clarify the point. In the case of embedded subtrees where the level of embedding is not restricted, node N_0 (left hand side of Fig. 5.11) from level 0 can be multiplied with r^1 number of nodes from level 1 and r^2 number of nodes from level 2, whereas when the δ is restricted to 1, node N_0 from level 0 can be multiplied with only r^1 number of nodes from level 1 (right hand side of Fig. 5.11).

Let us denote nodes at level i as N_i and the level of embedding between two nodes x & y as function $\Delta(x,y)$. Using this example, we can see that for each multiplication, the specified level of embedding constraint δ, has to be satisfied, i.e. $\Delta\ (N_i,N_j) \leq \delta$. From Fig. 5.11, in the case of embedded subtrees (left), there are $r^1 + r^2$ numbers of nodes available for selection whereas for the case of induced subtrees (right), there are only r^1 numbers of nodes available for selection when extending N_0 since only nodes $\Delta\ (N_0,N_1) \leq \delta$ whereas $\Delta\ (N_0,N_2) > \delta$. Consequently if we convert it into

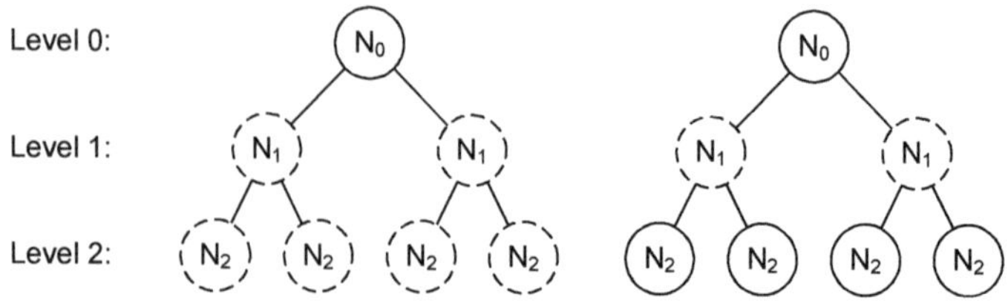

Fig. 5.11 Illustration of the scope of multiplication between N_1 and N_2 of embedded subtrees (left) and induced subtrees (right)

a combinatorial expression the embedded subtree case resolves to ${}_1C_1^0 \cdot {}_2C_1^1 \cdot {}_4C_1^2$ and the induced subtree case resolves to ${}_1C_1^0 \cdot {}_2C_1^1 \cdot {}_2C_1^2$.

Please note that equation 5.18 is basically equation 5.14 with fewer terms without the combinatorial multiplication between nodes at level λ_i and node(s) at level λ_j for $\Delta\ (\lambda_j, \lambda_i) > \delta$.

A similar deduction can be applied to equation 5.15 to obtain an equation for determining the number of all embedded 3-subtrees where maximum level of embedding is equal to 1 (i.e. induced 3-subtrees), and this is given by:

$$|C_3| = r^0 . {}_1C_1^0 \cdot {}_{r^1}C_2^1 + r^1 . {}_1C_1^0 \cdot {}_{r^1}C_2^1 + \ldots + r^{d-1} . {}_1C_1^0 \cdot {}_{r^1}C_2^1$$
$$+ r^0 . {}_1C_1^0 \cdot {}_{r^1}C_1^1 \cdot {}_{r^1}C_1^2 + \ldots + r^{d-2} . {}_1C_1^0 \cdot {}_{r^1}C_1^1 \cdot {}_{r^1}C_1^2$$

This can be simplified to give equation 5.19:

$$|C_3| = \left(\frac{1 - r^{(d-1)+1}}{1-r}\right) {}_1C_1^0 \cdot {}_{r^1}C_2^1 + \left(\frac{1 - r^{(d-2)+1}}{1-r}\right) {}_1C_1^0 \cdot {}_{r^1}C_1^1 \cdot {}_{r^2}C_1^2 \qquad (5.19)$$

With the introduction of the level of embedding constraint into the formula, the level of embedding between any two adjacent nodes needs to satisfy the specified constraint. Here we are considering only where the level of embedding is constrained to 1. In this case the Δ between two combinatorial terms must not be greater than 1. If the multiplication between two nodes from two different levels λ_i and λ_j is represented by the multiplication of two combinatorial terms ${}_{n_1}C_{p_1}^{\lambda_i} \cdot {}_{n_2}C_{p_2}^{\lambda_j}$ and the multiplication between three nodes from three different levels λ_i, λ_j and λ_k is represented by the multiplication of three combinatorial terms ${}_{n_1}C_{p_1}^{\lambda_i} \cdot {}_{n_2}C_{p_2}^{\lambda_j} \cdot {}_{n_3}C_{p_3}^{\lambda_k}$ then $\Delta(\lambda_j, \lambda_i)$ and $\Delta(\lambda_k, \lambda_j)$ must be 1.

Equation 5.20 is a count of all embedded k-subtrees that can be generated from a uniform tree *T(r,d)* for which level of embedding constraint is set to 1. Equation 5.20 has a similar form to equation 5.17, with the change consisting of the additional rule that enforces the level of embedding constraint δ:1 between adjacent nodes from different levels.

$$|C_k| = \left(\frac{1 - r^{(d-\lambda_\Omega+1)}}{1-r}\right) \left\{ \prod_{i=0}^{\Omega,\alpha,\beta} \left({}_{n^i}C_{p_i}^{\lambda_i}\right) \right\}, \Delta = \lambda_\Omega - \lambda_0 = \lambda_\Omega \qquad (5.20)$$

Where $n \geq p$ and Ω is a positive integer such that $\Omega \leq d$ and $p_0 + p_1 + \cdots + p_\Omega = k$ for $k = 1,\ldots,|T(r,d)|$. For equation 5.20, α is an update function such that $n_i = p_{i-1}r$. β is a constraint such that for two adjacent combinatorial multiplication ${}_{n_1}C^{\lambda_i}_{p_1} \cdot {}_{n_2}C^{\lambda_j}_{p_2}$, $\Delta(\lambda_j, \lambda_i) = 1$. Also, $\lambda_i < \lambda_{i+1}$. The following section will demonstrate the difference between the equations using an example to show the varying degree of complexity, as one progressively moves from induced to embedded subtrees.

5.4.2.3 Counting Induced/Embedded Subtrees

In this section, we will show how we can use the previous equations to count all embedded k-subtrees from a uniform tree *T(2,2)* as shown in Fig. 5.12. We will first set the maximum level of embedding constraint to 1 and then increase it to 2 to demonstrate how the level of embedding is used to control the complexity of generating embedded subtrees.

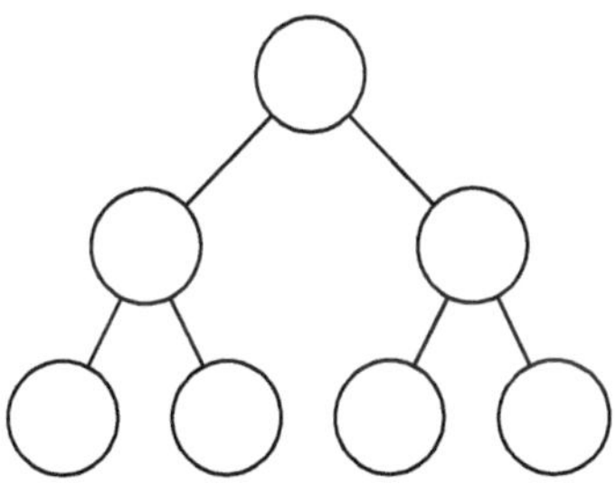

Fig. 5.12 A uniform tree *T(r,d)* with degree *r*:2 and depth *d*:2

The following shows the calculation of $|C_1|, |C_2|, \ldots, |C_7|$ of embedded subtrees that can be generated from *T(2,2)* for which the maximum level of embedding δ:*1*:

$$
\begin{aligned}
&|C_1| : 2^0.{}_1C_1 + 2^1.{}_1C_1 + 2^2.{}_1C_1 = 7\\
&|C_2| : 2^0.{}_1C_1.{}_2C_1. + 2^1.{}_1C_1.{}_2C_1 = 6\\
&|C_3| : 2^0.{}_1C_1.{}_2C_2 + 2^1.{}_1C_1.{}_2C_2 + {}_1C_1.{}_2C_1.{}_2C_1 = 7\\
&|C_4| : 2^0.{}_1C_1.{}_2C_1.{}_2C_2 + 2^0.{}_1C_1.{}_2C_2.{}_4C_1 = 6\\
&|C_5| : 2^0.{}_1C_1.{}_2C_2.{}_4C_2 = 6\\
&|C_6| : 2^0.{}_1C_1.{}_2C_2.{}_4C_3 = 4\\
&|C_7| : 2^0.{}_1C_1.{}_2C_2.{}_4C_4 = 1
\end{aligned}
$$

All the counted k-subtrees are generally known as induced subtrees. There are a total of 37 induced subtrees generated from uniform tree *T(2,2)*.

Next, we will increase the maximum level of embedding δ to 2, and this gives:

$$|C1| : 2^0._1C_1 + 2^1._1C_1 + 2^2._1C_1 = 7$$
$$|C2| : 2^0._2C_1 + 2^1._2C_1 + 2^0._1C_1._4C_1 = 10$$
$$|C3| : 2^0._1C_1._2C_2 + 2^1._1C_1._2C_2 + 2^0._1C_1._2C_1._4C_1 + 2^0._1C_1._4C_2 = 17$$
$$|C4| : 2^0._1C_1._2C_1._4C_2 + 2^0._1C_1._2C_2._4C_1 + 2^0._1C_1._4C_3 = 20$$
$$|C5| : 2^0._1C_1._2C_1._4C_3 + 2^0._1C_1._2C_2._4C_2 + 2^0._1C_1._4C_4 = 15$$
$$|C6| : 2^0._1C_1._2C_1._4C_4 + 2^0._1C_1._2C_2._4C_3 = 6$$
$$|C7| : 2^0._1C_1._2C_2._4C_4 = 1$$

As expected, with the increase in the maximum level of embedding δ there are more embedded subtrees counted with a total of 76. In fact, by setting the maximum level of embedding to 2 all the counted k-subtrees are generally known as embedded subtrees, since $\delta = d$ where d is the depth of the tree *T(n,d)*.

The graph in Fig. 5.13 shows the number of embedded subtrees of varying lengths generated from a uniform tree *T(2,2)* at two different maximum levels of embedding δ, 1 and 2.

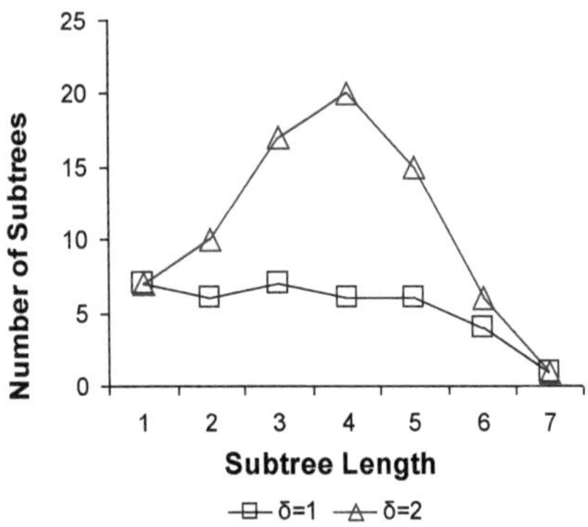

Fig. 5.13 Comparison of the numbers of embedded subtrees generated between δ:*1* (induced) and δ:*2* (embedded)

Throughout this section we have seen how the level of embedding constraint can be utilized to contain the complexity of mining embedded subtrees at a manageable level. If the maximum level of embedding δ is set to 1 then all the embedded subtrees obtained are in fact induced subtrees.

5.5 Experimental Evaluation and Comparisons

In this section, we will evaluate the proposed approach for mining frequent ordered subtrees by comparing them with other algorithms such as PatternMatcher (Zaki 2005), VtreeMiner (Zaki 2005) and FREQT (Asai et al. 2002). Among those algorithms, VTreeMiner is considered as the current state-of-the-art technique in the literature and is one of the first techniques proposed for solving the problem of mining frequent ordered embedded subtrees.

The section outline is as follows. Firstly, we will discuss the method of experimental comparison in Section 5.5.1. Then we will cover some implementation issues in Section 5.5.2. Experimental results and discussions will then be provided in Section 5.5.3.

5.5.1 The Rationale of the Experimental Comparison

In general, experimental comparisons are an important part of data mining algorithm research. In the area of mining frequent patterns from large database, experimental comparisons have been widely used as a means of confirmation, validation, and comparison of the proposed techniques against the theoretical framework, meeting certain correctness criteria, and the state-of-the-art techniques respectively.

Our performance criteria for assessing the algorithms are built upon the concepts of *efficiency*, *scalability*, *predictability*, and *extendibility*. The *efficiency* in the context of this work corresponds to the capability of an algorithm to complete the task in a reasonable amount of time. One way to measure efficiency is to benchmark against state-of-the-art algorithms in the same class. These algorithms have been considerably reviewed and acknowledged in the current literature and hence make good candidates for evaluation of the developed approach. In this sense, the efficiency is considered relative rather than absolute and is acceptable when an algorithm's performance is comparable to, or better than, the performance of current state-of-the-art algorithms in the same field.

Scalability of an algorithm can correspond to two aspects. One is when an algorithm shows linear performance on the datasets of different size and complexity of the underlying tree structure. Another way in which an algorithm can be considered scalable is if it shows linear performance when the number of frequent patterns to be extracted from a particular dataset is increased. In the context of tree mining, this corresponds to reducing the minimum support threshold σ, as with smaller σ, many more patterns will be considered as frequent. To evaluate the scalability, the comparisons of algorithms will be performed on a wider range of synthetically generated and real-world datasets of varying size and complexity, and the minimum support threshold σ will be varied for each dataset.

A framework is said to be predictable whenever its behaviour is measurable across broad characteristics (size, granularity, complexity) of data with different complexity, whether it is artificial data or real-world data. Also, the boundaries and limitations of the system are known and an alternative solution exists whenever such limitations are encountered.

An algorithm is considered as extendible in the sense that minimal effort is required to adjust the general framework so that different but related problems can be solved.

With the above definition of performance, we do not limit our criteria when measuring performance solely based on the timing variable. Timing is an important criterion to consider, however, it becomes less important if it does not handle other criteria gracefully. A framework that is faster than other frameworks only on data with certain characteristics, but very poor on different data, is a poor framework.

A framework that is fast and efficient on small data, but cannot deal with a large dataset is usually not considered useful. A framework that provides solutions for a problem but sees no way to tackle similar problems or sub-problems of the domain of interest, will have only a short lifetime and will not be extensively applied in both scientific and industrial domains.

Many researchers have previously used artificial datasets for making experimental comparisons. An advantage of using artificial datasets is that they give more freedom to embed different desired characteristics into the dataset to allow explanation of certain issues. It also allows us to design the dataset in a bottom-up approach that helps researchers to meet certain goals they want to achieve from such a dataset or to highlight certain characteristics of their approaches. However, artificial datasets can sometimes be oversimplified and might not encode the same level of complexity as does the real dataset. On the other hand, a real-world dataset would normally have certain complexity that is harder to predict and in many cases it can be very difficult to deal with. Therefore, it is always important that real datasets be used in any experimental comparisons. More importantly, most of the proposed algorithms are to be used for solving real-world problems and hence, it is always a good benchmarking mechanism to run it against the real-world dataset that is part of the real-world problems.

In our experimental comparison, we will utilize both artificial and real-world datasets. As performing comparisons on restricted datasets is not sufficient, and would have limited value for drawing reliable conclusions about the relative performance of algorithms, a variety of datasets, covering a broad range of characteristics that are important in the domain, are chosen for experimental comparison. Thus, for the performance evaluation of the proposed framework, the datasets are designed and chosen with different mixtures of attribute types (high fan-out, deep tree, mixed, long tree, sparse, dense, small labels, large labels, etc), different sizes of dataset, and different levels of complexity.

5.5.2 *Implementation Issues*

The theoretical aspects of an algorithm are important. On the other hand, an algorithm without an implementation has no real value. Implementation issues are seldom discussed when describing and comparing various algorithms. We have found that if some implementation issues are overlooked, the performance can be greatly affected. In fact, implementation can be the decisive factor when we are talking about the performance. In this section, we discuss some of the implementation issues that we consider to be both important and common when implementing tree mining algorithms.

5.5.2.1 Accelerating the Object Oriented (OO) Approach

The use of the OO development approach is one of the common practices today in software development and it is claimed that it results in a more manageable, extensible and easy-to-understand code. We have developed our algorithms in C++ making use of some of the OO features of the language such as class and function overloading. There are a few important things worth noting when implementing time-critical systems using the OO approach. Constructor calls on objects have overheads and one should keep them as low as possible. Inheritance between objects should be avoided whenever possible, if real-time performance is important as it degrades this. We found that performing equivalent computations through primitive objects such as array of integers, instead of thick objects such as vector objects or linked list objects, could improve the performance, especially when we need to construct a large number of objects in memory. Another reason not to use thick objects is that they may implement inheritance. An additional way to avoid expensive constructor calls is by storing only a hyperlink (Wang et al. 2004) or a kind of pointer to the existing object and using that pointer to access information from the same object. Another well known way to increase the performance is by passing an object through function by reference instead of by value. Moreover, when copying a block of memory from one location to another location, performing the operation through the use of a memory block copying routine such as memcpy in C library instead of using the *for loop,* could also be the next step in performance tweaking. Last but not least, one can consider writing inline functions when they are called very frequently. However, it is not always a good practice to create all functions in inline mode.

5.5.2.2 Choosing Hash Functions

In the context of mining frequent patterns, one of the most common approaches to perform frequency counting is to use a hash table. When using a hash table, it is very important to choose a good hash function. Unfortunately, choosing a hash function can be more than a trivial task (Jenkins 1997). The following are several known hash functions that were compared: Rotation Hash (RH), Additive Hash (AH), Bernstein Hash (BH), Zobrist Hash (ZH), and One-At-A-Time Hash (OH). Detailed descriptions of each hash function and a few other hash functions can be found in (Jenkins 1997). To discover the effects of using different hash functions, we ran experiments on the CSLogs data (Zaki 2005).

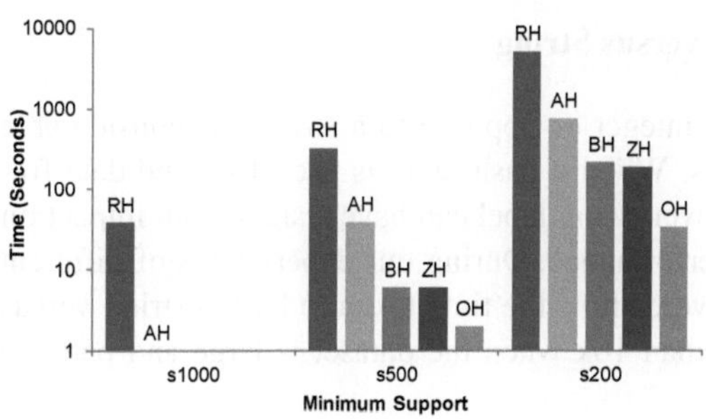

Fig. 5.14 Number of collisions plotted over different minimum supports

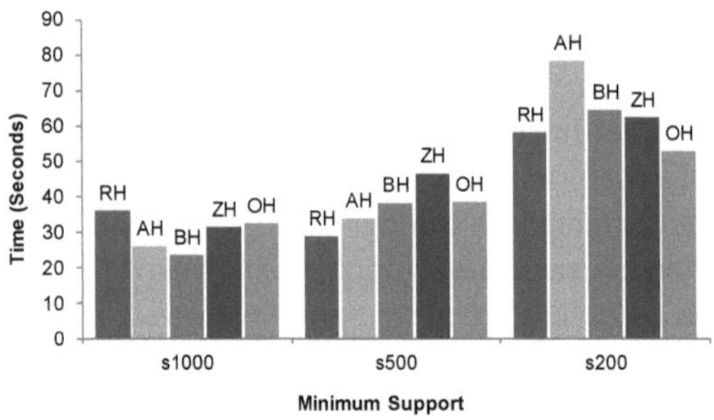

Fig. 5.15 Time performance plotted over different minimum supports

From Fig. 5.14 and Fig. 5.15 it is not immediately evident which function is truly the winner. When minimum support is set to 1000, BH is the fastest. At minimum support 500 and 200, RH and OH are the fastest respectively. In terms of the number of collisions produced, RH is consistently the worst for different minimum support and OH seems to be the best of all. One interesting point to note is that at minimum support 500, RH has a faster execution time than does RH at minimum support 1000, whereas the other functions seem to have increasing execution time when the minimum support threshold is decreased. There is no direct explanation for this but we can see an indication that certain functions can have varying performances at different minimum support thresholds. In general, simple functions such as AH and RH are not recommended as they do not handle collisions very well (Jenkins 1997). Our experiment supports this view. AH and RH are the two functions that produce the worst number of collisions (Fig. 5.14). From Fig. 5.14 and Fig. 5.15 we can infer that BH and OH are the two top performers. For all the experiments used in later sections, we use BH. BH can produce fewer collisions than a hash that gives a more truly random distribution if all 32 bits in the keys are used (Jenkins 1997). However, if not all the 32 bits are used, this function has detected flaws. There is a possibility that if a better hash function is used, further optimization can be attained.

5.5.2.3 Hashing Integer Array versus String

Representing labels of trees as an integer as opposed to a string has considerable performance and space advantages. When a hash table is used for candidate frequency counting, hashing integer over string label can have a significant impact on the overall candidates counting performance. During our experiments of different implementation choices, we discovered how the time taken to hash a string versus and integer could differ by more than 10x when the dataset is large and patterns become relatively long.

Label sensitivity Test

Let us consider a very large database of integer-labeled trees with large labels set. In this case, the labels can be a very big integer value. We performed a label sensitivity test and created four synthetic datasets by varying the maximum integer label values: 24; 24,000; 24,000,000; 240,000,000. It is important to see that the algorithms can handle databases of small and large integer-labeled trees.

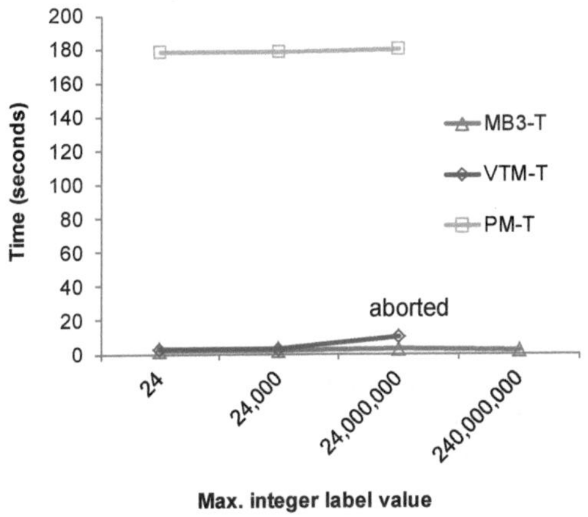

Fig. 5.16 Label sensitivity test

As we can see from Fig. 5.16, MB3 can handle both small and large integer-labeled trees very well. The performance of MB3 remains the same for all 4 different datasets. On the contrary, we find that both TreeMiner algorithms VTreeMiner (VTM) and PatternMatcher (PM) suffer performance degradation whenever the maximum value of the integer-labeled trees is increased. Surprisingly, for the last dataset with maximum integer-labeled value equal to 240,000,000 both VTM and PM were aborted. We observe that the implementation of the TreeMiner for generating 1 and 2-subtrees employs a perfect hashing scenario using array objects and using the label as the key for each cell in the array. What essentially happens with this implementation is that it is performance-optimized but space-inefficient. In the last scenario where the maximum label can reach up to 240,000,000, VTM and PM will unnecessarily allocate an array with 240,000,001 cells even when there is only one node with label equal to 240,000,000 in the tree database. Using a hash table, as opposed to using an array-based implementation, would be a better approach whenever the number of items to be counted is huge.

5.6 Experimental Results and Discussion

We divide our experimental results into two parts, to distinguish the two different versions we used when implementing the framework. The first experiment set shows the comparison of the MB3 (Tan et al. 2005a) and iMB3 Miner (Tan et al. 2006) with other approaches such as FREQT (Asai et al. 2002), VTreeMiner and PatternMatcher (Zaki 2005). The MB3 and iMB3 algorithms are implementations that utilize the *dictionary* and the *embedding list* (*EL*) structures, explained in Chapter 4. In addition, the set of full occurrence coordinates were stored in the *vertical occurrence list* (VOL) (see Chapter 4), rather than only the right-most path occurrences as explained in this chapter. In the second experiment set, we show experimental results of the most optimized versions of the TMG framework implementation for the mining of ordered induced/embedded subtrees (as explained in this chapter). For generating artificial datasets, we use the TreeGen program that was developed by (Zaki 2005). Throughout our book, we will describe all datasets generated by TreeGen using the following parameters:

|Tr|: Number of transactions
|T|: Number of nodes in a transaction
|D|: Depth
|F|: Fan-out factor
|N|: Number of items

Additionally, Table 5.1 describes the characteristics of all the datasets used for experimental comparisons throughout this chapter.

CSLogs dataset is a real-world data set previously used by (Zaki 2005) for testing VTM and PM using transaction-based support and since then has been used in many experimental studies (Chi et al. 2005; Tatikonda, Parthasarathy, & Kurc 2006; Wang et al. 2004). The CSLogs dataset contains one month of web access trees from the computer science department of the Rensselaer Polytechnic Institute. There are a total of 59,691 transactions and 13,209 unique vertex labels (corresponding to the URLs of the web pages). The average string encoding length for the data set is 23.3 (Chi et al. 2005). In (Zaki 2005), the experiments were conducted using transaction-based support. When used for occurrence-match support, the tested algorithms had problems in returning frequent subtrees including the TreeMiner algorithms. The dataset was then progressively reduced until interesting results appeared at 32,241 transactions. The Prions dataset describes a protein ontology database for Human Prion proteins in XML format (Sidhu et al. 2005). The remaining datasets are artificial datasets generated to test the algorithms on a variety of characteristics of tree structures, namely: deep trees, wide trees, mix (deep-wide-short-long), large dataset, dense, and sparse. The artificial datasets in general are sparser than the real-world CSLogs dataset.

5.6.1 Experiment Set I

We compare iMB3 Miner (iMB3), FREQT (FT) for mining induced subtrees and MB3-Miner (MB3), VTreeMiner (VTM) and PatternMatcher (PM) for mining embedded subtrees. We created an artificial database of trees with varying: max. size (s), max. height (h), max. fan-out (f), and number of transactions ($|Tr|$). Notation XXX–T, XXX-C, and XXX–F are used to denote execution time (including data

Table 5.1 A table describing the characteristics of the datasets used for experimental comparisons

Dataset	|N|	|Tr|	Avg |T|	Avg |D|	Avg |F|	Max |T|	Max |D|	Max |F|
CSlogs	10698	59,691	7.46	2.56	2.3	218	31	60
Prions	46,851	17,511	12.97	1	11.98	19	1	18
Deeptree	100	10,000	27.31	16.66	1.27	28	17	3
Deeptree2	9724	20,000	34.23	15.38	1.21	63	17	3
Widetree	10000	6,000	217.27	6.19	9.97	428	9	69
Widetree2	10001	10,000	217.245	6.19	9.97	428	9	69
Mixed1	10000	76,000	27.63	4.9	2.50	428	17	69
Mixed2	100	26,550	8.15	2.98	1.91	233	24	7
100,000s	50	100,000	6.98	2.37	1.94	10	3	4
500,000s	50	500,000	6.98	2.37	1.94	10	3	4
1Ms	50	1000,000	6.98	2.37	1.94	10	3	4
D40F40T100KM200N100	24	100,000	2.54	1.22	0.80	19	12	4
D40F40T500KM200N100	24	500,000	2.53	1.23	0.80	19	12	4
D40F40T1000KM200N100	24	1000,000	2.53	1.23	0.79	19	12	4

preprocessing, variables declaration, etc), number of candidate subtrees $|C|$, and the number of frequent candidate subtrees $|F|$ obtained from XXX approach respectively. Additionally, iMB3-(NP)-dx notation is used where x refers to the maximum level of embedding Φ and (NP) is optionally used to indicate that full pruning is not performed. The minimum support σ is denoted as (sxx), where xx is the minimum frequency. Occurrence-match support was used for all algorithms unless it is indicated that the transaction-based support is used. Experiments were run on 3Ghz (Intel-CPU), 2Gb RAM, Mandrake 10.2 Linux machine and compilation were performed using GNU g++ 3.4.3 with the –O3 parameter. We use the Kudo's FT implementation (Kudo 2003), which was also used by Chi (Chi et al. 2005) and disable the output mode so that it does not spend its execution time printing all the frequent subtrees when timing its performance.

5.6.1.1 Scalability Test

The datasets used for this experiment are 100Ks, 500Ks, and 1000Ks with σ set to 25, 125 and 250, respectively. From Fig. 5.17(a) we can see that all algorithms are scalable. MB3 outperforms VTM and PM for mining embedded subtrees and

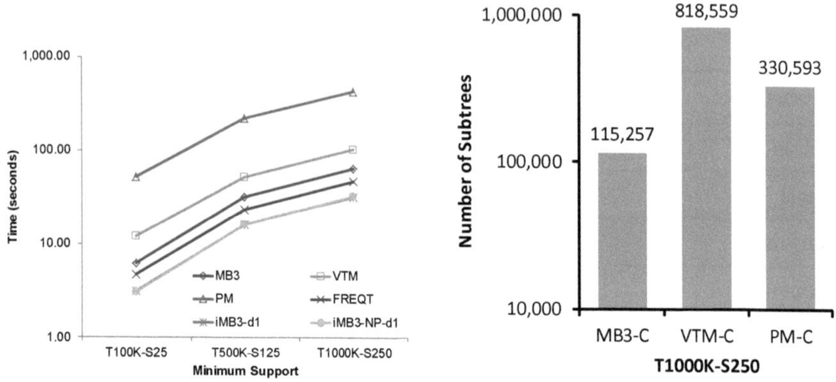

(a) (b)

Fig. 5.17 Scalability test: (a) time performance (b) number of subtrees $|C|$

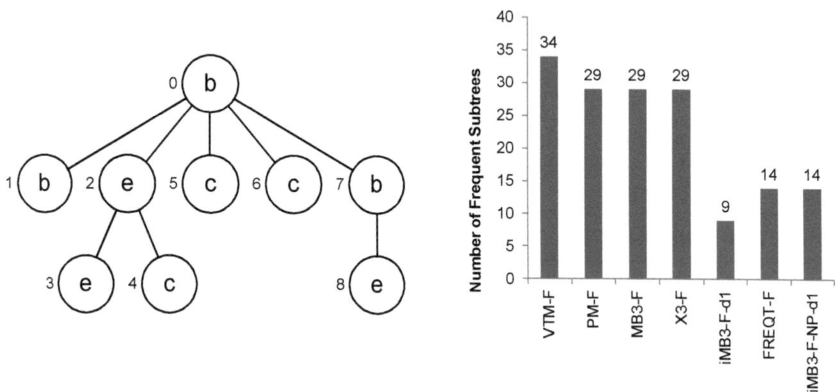

Fig. 5.18 Pseudo-frequent test: number of frequent subtrees $|F|$ for the example tree

iMB3 outperforms FT for mining induced subtrees. Fig. 5.17(b) shows the number of candidates generated by MB3, VTM and PM for $|T_r|$:1000K, σ:250. It can be seen that VTM and PM generate more candidates (VTM-C & PM-C) by using the join approach. The extra candidates are invalid; i.e. they do not conform to the tree model.

5.6.1.2 Pseudo-frequent Test

We created a simple tree as shown on the left of Fig. 5.18, to illustrate the importance of full pruning when occurrence-match support is used. We set σ to 2 and compared the number of frequent subtrees generated by various algorithms. The X3 algorithm (Tan et al. 2005b) is our initial implementation of the TMG framework to ensure the generation of valid subtree candidates. The main difference is that it processes XML documents directly without performing integer to label mapping in the pre-processing stage (see discussion in Section 5.5.2.3). As such its time performance is worse and in fact not compatible with the other algorithms discussed in this section, and we exclude it from further experimentation. From Fig. 8.6, we can see that the number of frequent subtrees detected by VTM employing depth-first search (DFS) is larger when compared with the number detected by PM and MB3 employing breadth-first search (BFS). The difference comes from the fact that the three BFS based algorithms perform full pruning whereas the DFS-based approach such as VTM relies on opportunistic pruning, which does not prune pseudo-frequent candidate subtrees. Fig. 5.18 shows that FT and iMB3-NP generate more frequent induced subtrees in comparison with iMB3. This is because they do not perform full pruning, and as such generate extra pseudo-frequent subtrees.

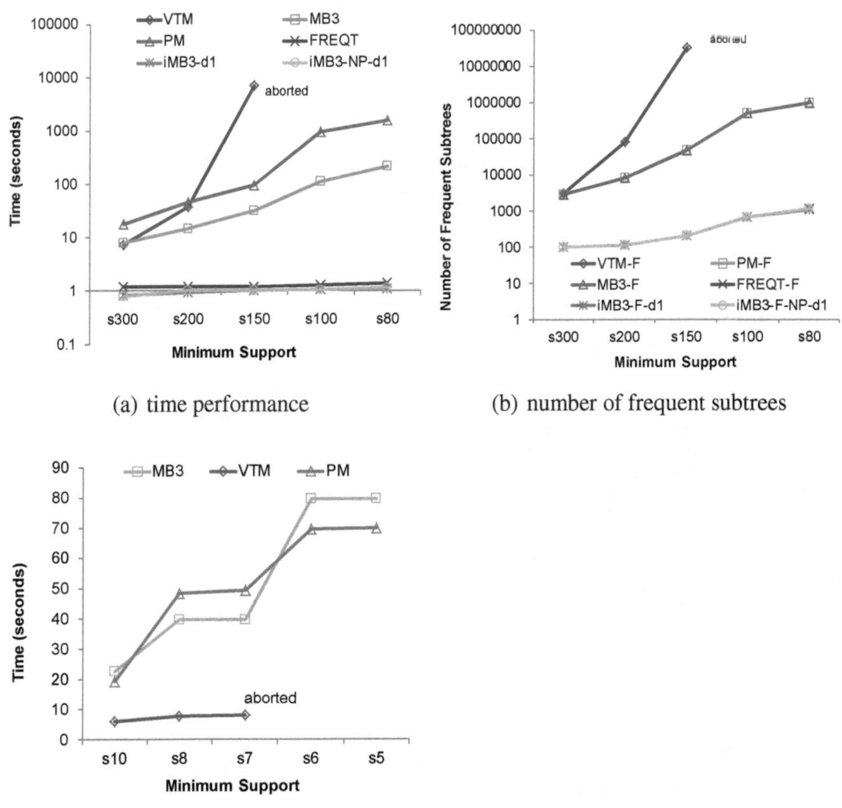

(a) time performance

(b) number of frequent subtrees

(c) wide tree time performance

Fig. 5.19 Deep and wide tree test

5.6.1.3 Deep Tree vs. Wide Tree Test

For the deep tree dataset, when comparing the algorithms for mining frequent embedded subtrees, MB3 has the best performance (Fig. 5.19(a)). The reason for VTM aborting when $\sigma < 150$ can be seen in Fig. 5.19(b) where the number of frequent subtrees increases significantly when σ is decreased. At σ:150, VTM generates a superfluous 688x more frequent subtrees compared with MB3 and PM. In regards to mining frequent induced subtrees, Fig. 5.19(a) shows that iMB3 has a slightly better time performance than does FT. At s80, FT starts to generate pseudo-frequent candidates.

For the widetree dataset, the DFS-based approach like VTM outperforms MB3 as expected. However, VTM fails to finish the task when $\sigma < 7$, due to the extra number of pseudo-frequent subtrees generated throughout the process. In general, the DFS- and BFS-based approaches suffer from, deep and wide trees respectively. In Fig. 5.19(c) we omit iMB3 and FT because the support threshold at which they produce interesting results is too low for embedded subtrees algorithms.

5.6.1.4 Mixed (Deep and Wide) Test

For this experiment, we utilized a mixed dataset. Since the DFS approach and BFS approach suffer from deep and wide trees respectively, it would be interesting to test the performance on a mixed data set, which is both deep and wide. When comparing algorithms for mining embedded subtrees MB3 has the best performance as is shown in Fig. 5.20. VTM gets aborted when $\sigma < 150$, and the drawback of opportunistic pruning is even more noticeable. With regards to mining induced subtrees, iMB3 performs better than does FT.

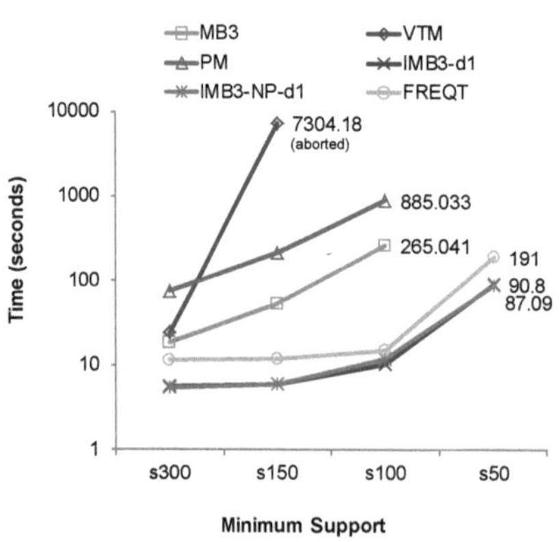

Fig. 5.20 Mixed data set

5.6.1.5 Prions Test

For this experiment, we use the prions dataset. This real-world data describes a protein ontology database for Human Prion proteins in XML format (Sidhu et al. 2005). For this dataset, we map the XML tags to integer-indexed labels similar to the format used in (Zaki 2005). The maximum height is 1. In this case, all candidate subtrees generated by all algorithms would be induced subtrees. Fig. 5.21(a) shows the time performance of different algorithms with varying σ. MB3 has the best time performance for this data.

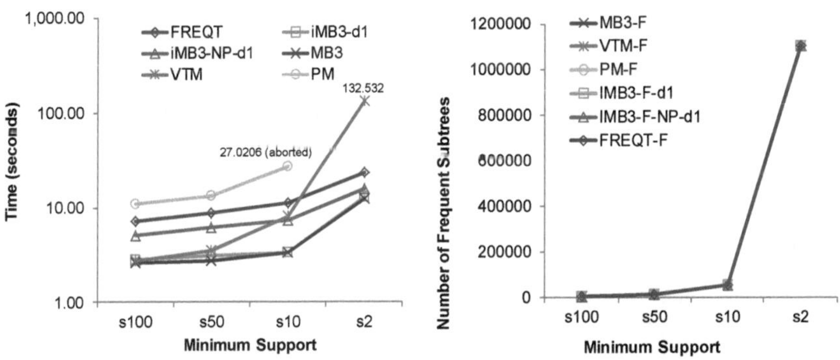

(a) time performance (b) number of frequent subtrees

Fig. 5.21 Prions protein data

Quite interestingly, with this dataset the number of frequent candidate subtrees generated is identical for all algorithms (Fig. 5.21(b)). Another observation is that when $\sigma < 10$, PM aborts and VTM performs poorly. The rationale for this could be that the utilized join approach enumerates additional invalid subtrees. Note that the original MB3 is faster than iMB3 due to additional checks performed to restrict the maximum level of embedding.

5.6.1.6 CSLogs Test

For this experiment, we use the CSLogs dataset. This data set was previously used by (Zaki 2005b) to test VTM and PM using transaction-based support. When used for occurrence-match support, the tested algorithms had problems in returning frequent subtrees. The dataset was progressively reduced and at $|T_r|$:32,241, interesting results appeared. VTM aborts when $\sigma < 200$ due to numerous candidates being generated. We demonstrate the usefulness of constraining the maximum level of embedding and provide some results between the algorithms when σ is varied.

From Fig. 5.22(b), we can see that the number of frequent subtrees generated by FT and iMB3-NP is identical. Both FT and iMB3-NP generate pseudo-frequent subtrees as they do not perform full pruning. Because of this, the number of frequent

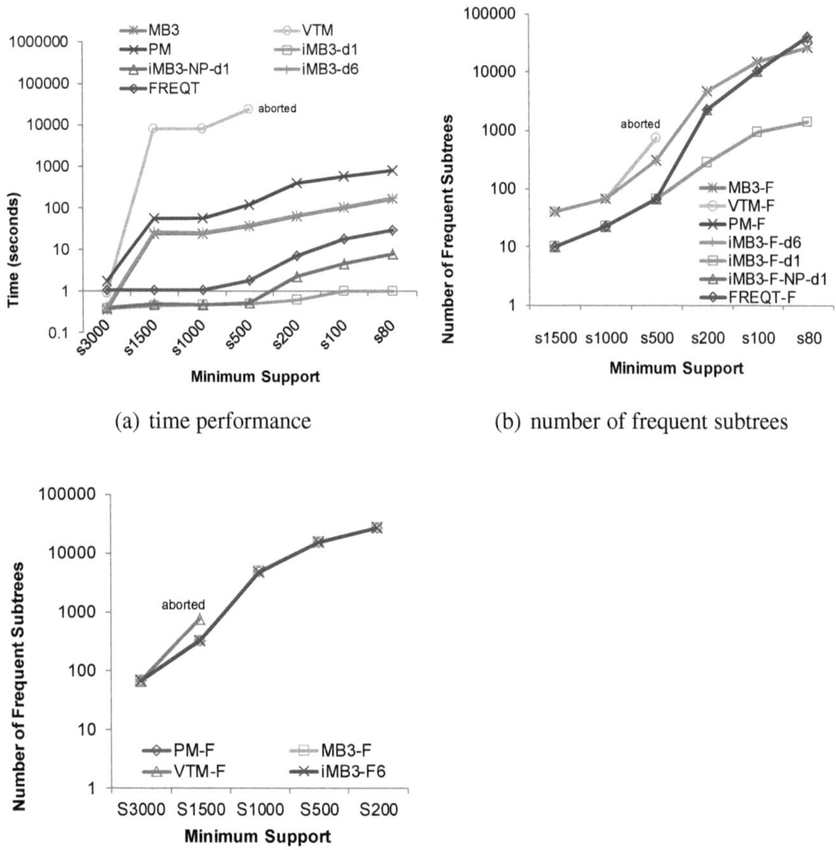

(a) time performance

(b) number of frequent subtrees

(c) number of frequent subtrees for unconstrained vs constrained approach

Fig. 5.22 Test on 54% transactions of original CSLogs data

induced subtrees detected by FT & iMB3-NP can unexpectedly exceed the number of frequent embedded subtrees found by MB3 & PM (Fig. 5.22(b), s80). Fig. 5.22(a) shows that both iMB3-NP and iMB3 outperform FT.

A large time increase for FT and iMB3-NP is observed at s200 as a large number of pseudo-frequent subtrees are generated (Fig. 5.22(b)). Secondly, we compare the results from VTM, PM and MB3 with the result obtained when the maximum level of embedding is restricted to 6 (iMB3-d6) (Fig. 5.22(c)). By restricting the embedding level, we expect to decrease the execution time without missing many frequent subtrees. The complete set of frequent subtrees was detected at $\sigma = 200$, while only less than 2% being missed with $\sigma < 200$. Overall, MB3 and its variants have the best performance.

5.6.1.7 Transaction-Based Support Test

In this experiment, we use the CSLogs data. In the previous sections, we have focused on the use of occurrence-match support. Previous experimental results show that, overall, our approach performs better than other techniques when occurrence-match support is considered. In the experiment in this section, the comparison is made using the transaction-based support. As discussed earlier, our framework is flexible and generic enough to consider different support definitions. From Fig. 5.23(a), it can be seen that MB3 performs better than does VTM, and when σ is lowered to 50, VTM aborts.

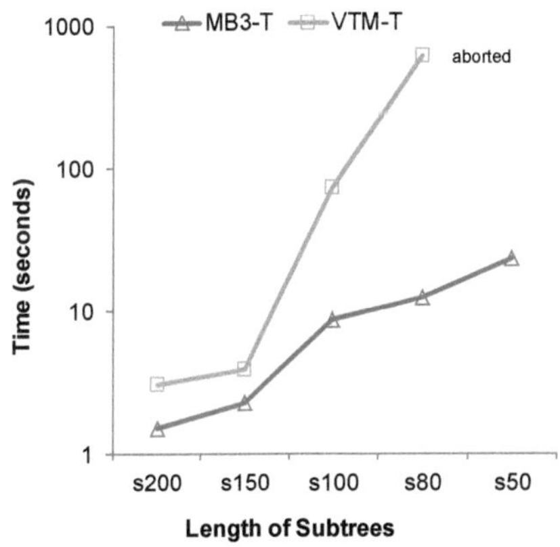

(a) time performance over different minimum support

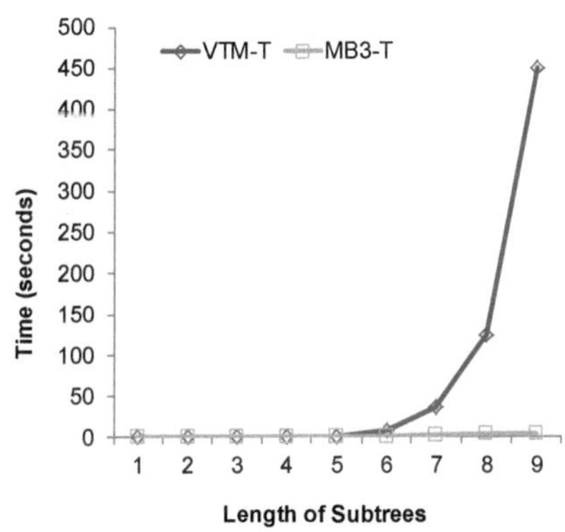

(b) time performance over subtree length (σ:80)

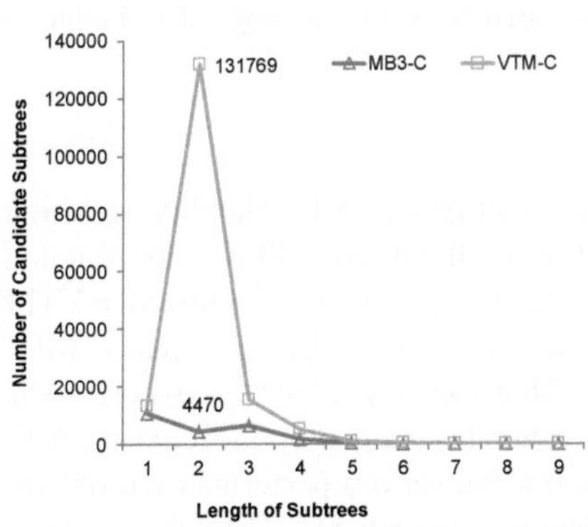

(c) number of candidate subtrees over subtree length (σ:80)

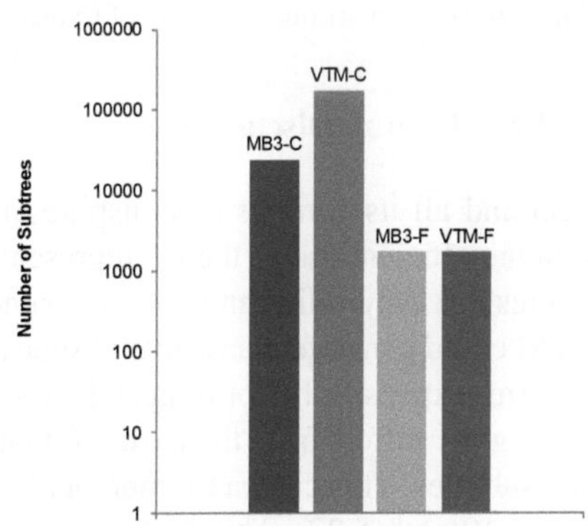

(d) total number of generated subtrees (σ:80)

Fig. 5.23 Benchmarking the usage of transaction-based support for mining embedded subtrees

VTM performance degrades with the increase in subtree length, as is shown in Fig. 5.23(b). In Fig. 5.23(c), we can see a spike of the total number of candidate 2-subtrees generated by the VTM. VTM generates 131,769, whereas MB3 generates only 4,470 candidate 2-subtrees. For generation of candidate 2-subtrees alone, VTM generates 29.47852 times more candidates in comparison with MB3. However, the total number of frequent subtrees produced by VTM and MB3 is identical, as evident from Fig. 5.23(d). The problem of generating pseudo-frequent subtrees, which was a major issue in our previous experiments, is eliminated here because the transaction-based support is considered. The flexibility inherent in our framework allows MB3 to swap from occurrence-match support to transaction-based support without a noticeable performance penalty. The performance comparison for induced subtrees case can be seen from Fig. 5.24. Overall, iMB3 performs slightly better than does FREQT.

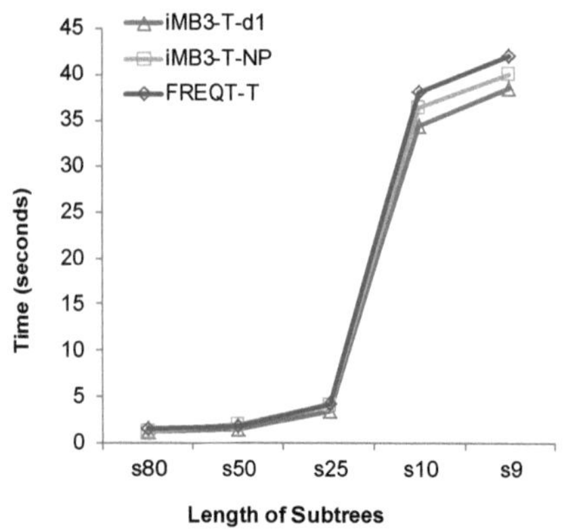

Fig. 5.24 Benchmarking the usage of transaction-based support for mining induced subtrees

5.6.1.8 Overall Discussion

MB3 and all its variants demonstrate high performance and scalability, resulting from the efficient use of the *EL* representation and the optimal TMG approach that ensures that only valid candidates are generated. The join approach utilized in VTM & PM could generate many invalid subtrees which degrades the performance. MB3 performs expensive full pruning, whereas VTM utilizes less expensive opportunistic pruning but suffers from the trade-off that it generates many pseudo-frequent candidate subtrees. This can cause memory blow-out and serious performance problems (Fig. 5.19(a) & 5.22). This problem is evident in cases where VTM failed to finish for lower support thresholds. Some domains aim to acquire knowledge about exceptional events such as fraud, security attacks, terrorist detection, unusual responses to medical treatments and others. Often exceptional cases are one of the means by which the current common knowledge is extended in order to explain the irregularity expressed by the exception. In order to find the exceptional patterns, the user needs to lower the support constraint, as these patterns are exceptional in the sense that they do not occur very often. It is therefore preferable that a frequent subtree mining algorithm be well scalable with respect to varying support.

In the context of association mining, regardless of which approach is used, for a given dataset with minimum support σ, the discovered frequent patterns should be identical and consistent. Assuming pseudo-frequent subtrees are infrequent, techniques that do not perform full pruning would have limited applicability to association rule mining. When representing subtrees, FT (Asai et al. 2002) uses string labels. VTM, PM, and MB3 (and its variants) use integer-indexed labels. As mentioned earlier, when a hash table is used for candidate frequency counting, hashing integer labels is faster than hashing string labels especially for long patterns. As we can see, iMB3 & iMB3-NP always outperform FT. When experimenting with the maximum level of embedding constraint (Fig. 5.22(c)), we have found that restricting the maximum level of embedding at a particular level leads to speed increases at the low cost of missing a very small percentage of frequent subtrees. This indicates that when dealing with very complex tree structures where it would be unfeasible to generate all the embedded subtrees, a good estimate could be found by restricting the maximum level of embedding.

The flexibility inherent in our framework allows the MB3 and iMB3 algorithms to consider the transaction-based support with only a slight change to the way subtree occurrences are counted. Despite the fact that the implementation was not tailored for transaction-based support, our algorithms still exhibit the best performance when compared with other algorithms (Fig. 5.23, Fig. 5.24).

5.6.2 Experiment Set II

We mentioned earlier that the framework described in this chapter is the most optimized one with respect to the implementation choices described in Chapter 4 for the TMG framework. Among the sets of experiments provided in this section, we will also demonstrate the difference in performance between the different implementation versions of the TMG framework. The main two optimizations occurred by storing only the right-most path coordinates of a subtree (referred to as the *RMP* optimization), and through the use of the *recursive list* structure rather than the *embedding list* (referred to as the *RL* optimization). Please refer to Chapter 4 for more detail about these differences. Hence, the following notations are now used when we compare iMB3 variants (iMB3 Miner (IM), iMB3 with *RMP* optimization (IMR), iMB3 with *RMP* & *RL* extension (IML)) and FREQT (FT) for mining induced subtrees and MB3 variants (MB3-Miner (M), MB3 with *RMP optimization* (MR), MB3 with the *RMP* & *RL* optimization (ML)), VTreeMiner (VTM) and PatternMatcher (PM) for mining embedded subtrees. We will refer to each algorithm in this section using its abbreviation as indicated within the brackets. Additionally, IM(NP)-dx notation is used where x refers to the *maximum level of embedding* δ and (NP) is optionally used to indicate that full pruning is not performed. Whenever IM is mentioned without '–dx' suffix it is assumed that the *maximum level of embedding* constraint is 1. The minimum support σ is denoted as (sxx), where xx is the minimum frequency. Occurrence-match support was used for all algorithms. The -u parameter is used for VTM and PM. Experiments were run on a 3Ghz (Intel-CPU), 2Gb RAM, Mandrake 10.2 Linux machine and compilations were performed using the GNU g++ 3.4.3 with the –O3 parameter. We use the Kudo's FT implementation (Kudo 2003) and disable the output mode so that it does not spend its execution time printing all the frequent subtrees when timing its performance.

5.6.2.1 RMP Coordinate Test

We use the widetree and deeptree dataset for this experiment. This set of experiments was performed to demonstrate the space efficiency obtained by storing only the *RMP* coordinates of a subtree. Fig. 5.25 shows the space required by the M and MR algorithms for storing the candidate subtree information for subtrees of different lengths. As the graphs clearly show, storing the *RMP* coordinates of a subtree (MR),

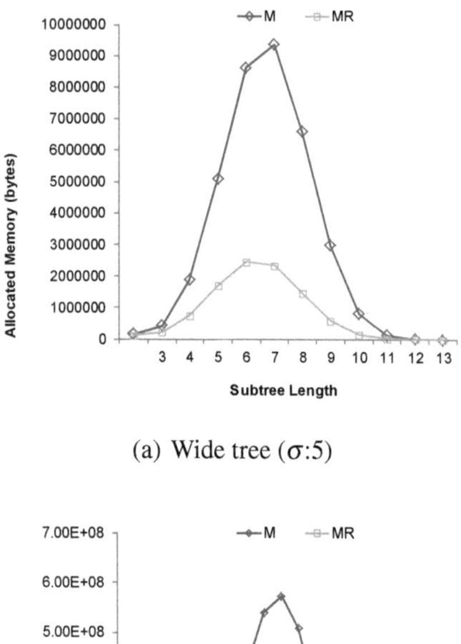

(a) Wide tree (σ:5)

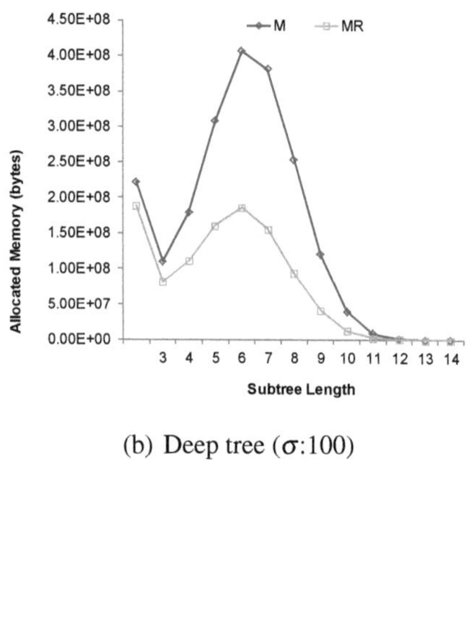

(b) Deep tree (σ:100)

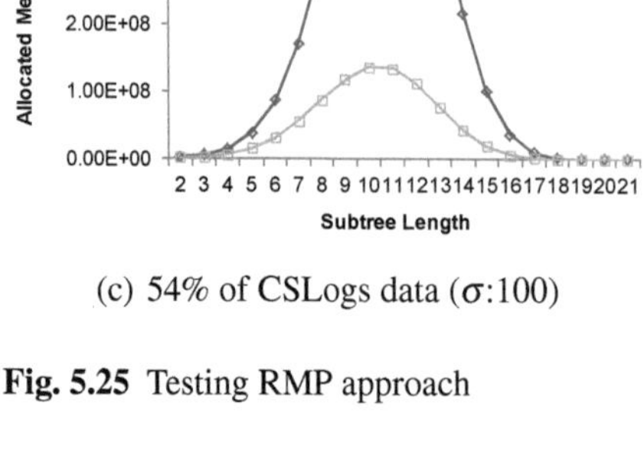

(c) 54% of CSLogs data (σ:100)

Fig. 5.25 Testing RMP approach

instead of the whole occurrence coordinate, greatly reduces the space requirement of the approach. The major improvement occurred for the mid-length subtrees since they are 'sufficiently' long and high in number. Sufficiently long refers here to the fact that they are long enough so that by storing only the RMP information, space requirement is greatly reduced.

5.6.2.2 Recursive List Test

The aim of these experiments is to demonstrate the time and space efficiency obtained by using the *RL* structure for implementing the *TMG* enumeration approach. Fig. 5.26(a-c) shows the time performance of the algorithm variants and, as can be seen, the candidate enumeration implementation using the *RL* structure has better time performance in all cases. Since the difference is not always easy to observe from the graphs, we have accompanied each graph with a table that shows the corresponding time differences among algorithm versions for the decreasing support values.

For some datasets, the difference was small, and this is because of varying tree structures of the datasets. Furthermore, the *RL* approach is more memory efficient as storing the *embedding list* occupies large amounts of memory for some datasets.

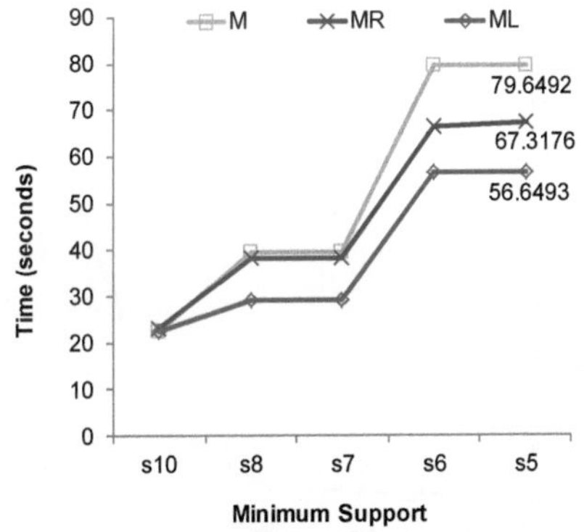

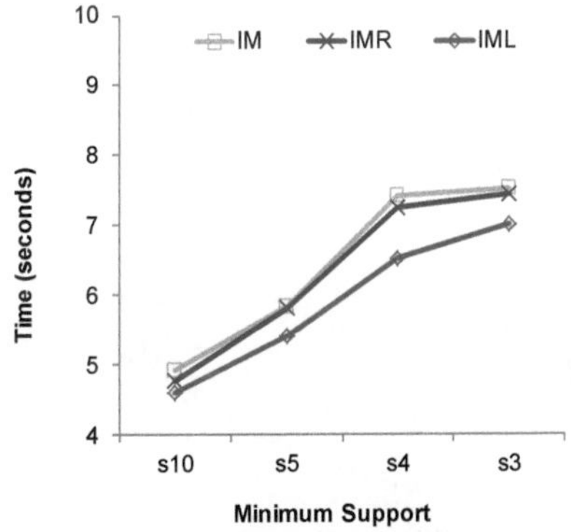

M	MR	ML
22.5277	23.1102	22.5343
39.4723	38.3161	29.2169
39.5097	38.3752	29.082
79.5588	66.3176	56.4371
79.6492	67.3176	56.6493

IM	IMR	IML
4.92006	4.77888	4.61213
5.83568	5.82475	5.41272
7.40688	7.23424	6.52615
7.52302	7.42466	6.99996

(a) Widetree2 dataset

Fig. 5.26 Testing *RL* optimization

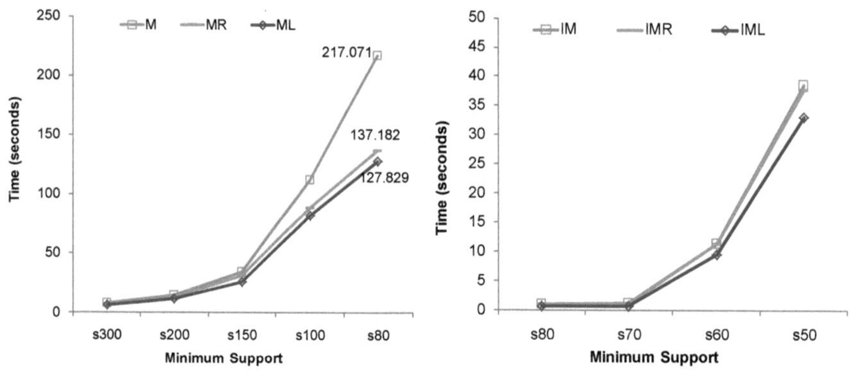

M	MR	ML
7.7	6.51206	5.5954
14.3001	12.5078	11.1789
33.6742	31.0909	25.7522
111.881	88.7228	81.6308
217.071	137.182	127.829

IM	IMR	IML
1.07	1.03	0.70416
1.09813	1.05359	0.716147
11.4172	11.2934	9.4472
38.5614	37.7936	33.0502

(b) Deeptree2 data set

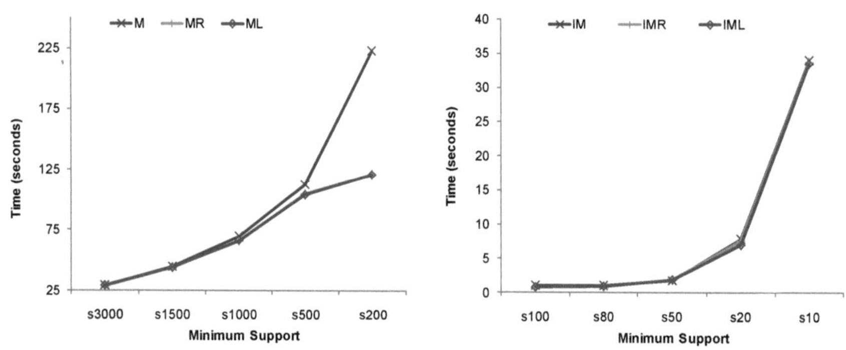

(c) 54% of CSLogs data

Fig. 5.26 (*continued*)

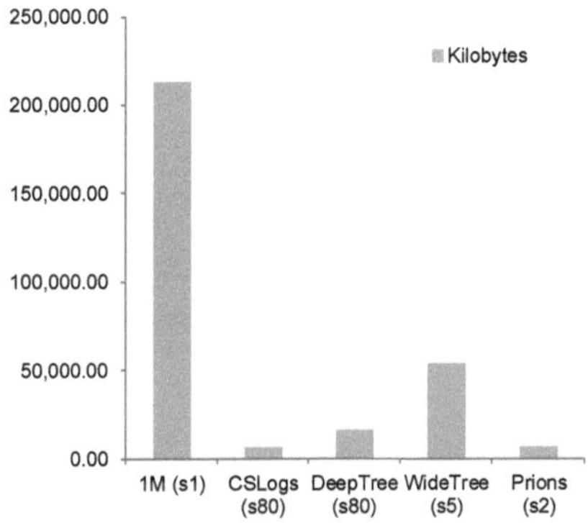

Dataset	Kilobytes
1M (s1)	212,698.00
CSLogs (s80)	6,421.75
DeepTree (s80)	15,771.00
WideTree (s5)	53,090.30
Prions (s2)	6,150.44

(d) Memory usage of Embedding List for different datasets

Fig. 5.26 *(continued)*

This is shown in Fig. 5.26(d) which also corresponds to the total amount of memory saved for each data set, by implementing the *RL optimization*. The 1M dataset is a synthetic data set consisting of 1M transactions with characteristics described in Table 5.1.

5.6.2.3 Scalability Test

For this experiment, we generated an artificial tree database where |Tr|was varied to 500K, 1000K, 1500K and 2000K and |D|:40 and |F|:40 using TreeGen with σ set to 5, 10, 15 and 20, respectively. From Fig. 5.27 we can see that all algorithms are well scalable. ML outperforms VTM for mining embedded subtrees and IML outperforms FT for mining induced subtrees.

5.6.2.4 Deep Tree Test

For this experiment, we utilize the deeptree2 dataset. When comparing the algorithms for mining frequent embedded subtrees, ML produces the best performance (Fig. 5.28(a)). Fig. 5.28(b) displays the number of frequent subtrees generated by all algorithms. As can be seen, the VTM approach detects more subtrees as frequent whose number can significantly grow with a decrease in σ (σ:260 - σ:220).

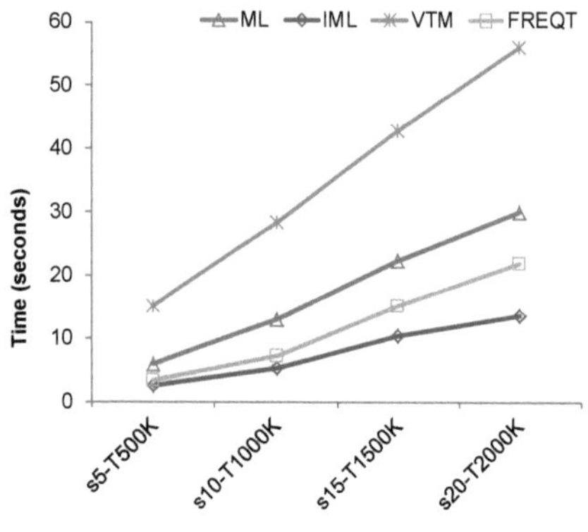

Fig. 5.27 Scalability test - time performance

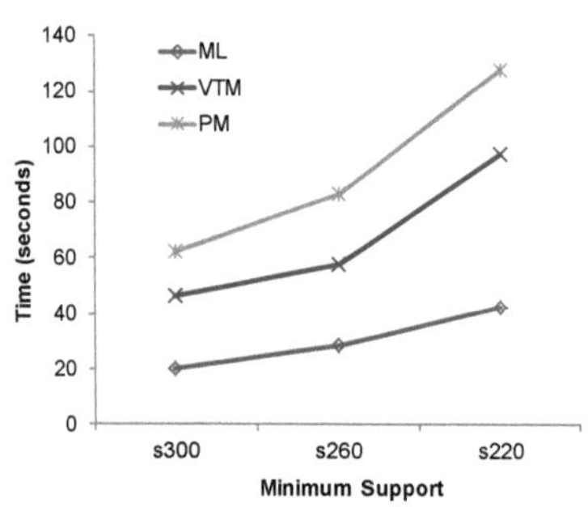

(a) time performance

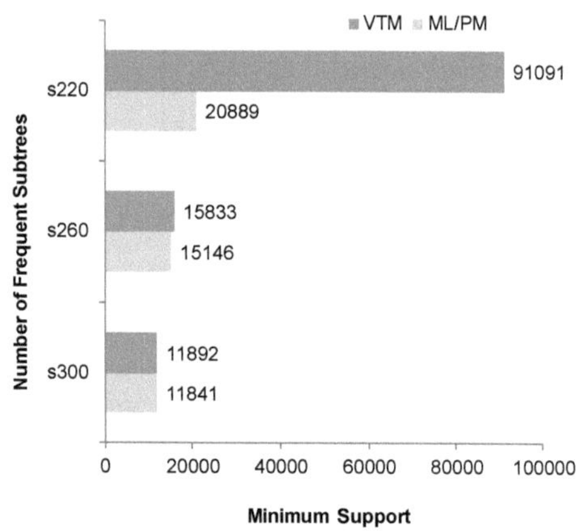

(b) number of frequent subtrees

Fig. 5.28 Deep tree test for embedded subtrees

The reason is that VTM cannot guarantee the absence of pseudo-frequent subtrees. When comparing algorithms for mining induced subtrees, Fig. 5.29 shows that IML performs better than FT.

5.6.2.5 Wide Tree

In this experiment, we use the widetree2 dataset. Fig. 5.30 shows the comparison of algorithms for mining embedded subtrees. ML has the best time performance, as is shown in Fig. 5.30(a). Fig. 5.30(b) shows the number of frequent subtrees detected by the compared algorithms, and for this particular dataset, the number of pseudo-frequent subtrees detected by VTM as frequent is much smaller in comparison with the wide tree dataset.

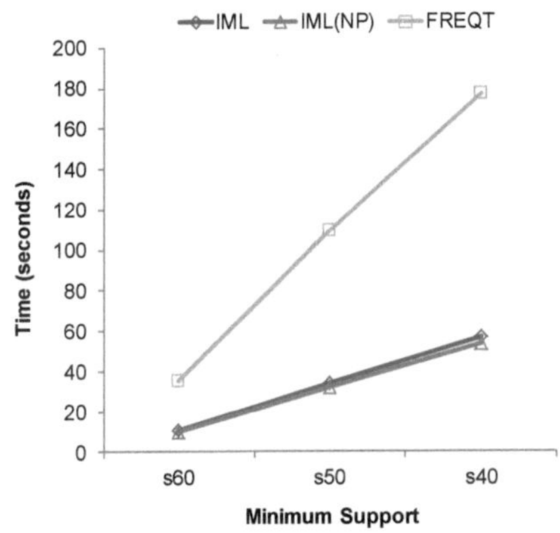

Fig. 5.29 Deep tree test for induced subtrees

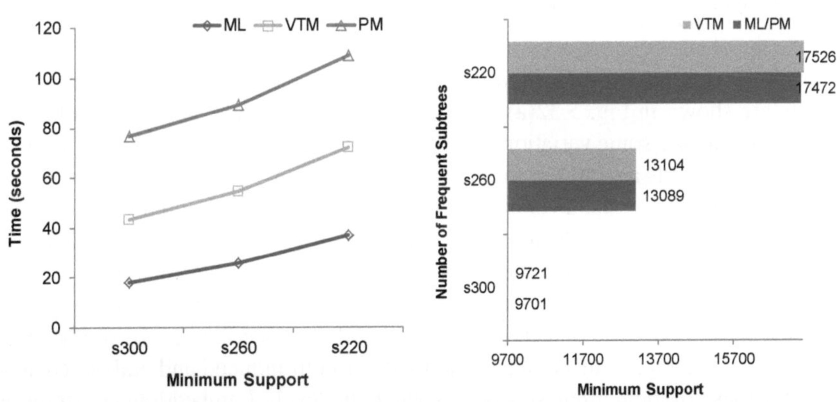

(a) time performance (b) number of frequent subtrees

Fig. 5.30 Wide tree test for embedded subtrees

Fig. 5.31(a) shows the time performance comparison between the IML and FT algorithms for mining frequent induced subtrees. As can be seen, the IML algorithm completes the task in a shorter amount of time.

The IML(NP) version was developed where full pruning is not performed and hence, the small difference in time performance between IML and IML(NP). However, since both IML(NP) and FT do not perform full pruning, extra pseudo-frequent subtrees are detected as frequent which is shown in Fig. 5.31(b).

5.6.2.6 Constraining the Level of Embedding Test

These experiments were performed in order to demonstrate the reduction in complexity that is achieved by constraining the level of embedding. ML, IML, and FT

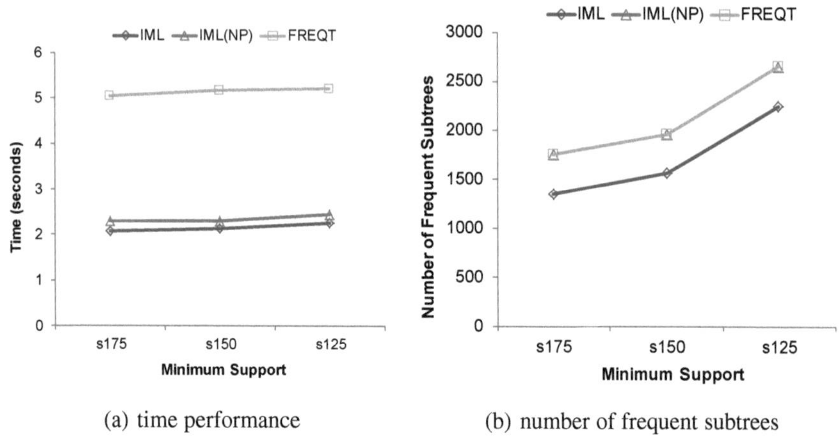

Fig. 5.31 Wide tree test for induced subtrees

algorithms were run on the widetree2 and deeptree2 datasets and the experimental results are shown in Fig. 5.32(a) & Fig. 5.32(b), respectively. On the left of those figures, we can see some variation in the total time taken by the IML algorithm when different thresholds are chosen for the level of embedding constraint (δ). As can be seen, any reduction in the level of embedding results in a reduction in the total time taken to perform the task. Similarly, on the right of Fig. 5.32(a) & Fig. 5.32(b), it can be seen that a variation in the allowed level of embedding results in a variation of the total number of subtrees detected as frequent. The graphs also display the total number of subtrees detected by FT and IML (induced), and ML (embedded) algorithms. The major difference in the totals of both induced and embedded subtrees for these particular datasets occurs when the level of embedding constraint is set to approximately 8 as this is far away from the induced embedding level and the depth of the database tree. However, this is highly dependent on the structure of the database tree and the frequency distribution of its substructures.

5.6.2.7 Hybrid Support Test

In this section we indicate the run time and the number of frequent subtrees extracted from the Prions dataset when the IML algorithm is applied using hybrid support definition. There are currently no algorithms that use the same hybrid support definition and hence providing comparisons with other tree mining approaches was not possible for this support definition. However, from Fig. 5.33 we can see that the approach is well scalable and the efficiency of our general TMG approach to ordered subtree mining is preserved when hybrid support definition is integrated.

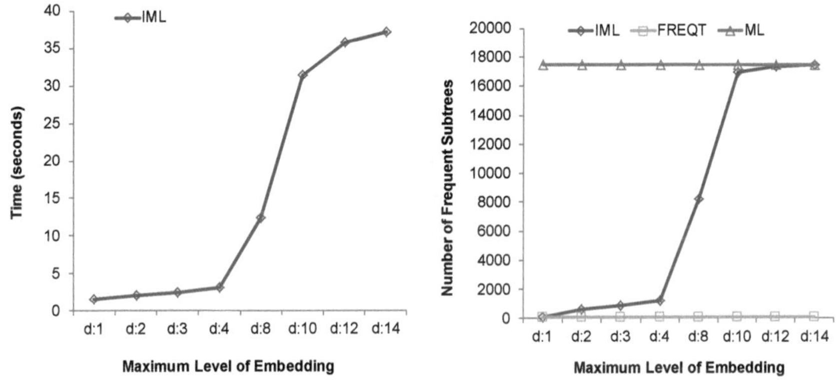

(a) Widetree2 dataset

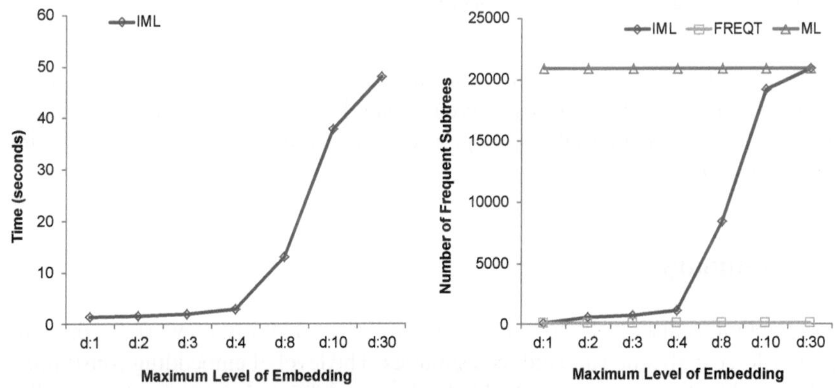

(b) Deeptree2 dataset

Fig. 5.32 Restricting the level of embedding

5.6.2.8 Overall Discussion

Throughout the experiments, both ML and IML demonstrate high performance which generally comes from the optimal TMG candidate generation approach which ensures that only valid candidates are generated. The *RMP* optimization proved very useful for memory conservation, especially for mid-length patterns, which are large in number and 'sufficiently' long (Fig. 5.25). This has also resulted in a faster algorithm (Fig. 5.25(a-c)). Furthermore, the ML and IML approach exhibits a faster performance and is much more space efficient since in our previous implementation *embedding list* structure occupied lots of memory (Fig. 5.26(d)). When experimenting with the level of embedding constraint (Fig. 5.32), it was found that restricting the *level of embedding* could lead to speed increases at the low cost of missing a small percentage of frequent subtrees. In many domains, subtrees with more than a

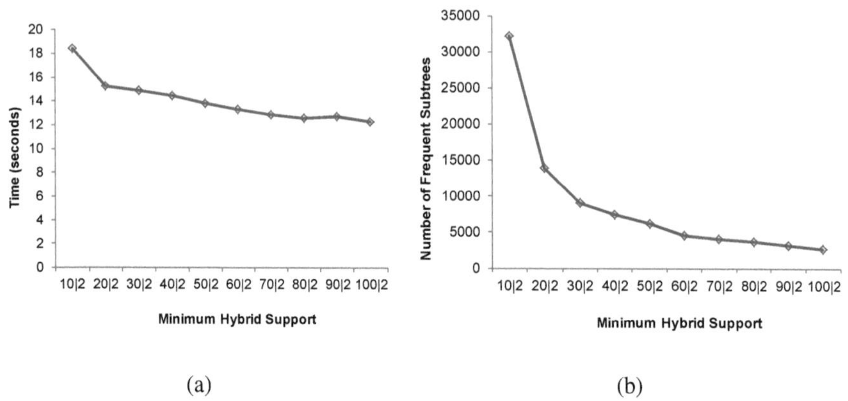

Fig. 5.33 Total time taken (a) and number of frequent subtrees extracted (b) from Prions dataset with varying hybrid support thresholds

certain level of embedding may not be meaningful. This indicates that when dealing with very complex tree structures where it would be computationally impractical to generate all the embedded subtrees, a good estimate could be provided by restricting the *level of embedding*.

5.7 Summary

This chapter has discussed the implementation of the TMG framework for the mining of ordered embedded and induced subtrees. The level of embedding constraint is used to enable the mining of embedded subtrees with varying levels of embedding between the nodes, whereas to mine induced subtrees, the constraint is set to one. A mathematical analysis of the worst case complexity was provided for the TMG candidate enumeration strategy, and for the complexity difference when the level of embedding is varied in the task. To be able to obtain a mathematical formula for worst case analysis of *TMG* candidate enumeration, we had to assume a uniform tree with support set to one. It was seen that for relatively small databases, the process can become unfeasible if certain structural characteristics are present. In the real world, the tree databases would have varying depths and degrees, and support thresholds would be set higher, and hence the infeasibility reflected in the analysis would not necessarily occur in practice. The number of candidate subtrees to be enumerated would be reduced since many infrequent candidates would be pruned throughout the process. As the frequency distribution can generally not be known beforehand, the support threshold could not be integrated into the TMG mathematical formula. Hence, despite the large complexity indicated by the formula, the developed tree mining algorithms are still well scalable for large databases of varying depths and degrees, as was demonstrated by the experiments presented in this chapter and in chapters to follow. The rationale behind experimental comparison was explained together with the justification of the evaluation strategy used. A number of important implementation issues were discussed and demonstrated through experiments using different implementation choices. We then conducted a number of experiments using both real-world and synthetic data to compare the TMG framework with the existing ordered subtree mining approaches, and highlighted the important differences.

References

1. Agrawal, R., Srikant, R.: Fast Algorithms for Mining Association Rules. In: Proceedings of the 20th International Conference on Very Large Data Bases (VLDB), Santiago de Chile, Chile, Septemebr 12-15, pp. 487-499 (1994)
2. Asai, T., Abe, K., Kawasoe, S., Arimura, H., Sakamato, H., Arikawa, S.: Efficient substructure discovery from large semi-structured data. Paper presented at the Proceedings of the 2nd SIAM International Conference on Data Mining (SIAM 2002), Arlington, VA, USA, April 11-13 (2002)
3. Chi, Y., Muntz, R.R., Nijssen, S., Kok, J.N.: Frequent Subtree Mining - An Overview. Fundamenta Informaticae, Special Issue on Graph and Tree Mining 66(1-2), 161–198 (2005)
4. Chodorow, M., Klavans, J.L.: Locating Syntactic Patterns in Text Corpora. Lexical Systems Project IBM T.J. Watson Research Center, Yorktown Heights, N.Y (1990)
5. Jenkins, B.: Hash Functions. Dr. Dobb's Journal (1997)
6. Kudo, T.: An implementation of FREQT (2003)
7. Neff, M.S., Roy, J.B., Omneya, A.R.: Creating and querying hierarchical lexical data bases. Paper presented at the Proceedings of the 2nd Conference on Applied Natural Language Processing, Austin, Texas, USA, February 9-12 (1998)
8. Shapiro, B., Zhang, K.: Comparing multiple RNA secondary structures using tree comparisons. Computer Applications in Biosciences 6(4), 309–318 (1990)
9. Sidhu, A.S., Dillon, T.S., Chang, E., Sidhu, B.S.: Protein ontology: vocabulary for protein data. Paper presented at the Proceedings of the 3rd International Conference on Information Technology and Applications (ICITA 2005), Sydney, Australia, July 4-7 (2005)
10. Tan, H., Hadzic, F., Feng, L., Chang, E.: MB3-Miner: mining eMBedded subTREEs using tree model guided candidate generation. In: Proceedings of the 1st International Workshop on Mining Complex Data in conjunction with ICDM 2005, Houston, Texas, USA, November 27-30, pp. 103–110 (2005a)
11. Tan, H., Dillon, T.S., Feng, L., Chang, E., Hadzic, F.: X3-Miner: Mining Patterns from XML Database. Paper Presented at the Proceedings of the 6th International Conference on Data Mining, Text Mining and their Business Applications, Skiathos, Greece, May 25 (2005b)
12. Tan, H., Dillon, T.S., Hadzic, F., Chang, E., Feng, L.: IMB3-Miner: Mining Induced/Embedded Subtrees by Constraining the Level of Embedding. In: Ng, W.-K., Kitsuregawa, M., Li, J., Chang, K. (eds.) PAKDD 2006. LNCS (LNAI), vol. 3918, pp. 450–461. Springer, Heidelberg (2006)
13. Tatikonda, S., Parthasarathy, S., Kurc, T.: TRIPS and TIDES: new algorithms for tree mining. Paper presented at the Proceedings of the 15th ACM International Conference on Information and Knowledge Management (CIKM), Arlington, Virginia, USA, November 6-11 (2006)

14. Wang, J.T., Zhang, K., Jeong, K., Shasha, D.: A System for Approximate Tree Matching. IEEE Transactions on Knowledge and Data Engineering 6(4), 559–571 (1994)
15. Wang, C., Hong, M., Pei, J., Zhou, H., Wang, W., Shi, B.: Efficient Pattern-Growth Methods for Frequent Tree Pattern Mining. In: Dai, H., Srikant, R., Zhang, C. (eds.) PAKDD 2004. LNCS (LNAI), vol. 3056, pp. 441–451. Springer, Heidelberg (2004)
16. Zaki, M.J.: Efficiently Mining Frequent Trees in a Forest: Algorithms and Applications. IEEE Transactions on Knowledge and Data Engineering 17(8), 1021–1035 (2005)

Chapter 6
TMG Framework for Mining Unordered Subtrees

6.1 Introduction

This chapter describes the extension of the TMG framework for the mining of unordered induced/embedded subtrees. While in online tree-structured documents such as XML the information is presented in a particular order, in many applications the order among the sibling-nodes is considered unimportant or irrelevant to the task and is often not available. If one is interested in comparing different document structures, or the document is composed of data from several heterogeneous sources, it is very common for the order of sibling nodes to differ, although the information contained in the structure is essentially the same. In these cases, mining of unordered subtrees is much more suitable as a user can pose queries and does not have to worry about the order. All matching sub-structures will be returned with the difference being that the order of sibling nodes is not used as an additional candidate grouping criterion. Hence, the main difference when it comes to the mining of unordered subtrees is that the order of sibling nodes of a subtree can be exchanged and the resulting tree is still considered the same.

To briefly illustrate this aspect, consider the simple example given in Fig. 6.1.

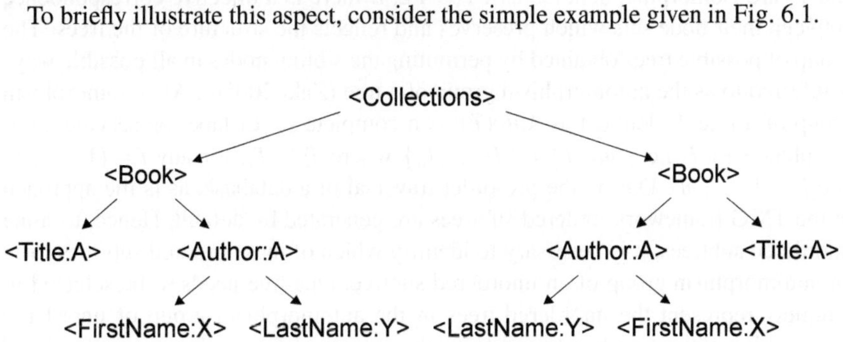

Fig. 6.1 Example of data where mining unordered subtrees is more desirable than mining ordered subtrees

F. Hadzic, H. Tan, and T.S. Dillon: Mining of Data with Complex Structures, SCI 333, pp. 139–174.
springerlink.com

When the order among siblings is taken into consideration, some subtrees that are actually automorphic would be considered to have different semantics. Sometimes this becomes too restrictive. For example, the following XML subtree fragment <Book><Title:A/><Author:A/> from Fig. 6.1 will be considered as a different subtree from subtree <Book><Author:A/><Title:A/>. Also, if we specify that the minimum support threshold is equal to 2, none of the above subtrees discovered from Fig. 6.1 would be frequent.

Often, we would like to relax the order among siblings such that the XML tree fragment <Book><Author:A/><Title:A/> (assume this is the representative form of automorphism group) would be considered equivalent with the XML tree fragment <Book><Title:A/><Author:A/>. With such an assumption, we would now find that the XML tree <Book><Author:A/><Title:A/> is frequent because its support count is $\geq$ minimum support specified, 2. Furthermore, sometimes we want to allow a certain degree of embedding to be present on the discovered subtrees. For example, we might want to consider the following XML subtree fragment <Book> <FirstName:X/><LastName:Y/>, which is an embedded subtree not an induced subtree. If we consider order among siblings, a similar case will occur; that is, it becomes too restrictive such that the subtree <Book><FirstName:X/><LastName:Y/> becomes infrequent. Again, it is more logical to relax the order notion among the siblings so that the XML subtree fragment <Book><FirstName:X/><LastName:Y/> is considered equivalent to the other XML subtree fragment <Book><LastName:Y/>< FirstName:X/>.

In the previous chapter, when we described the technique for ordered subtree mining, the order of sibling nodes is automatically preserved. This is because the candidates are enumerated as they are found in the tree database, and every variation of a subtree with respect to the order of its sibling nodes is considered as a separate candidate subtree. In unordered subtree mining, on the other hand, a subtree that occurs with different order among its sibling nodes needs to be considered as the same candidate. As mentioned in Chapter 2 and 3, this produces the problem of determining whether two trees are isomorphic to one another. Two trees T_1 and T_2 are isomorphic, denoted as $T1 \cong T2$, if there is a bijective correspondence between their node sets which preserves and reflects the structure of the trees. The group of possible trees obtained by permuting the sibling nodes in all possible ways is referred to as the automorphism group of a tree (Zaki 2005b). An automorphism group of a tree T, denoted as *Auto*(T), is a complete set of label preserving automorphisms of T, i.e. *Auto*(T) = $\{T_1,\ldots,T_n\}$ where $T_i \cong T_j$, for any $i = (1, \ldots, n)$ and $j = (1, \ldots, n)$. During the pre-order traversal of a database, as is the approach in the TMG framework, ordered subtrees are generated by default. Hence, to mine unordered subtrees, it is necessary to identify which of these ordered subtrees form an automorphism group of an unordered subtree. One tree needs to be selected to uniquely represent the unordered trees in the automorphism group of unordered trees. This unique tree is referred to as the canonical form (CF) of an unordered tree. The problem of candidate generation is then to enumerate subtrees according to the canonical form so that the frequency of a candidate unordered subtree T, is correctly determined as the frequency of the members of *Auto(T)*. Hence, additional steps are required to identify ordered subtrees that form an automorphism group. Due to this extra processing, the mining of frequent unordered subtrees is more difficult than the mining of frequent unordered subtrees. On the other hand, when the notion of order in data is considered, the processing complexity can increase due to potentially numerous combinations of entities (candidate subtrees) in data.

This chapter starts by explaining the aspects of CF ordering in Section 6.2 and describes the properties of the CF according to which the candidate subtrees will be ordered in the TMG framework. We then give a top level description of the framework as a whole in Section 6.3. Section 6.4 explains the details of the steps involved in the TMG framework for unordered subtree mining. It starts with an explanation of the way in which CF ordering is performed at the implementation level, and then proceeds by explaining the whole process, focusing on the parts that are different from the way ordered subtrees are handled, as explained in the previous chapter. A number of experiments are then provided in Section 6.5 to confirm the conceptual and theoretical formulations and these are used to provide a comparison with the current state-of-the-art algorithms for unordered subtree mining.

6.2 Canonical Form Used in TMG Framework

The correct enumeration of unordered subtrees in the TMG framework comes down to choosing an appropriate canonical form (CF) according to which all the enumerated candidate subtrees are to be ordered prior to them being hashed for frequency calculation. With respect to the CF chosen, the aim should be that the enumerated candidate subtrees will require less sorting on average, since sorting the tree encodings usually becomes one of the performance bottlenecks. Since the generally developed TMG framework utilizes vertical relationships (level of embedding) in order to tackle both induced and embedded subtrees, the depth-first CF (DFCF) proposed in (Chi, Yang & Muntz 2004a) was used which uniquely maps each subtree. To obtain the DFCF, the sorting of subtrees is done based upon the alphabetical order of the node labels of the subtree. The nodes are sorted at each level of the subtree in a bottom-up fashion (i.e. starting from the leaf nodes), and the nodes with labels that sort lexicographically smaller are placed to the left of the subtree. If the node labels are the same, an additional criterion utilized is the labels of the descendants of the node (if it has any) encountered during the pre-order (depth-first) traversal of the subtree rooted at that node. If one node has no descendants, i.e. there is no subtree rooted at it, then it is automatically considered smaller than a node with the same label that has some children. Otherwise, the pairs of node labels encountered at each step of the pre-order traversal of both subtrees are compared. Furthermore, if one needs to backtrack during the pre-order traversal of one subtree, but not during the traversal of the subtree with which it is being compared, and there are more nodes in each of the subtrees, then the former will be considered smaller than the latter. For example, let us say that we are comparing the pairs of nodes encountered in the pre-order traversal of two subtrees *S1* and *S2*. If we had to backtrack in S_1 but not in S_2 and more nodes remain in the traversal of S_2, then S_1 is considered as being smaller than S_2. This aspect is usually enforced by using a special symbol to indicate that a backtrack has occurred in the pre-order traversal of a subtree. To enforce the additional condition, the backtrack symbol is considered to be smaller than any other node label.

At the implementation level, the ordering is done with respect to the way that subtrees are represented and the process will be formally explained in the next section.

The DFCF ordering scheme as proposed in (Chi, Yang & Muntz 2004a), can be formally explained as follows:

Definition: Given two trees T_1 and T_2, with *root[T_1]* = r_1 and *root[T_2]* = r_2, let *descendants(r_1)*: $\{r_1d_1,\ldots,r_1d_m\}$ be the set of descendants of node r_1 and *descendants(r_2)*: $\{r_2d_1,\ldots,r_2d_n\}$ be the set of descendants of node r_2, ordered according to the pre-order traversal. Note that these sets can also contain the special backtrack symbol to indicate the backtracking during the traversal of the descendant nodes. Let *$ST_x(r)$* denote the subtree of tree T_x with root node r, and $|descendants(r)|$denote the number of descendants of node r. We define *label(r_1d_i)* < *label(r_2d_j)* if *label(r_1d_i)* lexicographically sorts smaller than *label(r_2d_j)*, and so $T_1 < T_2$ iff:

(a) *label(r_1)* < *label(r_2)* or,

(b) if *label(r_1)* = *label(r_2)* and $|descendants(r_1)| = m$, $|descendants(r_2)| = n$, then either:

1. $1 \leq i < j \leq \min(m,n)$ $\exists\, j$ such that $ST_1(r_1d_i) = ST_2(r_2d_i)$ and $ST_1(r_1d_j) < ST_2(r_2d_j)$
2. $m \leq n, \forall\, 1 \leq i \leq m$ and $m \leq n$, $ST_1(r_1d_i) = ST_2(r_2d_i)$

Within the described ordering scheme above, the smallest subtrees are placed to the left of the original tree. As an example, consider Fig. 6.2 which shows the steps of sorting a tree T into its canonical form (T_{CF}) in a bottom-up manner. The bolded nodes indicate that the ordering scheme has just been applied to those nodes (i.e. all the nodes at that level). At the implementation level, the ordering may be performed differently depending on the way in which a tree is represented, and the choice that will lead to performing the task in the shortest time possible. The ordering process as it occurs at the implementation level is explained in Section 6.4.1.

6.3 Overview of the TMG Framework for Mining Unordered Induced/Embedded Subtrees

A high level description of the general steps taken by the proposed approach is provided, and in the following section each step is explained at a greater level of detail taking implementation considerations into account. The basic steps taken by the algorithm are presented in the flowchart of Fig. 6.3. The tree database is first transformed into a database of rooted integer-labeled trees. It is ordered according to the canonical form ordering scheme described earlier. This step will reduce the

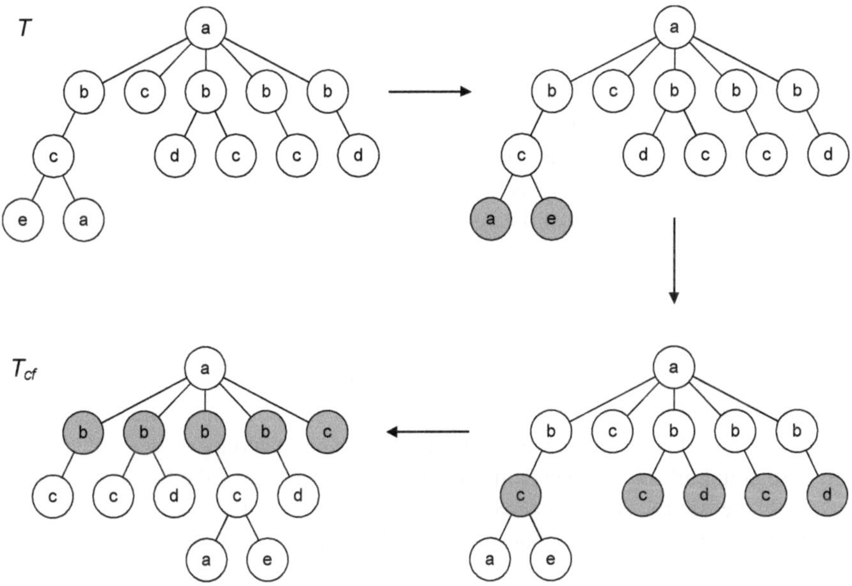

Fig. 6.2 Converting a tree T (top left) into its canonical form T_{cf} (bottom left)

average number of generated candidate subtrees that need to be ordered into their canonical form. The *recursive list* (*RL*) is constructed which is a global sequence of encountered nodes in the pre-order traversal together with the necessary node information. During this process, the node labels are hashed to obtain the set of frequent 1-subtrees (F_1) and then the set of frequent 2-subtrees is also obtained. TMG candidate generation using the *RL* structure takes place and the string encodings of candidate subtrees are hashed to the C_k (candidate k-subtrees) hash table and the right-most path (*RMP*) occurrence coordinates are stored in the *vertical occurrence list* (*VOL*). Prior to hashing the string encoding of each candidate subtree, it is first ordered into its canonical form unless the enumerated encoding was already in the canonical form. During C_k construction, each frequent (k-1)-subtree is extended one node at a time, starting from the last node of its *RMP* (right most node), up to its root. The whole process is repeated until all frequent k-subtrees are enumerated.

As can be seen, the process is somewhat similar to the process described in the previous chapter for ordered tree mining. The main difference at the top level comes in the canonical form ordering performed on the database at the start and on candidate subtrees before they are hashed. Some further implications and differences occur at the implementation level, which are explained next.

6.4 Detailed Description of TMG Framework for Unordered Subtree Mining

While the general steps have been explained in the previous section, in this section each step is explained in more detail, taking implementation considerations into account. Even though some of the steps are very similar to the steps described in the previous chapter, some of the explanation will be replicated here for clarity and to make the necessary changes easier to understand. The way that a subtree is represented at the implementation level was described in the previous chapter and it remains the same. However, since the canonical form ordering process can, to some extent, be seen as a separate process that needs to occur to the database tree at the start, and also to each of the generated candidate subtree, it is explained first in Section 6.4.1. We will make use of the example tree database in Fig. 6.4 as a running example. As usual, the pre-order position, i.e. occurrence coordinate (*oc*) of each node from T_{db} is shown on the left of the node. For example, if unordered induced subtrees are mined, the subtree 'st_1' with $\varphi(st_1)$:'*b c / e*', occurs in T_1 with *oc*:1 2 5 and in T_2 with *oc*:'1 3 2'. In T_2 the nodes 'e' and 'c' of the subtree 'st_1' occur in different order from the encoding but it is still considered the same as we are mining unordered subtrees. So, if we have a string encoding of '*b c / e*' and the aim is to count all the subtrees with that encoding, then when the transaction T_2 is traversed, the subtree is encountered in a form different from the string encoding. It is encountered with *oc*:'1 2 3', but since the sibling order can be exchanged, i.e *oc*:'1

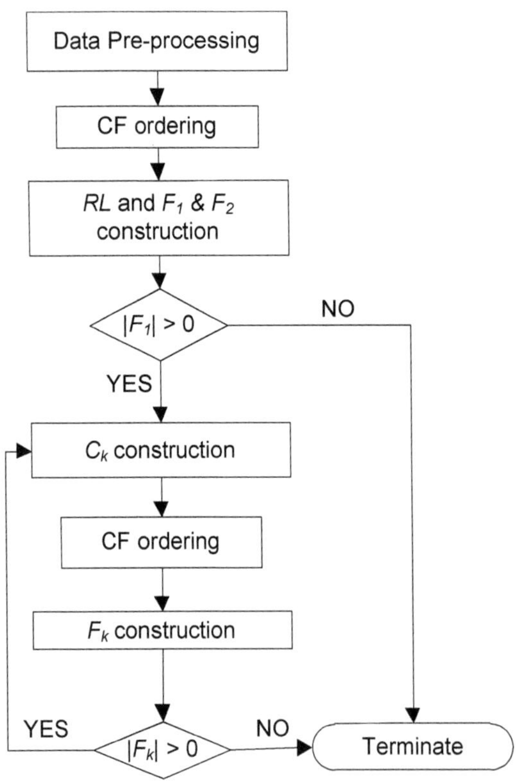

Fig. 6.3 General description of the steps taken

3 2', it still needs to be recognized as the same subtree. This is the main difficulty of unordered tree mining in comparison to ordered tree mining and the way it is handled is explained next.

6.4.1 Canonical Form Ordering

Generally speaking, when all permutations of subtrees rooted at any node in a particular tree *T* are performed, the resulting trees are non-isomorphic ordered trees and isomorphic unordered trees to *T* and among themselves. We could say that they form an unordered automorphism group of *T*. This aspect is illustrated in Fig. 6.5 where all possible unordered automorphisms of subtree st_2 from Fig. 6.4 are shown. They are all isomorphic unordered trees since each one can be mapped into another by permuting the children of vertices.

To illustrate this aspect further, consider the example from Fig. 6.4 when unordered embedded subtrees are mined. The subtree 'st_2' as displayed at the bottom right of the Fig. 6.4 would have the string encoding of $\varphi(st_2)$:'a e / c c / e'. In T_{db}, it occurs at 9 positions and from these 9 occurrences, the subtree st_2 is encountered four times in T_2 with sibling nodes in the same order, i.e. *oc*:'0 2 5 6 7', *oc*:'0 4 5 6 7', *oc*:'0 2 5 6 8', and *oc*:'0 4 5 6 8'. In T_1, the subtree 'st_2' is not encountered with the same sibling node order as the one displayed on the bottom right of Fig. 6.4. The corresponding string encodings of the subtree 'st_2' are different with respect to their

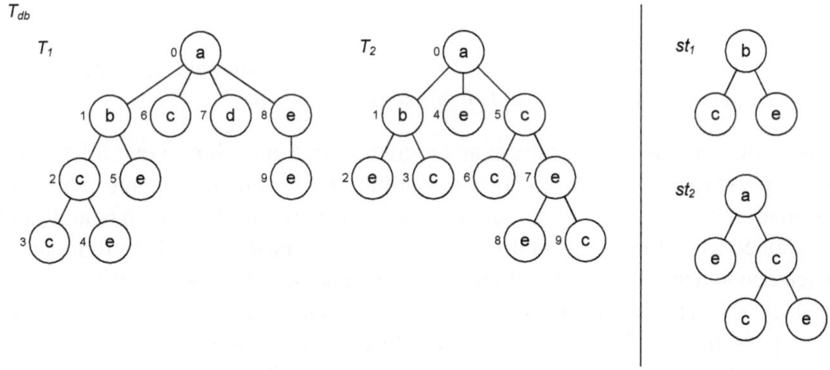

Fig. 6.4 Example tree database (T_{db}) with two subtrees st_1 and st_2

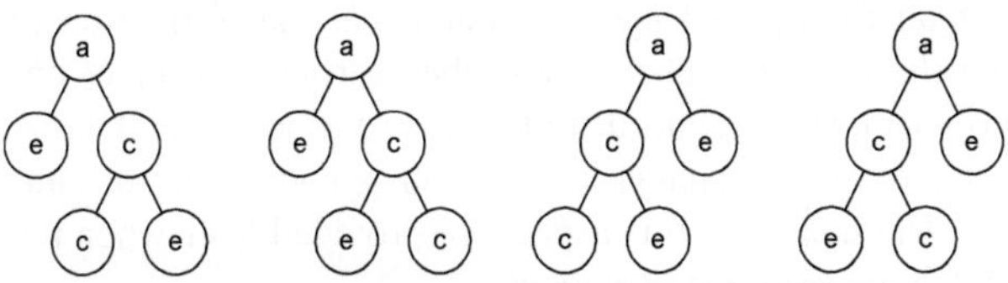

Fig. 6.5 Possible subtree permutations of subtree st_2 (Fig. 6.4)

sibling node order. For all the occurrences of subtree st_2 in T_1, *oc*:'0 5 2 3 4', *oc*:'0 8 2 3 4' and *oc*:'0 9 2 3 4', the string encoding as encountered during the pre-order traversal would be 'a c c / e / / e'. Similarly, the string encoding of st_2 for *oc*:'0 2 5 6 7', *oc*:'0 4 5 6 7', *oc*:'0 2 5 6 8', and *oc*:'0 4 5 6 8'. in T_2, would be 'a e / c c / e', while the string encoding for *oc*:'0 2 5 9 8' and *oc*:'0 4 5 9 8' would be 'a e / c e / c'. These encodings would all be representatives of the automorphism group of the subtree st_2 (shown in Fig. 6.5) together with the encoding 'a c e / c / / e' which did not occur in T_{db} from Fig. 6.4. One needs to be selected as a representative of the CF for that automorphism group.

Whenever a subtree is encountered, it is ordered according to this CF so that all the occurrences of the representatives of an automorphism group are considered as one unordered subtree. The depth-first canonical (DFCF) as proposed in (Chi, Yang & Muntz 2004a) was explained earlier. At the implementation level, the string encoding is used to represent a subtree and the same ordering rule is enforced by sorting encodings based upon the alphabetical order of the labels and considering the special backtrack ('/') symbol as being smaller than any other label. This ordering can be formally explained as follows:

Given two trees T_1 and T_2, with $root[T_1] = r_1$ and $root[T_2] = r_2$, let $D(r_1)$ and $D(r_2)$ denote the descendant sets of r_1 and r_2, respectively. Let $|D(r_x)|$ x = 1 or 2) denote the number of descendants of the node r_x. Further, let $\varphi(T_x)_k$ (x = 1 or 2) denote the kth element of the pre-order string encoding of tree T_x (this can be either a node label or the special backtrack ('/') symbol which, as mentioned earlier, is considered smaller than any other label). The length or size of encoding $\varphi(T_x)$ is denoted as $|\varphi(T_x)|$. T_1 is considered smaller than T_2 iff either:

1. $label(r_1) < label(r_2)$, or
2. $label(r_1) = label(r_2)$ and either $|D(r_1)| = |D(r_2)|$ and $\varphi(T_1)_k = \varphi(T_2)_k \forall 1 \leq k \leq |\varphi(T_1)|$, or $\varphi(T_1)_k < \varphi(T_2)_k$ for some $1 \leq k \leq |\varphi(T_1)|$.

These rules are used to transform an ordered subtree encoding so that it becomes a CF representative of the automorphism group of an unordered subtree. This transformation process is described in the pseudo code displayed in Fig. 6.6, and can be explained as follows. The subtree encoding is traversed from left-to-right and for each encountered label l in $\varphi(T)$ if l is not a backtrack symbol ('/'), it is pushed to a stack S. Every time a backtrack symbol is encountered (i.e. l = '/'), the top node is popped from S. The encoding of the subtree rooted at this top node is appended to the sibling list of the parent of this top node. The sibling list is then sorted from left-to-right so that smaller subtree encodings according to the CF ordering scheme described in the previous section, are placed on the left of the list.

The following illustrates an example of applying canonical transformation to an ordered subtree T using the ordering scheme described earlier. The pivot position (the position of the right-most-node) needs to be tracked properly so that the TMG enumeration is unaffected by the change in the sibling node order of enumerated candidate subtrees. The details of this aspect are provided later when the implementation of the TMG enumeration is explained.

```
L:OrderedSubtree-Encoding
S:stack;
pos := 1;
pivot-pos = pos of right-most-node;

CFSubtree ComputeCanonicalSubtree(L, in/out pivot-pos) {
    Push(S,L[0]); // push L[0]to stack S
    while(Size(S) > 0 && pos < Length(L)) {
        if( L[pos] != {backtrack}) {
            Push(S,L[pos]); // push L[pos]to stack S
        }
        else {
            prevtop := Pop(S);
            curtop := Top(S); // peek the top of the stack
            E(curtop) := E(curtop) union {label(prevtop) + child-encoding(prevtop)};
            pivot-pos := Sort(E(curtop), pivot-pos));

            foreach( encoding e in E(curtop) ) {
                child-encoding(curtop) += e + {backtrack};
            ]
        }
        pos++;
    }
    top = Pop(S);
    return (label(top) + child-encoding(top));
}
```

Fig. 6.6 Pseudo-code to transform an encoding of a subtree to its canonical form.

From the example in Fig. 6.7, the labels are read from left to right from *φ(T)*. To distinguish the same node labels that occur at a different position in the tree, when referring to nodes, an integer will be added into the brackets that corresponds to the pre-order position of that node in *T*. Each node in the encoding may have a sibling list associated with it and an encoding which in this case corresponds to the encoding of the whole subtree rooted at that particular node. To assist the explanation, we show the states of the stack at different times of the enumeration process, but for simplicity we do not show the encodings associated with nodes

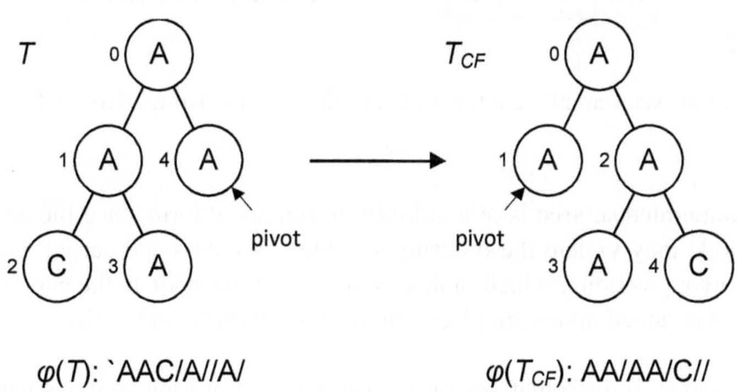

Fig. 6.7 Illustration of transformation of tree T into its DFCF T_{CF}

(Fig. 6.8). The process starts by pushing the first 3 nodes A(0), A(1), C(2) to the stack *S* (i.e. ST_1 in Fig. 6.8). After the first backtrack is read, node C(2) is popped and the subtree encoding rooted at node C(2) (which is simply 'C') is appended to the sibling list of node A(1) (ST_2 in Fig. 6.8). Since node C(2) is the only child then no sorting needs to be performed on the sibling list of node A(1). The process continues by reading the next label, A(3), and pushing A(3) to the stack *S* whose current snapshot is A(0), A(1), A(3) (ST_3 in Fig. 6.8). After the next backtrack is read, node A(3) is popped from S and its encoding is appended to the sibling list of its parent node (A(1)) (ST_4 in Fig. 6.8). Previously, the encoding of node C(2) was appended to the sibling list of node A(1), and so the A(1) sibling list now contains ('C','A'). We perform sorting on this list and since A is lexicographically smaller than C it is placed to the left, i.e. ('A','C') (ST_5 in Fig. 6.8). The encoding of node A(1) is then updated and it becomes 'AA/C'. The next label read from φ(T) is a backtrack. At this stage, the stack snapshot becomes A(0), A(1). The stack is popped and A(1) encoding 'AA/C' is obtained and appended to the sibling list of A(0) (ST_6 in Fig. 6.8). No sorting of the A(0) sibling list is required yet at this stage. Next, node A(4) is read and its encoding appended to the sibling list of A(0) (ST_7 in Fig. 6.8). The sibling list is then sorted and it becomes {'A','AA/C/'} (ST_8 in Fig. 6.8). Finally, we obtain the DFCF of subtree rooted at A(0) as 'AA/AA/C//'.

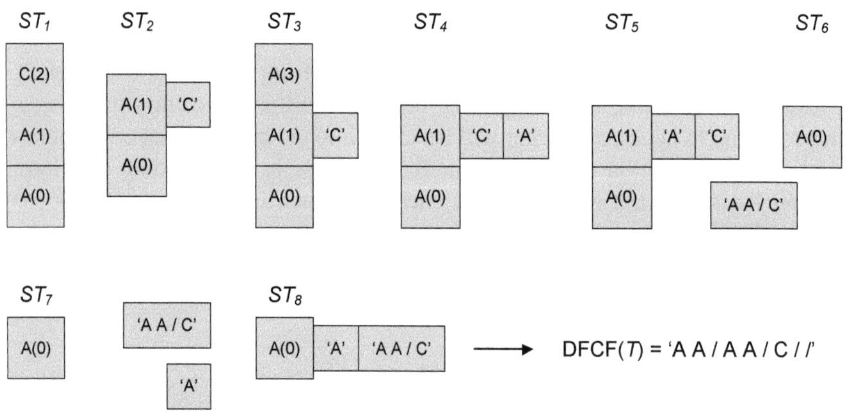

Fig. 6.8 The states of the stack at different times of canonical form ordering of tree *T* from Fig. 6.7

Each newly enumerated subtree is ordered into its canonical form since the addition of a new node may violate the ordering scheme. As mentioned earlier, we keep track of the pivot position ς which indicates the right-most node in the encoding, so that any newly added nodes are placed in the correct place within the string encoding.

Performing canonical transformations on a large number of long subtrees can be very expensive, due to the expensive traversal of the string encodings in order to determine and compare the sibling nodes. Due to the fact that the whole tree database is first sorted into its CF and each previously enumerated (k-1)-subtree is ordered, many subtrees may already be in their CF when k-subtree enumerations are performed for $k > 2$. Hence, within the CF ordering scheme, we have determined a few preconditions that indicate that a subtree is already in its CF and that no ordering is required. These preconditions occur when we are appending a new node 'n' to the right-most node 'r' of the currently expanding subtree, or in other words, the pivot position ς points to the rightmost node of the currently expanding subtree.

Precondition 1: $parent(n) = r$;

Precondition 2: Let the left sibling of n be ln, then $children(ln)$ = null and $label(ln)$ = $label(n)$, OR $label(ln) < label(n)$.

If any of the above conditions are met, the ordering can be omitted which results in a run-time reduction as will be demonstrated later in the experiments.

6.4.2 Enumerating Unordered Subtrees Using TMG Framework

6.4.2.1 Data Pre-processing

The tree database is first transformed into a database of rooted integer-labeled ordered trees where each node label is mapped to a unique integer. The whole underlying tree structure of the database (i.e. the string encoding representing the whole T_{db}) is then ordered into its canonical form. During the candidate generation phase that will be explained later, each enumerated k-subtree from the tree database will need to be in its canonical form prior to being hashed to the frequent k-subtree set (F_k). If the underlying tree structure of the database is already in its canonical form, then one would expect that the candidate subtrees enumerated throughout the process, are more likely to already be in their canonical form. This would on average reduce the number of candidate subtrees that need to be ordered and hence will reduce the processing time. This will be demonstrated in the experiments provided later in the chapter.

6.4.2.2 Construction of Recursive List, F_1 and F_2

This step remains exactly the same as in the ordered case explained in the previous chapter. This is because, for subtrees consisting of 1 or 2 nodes, the canonical form ordering process does not need to be performed. Hence, the tree database, T_{db}, is traversed using the pre-order traversal, and *position*, *label*, *scope*, and *level* information are stored for each encountered node. The scope of a node refers to the position of its right-most leaf node or its own position if it is a leaf node itself (Zaki 2005a, Tan et al. 2006), whereas the level refers here to the level in the T_{db} tree, where this node occurs. An item in *RL* at position i is referred to as *RL[i]*. Every time a node is inserted into the *RL*, we generate a candidate 1-subtree. Its label is hashed to the C_1 hash table and the occurrence coordinates of that particular node are stored in the *VOL*. Depending on the support definition (as was explained in previous chapters and will be explained again later) its support is worked out from the size of the *VOL* or the number of unique transaction identifiers in *VOL*. If a subtree's support count is $\geq \sigma$ (user-specified minimum support count), we insert the candidate 1-subtree to the frequent 1-subtree set, F_1. For every item in F_1, the set of its occurrences is obtained which corresponds to the entries in the *RL* structure. For each entry in *RL*, a number of candidate 2-subtrees are generated by appending the label of the node contained in the *RL* entry to each of the labels of the nodes (*RL* entries) that are within the scope of the node being processed. For example, let us say that 1-subtree with label 'a' has an occurrence at position p in *RL*, where the scope is equal to $p + n$. The set of candidate 2-subtrees that can be obtained by extending this particular node, is formed by appending the label 'a' to each of the node labels from all the positions x in *RL*, where $p < x \leq p + n$. The resulting encoding is then hashed to the C_2 hash table and the occurrences for each candidate are stored in the *VOL* structure. Similarly, if a subtree's support count in C_2 is $\geq \sigma$, we insert the candidate 2-subtree to the frequent 2-subtree set, F_2. The process of enumerating the remaining k-subtrees ($k > 2$) follows the same procedure and is explained next.

6.4.2.3 Candidate k-Subtree (C_k) Generation

In what follows, the process of generating all candidate k-subtrees is described. This is the core process of the TMG framework and was explained in detail in the previous chapter. Hence, in this chapter we will focus on the slight modifications necessary for generating unordered subtree candidates. As previously, the right-most path occurrence coordinates (*RMP-oc*) are used to distinguish the different instances of a particular subtree in the database. To enumerate all k-subtrees from a (k-1)-subtree, the TMG enumeration approach extends one node at a time for each node in the right most path (*RMP*) of a (k-1)-subtree. Suppose that nodes in the *RMP* of a subtree are defined as extension points and the level of embedding between two nodes at position n and t is denoted by $\Delta(n,t)$. The TMG can be formulated as follows. Let $\Psi(T_{k-1})$:$[e_0,e_1,\ldots, e_j]$ denote the *RMP-oc* of a frequent (k-1)-subtree T_{k-1}, Φ the scope of the root node e_0 and δ the maximum level of embedding constraint. TMG generates k-subtrees by extending each extension point $n \in \Psi(T_{k-1})$ with a node with *oc* t for which the following conditions are satisfied: (1) $n < t \leq \Phi$, (2) $\Delta(n,t) \leq \delta$. Suppose that the encoding of T_{k-1} is denoted by $\varphi(T_{k-1})$ and $l(e_j,t)$ is a labelling function for extending extension point n with a node at position t. $\varphi(T_k)$ would be defined as $\varphi(T_{k-1}) + l(e_j,t)$, where $l(e_j,t)$ determines the number of backtrack symbols '/' to be appended before the label of the new node is added to $\varphi(T_k)$. The number of backtrack symbols is calculated as the shortest path length between the extension point n and the right-most node r, (notation $pl(n,r)$). For an illustrative example of this, please refer to Section 5.3.2 and Fig. 5.4 from Chapter 5, as this process remains unchanged, except for the use of a 'pivot position' that is explained next.

In the case of unordered subtrees, the right-most node may not always correspond to the last node (tail position) in the encoding as it does for the ordered subtree case. We refer to this case as non-tail expansion. This calls for an adjustment of our general TMG framework in order to enumerate unordered subtrees. The notion of pivot position ς is used to denote the position in the subtree encoding that corresponds to the right-most-node (i.e. the extension point). Each *RMP-oc* of a subtree will store an integer indicating the pivot position ς in the encoding for that particular occurrence of the subtree. Hence, for a non-tail expansion of a subtree T_{k-1}, if we are appending a new node with label l and *oc* t, rather than appending the backtrack symbols (if any) and l to the last node in $\varphi(T_{k-1})$, it will be appended to the pivot position ς by the function $l(\varsigma, t)$, in order to obtain $\varphi(T_k)$. The function $l(\varsigma, t)$ computes the length of the shortest path between the extension point ς and the new node with *oc t*. The length of the shortest path corresponds to the number of backtrack symbols that need to be appended with l. Please note that if there are b backtrack symbols to be appended with l and there were already some backtrack symbols after the pivot position ς in $\varphi(T_{k-1})$, then l will be appended after the b_{th} backtrack symbol. Furthermore, an additional backtrack symbol will be appended after the position in the encoding where l has been appended.

To illustrate this, please consider the subtree st_2 from T_2 in Fig. 6.4, with *oc*:[0, 5, 6, 7, 4], $\Psi(st_2)$:[0, 5 ,7] and $\varphi(st_2)$:'a c c / e / / e'. Please note that the order of *oc* and $\Psi(st_2)$ does not follow the pre-order traversal of T_2, since they are ordered according to the DFCF explained in section 6.4.1. As can be seen, the right-most node does not correspond to the last node 'e' in the encoding with *oc*:4, but rather to node 'e' with *oc*:7. Therefore, if we are extending st_2 from extension point node 'e' (*oc*:7) with node 'c' (*oc*:9) then $l(7,9)$ will append the label 'c' to $\varphi(st_2)$ at pivot position ς and add '/' after 'c'. The new encoding becomes 'a c c / e c / / / e'.

6.4.2.4 Pruning

As is the case in ordered subtree mining, full (k-1) pruning must be performed to make sure that all generated subtrees do not contain infrequent subtrees. This ensures that the downward-closure lemma (Agrawal & Srikant 1994) is satisfied and in the case of occurrence match support, the method will not generate any pseudo-frequent subtrees (see Section 2.6.1 of Chapter 2). Hence, at most (k-1) numbers of (k-1)-subtrees need to be generated from the currently expanding k-subtree, and when mining induced subtrees (i.e. δ = 1), we need to generate only n numbers of (k-1)-subtrees where n < (k-1) and equal to the number of leaf nodes in the k-subtree. When the removal of root node of k-subtree does not disconnect the subtree, then an additional (k-1)-subtree is generated by removing the root node from the expanding k-subtree. Each of these generated (k-1)-subtrees is checked for existence in the F_{k-1} set. This is where the difference occurs in the pruning process in comparison with ordered tree mining explained in the previous chapter. Since each of the subtrees in F sets are in their canonical form, all the generated (k-1)-subtrees also need to be in their canonical form prior to checking for their existence in F set. This is because the removal of a node may cause the subtree to no longer be in its canonical form, and hence, ordering needs to take place again. Therefore, if any of these (k-1)-subtrees cannot be found in the F_{k-1} set (i.e. it is infrequent), then the expanding k-subtree is pruned. Otherwise, it is added to the F_k set. Performing full (k-1) pruning is even more time consuming and expensive in the case of unordered subtree mining, since each of those generated (k-1)-subtrees has to be in its CF prior to it being looked up in the F_{k-1} set, and ordering takes place when necessary. A caching technique is used to check whether a k-subtree candidate is already in the frequent k-subtree set (F_k), as then it is known that all its (k-1)-subtrees are frequent, and only one comparison is made.

6.4.3 Frequency Counting of Candidate Subtrees

The candidate enumeration just explained makes use of the encoding of a subtree and the *RMP-oc* that represent the different instances of the subtree with that encoding. Hence, we need a means of storing all the information about a candidate subtree, that allows us to determine its support and expand the frequent subtrees with suitable nodes, as well as update the new encoding in the proper manner using the pivot position. The *vertical occurrence list* (*VOL*) is used to store all the *RMP-oc* of a particular subtree with its encoding and to determine the support of a subtree using any of the current support definitions. For occurrence-match support, the notion of the transaction id (*tid*) associated with each *RMP-oc* is ignored, while for transaction-based and hybrid support, the notion of *tid* of each occurrence coordinate is accounted for when determining the support. In Fig. 6.9 we show an example *VOL*, and one can see that besides the first column that keeps the transactional identifiers, the second column stores the pivot position ς so that it is known where we are to append a new node label and any backtracks, as explained earlier. Furthermore, to correctly determine the *RMP-oc* for the next set of k-subtrees, an integer is stored for each *RMP-oc* that indicates its size $|RMP\text{-}oc|$. The way that this information is used to enumerate (k+1)-subtrees from a current k-subtree was explained earlier in the TMG enumeration subsection. Please note that in the figure we use the pre-order string encoding (last column of *VOL* in Fig. 6.9) to keep the discussion consistent, but as we stated earlier, hashing of integers is faster than hashing of string labels. Hence, the labels in the string encoding are at the implementation level replaced by mapped integers, while the backtrack ('/') symbols are replaced by a negative one (-1).

In this example, we have assumed that unordered embedded subtrees were mined. When the occurrence-match support is used, the frequency of ordered embedded subtree st_2 of tree database T_{db} (Fig. 6.4) with encoding ordered in canonical form 'a c c / e / / e', is equal to the size of the *VOL*, i.e. 9 (Fig. 6.9). When transaction-based support is used, the support of 'st_2' is 2 since there are two transactions (*tid*:1, *tid*:2) that support subtree 'st_2'. When hybrid support definition is considered, two values are taken into account so that, given a hybrid support of $x|y$, a subtree will be considered frequent if it occurs at least y times in x transactions. At the implementation level, a transaction t will be considered to support a subtree if t occurs at least y number of times in the first column of *VOL*. In other words, a subtree will be considered frequent if there are at least x unique identifiers in *VOL* which are repeated at least y times within the *VOL*. Thus, from Fig. 6.9, the hybrid support of the subtree 'st_2' with encoding 'a c c / e / / e' is equal to 2|3 since there are two unique transaction identifiers and they both repeat at least 3 times.

tid	pivot pos ς	$\lvert\Psi(st_2)\rvert$	$\Psi(st_2)$		
1	7	2	0	5	
1	7	2	0	8	
1	7	2	0	9	
2	4	3	0	5	7
2	4	3	0	5	7
2	4	3	0	5	8
2	4	3	0	5	8
2	2	3	0	5	9
2	2	3	0	5	9
'a c c / e / / e'					

Fig. 6.9 VOL of subtree st_2 from Fig. 6.4

If induced subtrees were mined by setting the maximum level of embedding constraint to 1, then the list would consist of only one occurrence of the subtree with *oc*:[0, 5, 6, 7, 4] in T_2 from Fig. 6.4. It would correspond to the 4th (or 5th) entry of the *VOL* from Fig. 6.9. The reason for the 4th and the 5th entry in *VOL* being the same is that their *RMP* is the same and the difference is in the occurrence of the node 'e' at the left of the subtree. In one of the cases, the node occurred at position 2 in T_2 and in the other case at position 4 in T_2. Since the level of embedding between the root node and the node at position 2 is > 1, this occurrence would not be stored when induced subtrees are mined.

6.4.4 Pseudo Code of the Algorithm

The algorithm as a whole can be described by the pseudo code outlined in Fig. 6.10. While in our earlier discussion we explained the canonical form transformation step of the whole tree database and the *RL* construction step as separate processes, in the pseudo code they are considered as one step. This is because the canonical form transformation is considered as one of the pre-processing steps. However, the final *RL* that is to be used by the TMG candidate enumeration phase is first constructed during the database scanning stage when the set of frequent 1-subtrees is simultaneously obtained. The canonical form transformation then occurs on the constructed *RL* and the set of frequent 2-subtrees is obtained as explained earlier.

```
Inputs: Tdb(tree database),σ(min.support), δ(max. level of embedding)
Outputs: Fk(frequent subtrees), RL(recursive List)
{RL, F1, F2}: CfOrderingRLConstruction(Tdb, δ)
k = 3

while(|Fk| ≥ 0) {
  Fk = GenerateCandidateSubtrees(Fk-1, δ)
  k = k + 1
}

GenerateCandidateSubtrees(Fk-1, δ):
for each frequent k-subtree tk-1 ∈ Fk-1 {
Lk-1 = GetEncoding(tk-1)
VOL-tk-1 = GetVOL(tk-1)
for each RMP occurrence coordinate RMP-ock-1 (r:[m,…n])∈ VOL-tk-1 {
  for (j = n + 1 to scope(r)) {
    {extpoint,slashcount} = CalcExtPointAndSlashcount(RMP-ock-1, j);
    Lk = Lk-1 + append('/', slashcount) + Label(j)
    if(Precondition1(Lk) == false && Precondition2(Lk) == false) {
      {Lk, ς} = CanonicalTransform(Lk, ς)
    }
    if(EmbeddingLevel(extpoint, j) ≤ δ) {
      RMP-ock = TMG-extend(RMP-ock-1, j)
      if(Contains(Lk, Fk))
        Insert(h(Lk), ς, RMP-ock, Fk )
    else
      if(AllSubpatternFrequent(Lk)) // all k-1 patterns frequent?
        Insert(h(Lk), ς, RMP-ock, Fk )
      }
    }
  }
}
return Fk

TMG-extend(RMP-ock-1, j):
// right-most-path computation
|RMP-ock|= |RMP-ock-1| + 1 - slashcount; // compute size of RMP coordinate
α = |RMP-ock| - 1; // update the tail index α
RMP-ock[α] = j; // store the new node pos at tail index α
return RMP-ock
```

Fig. 6.10 Pseudo code of the proposed algorithm

6.5 Experimental Comparison with Existing Approaches for Unordered Subtree Mining

In this section, a variety of experiments are performed in order to evaluate the developed framework for mining unordered induced/embedded subtrees. It also serves the other purpose of demonstrating some of the important issues mentioned throughout the book with respect to tree mining. In the last section of the chapter, we will summarize some general findings and indicate some directions for future work.

The existing algorithms for mining unordered subtrees mine either induced or embedded subtrees. In the developed framework, the induced subtrees are mined when the maximum level of embedding constraint is set to 1, and embedded when no such constraint is imposed. The former case where the developed framework is used for mining unordered induced subtrees will be referred to as the UNI3 algorithm (Hadzic, Tan and Dillon 2007), while the latter is referred to as the U3 algorithm (Hadzic, Tan & Dillon 2008) in the experiments provided in this section. The section concludes with experiments for evaluating the approach when the hybrid support definition is used. In a number of experiments we make use of the CSLogs dataset (Zaki 2005a) which is a representative of real-world Web log data represented in tree-structured form. It describes a set of web logs files collected at a computer science department. In total, there are 59,691 subtrees (transactions) describing user browsing of the computer science department website (Zaki 2005a). Even though the CSLogs datasets is a real-world file in its current form, it is an integer labeled tree, which prevents us from analyzing the patterns in a meaningful manner. However, throughout our experiments, we demonstrate the difference in the number of patterns extracted when different subtree types are mined and/or support definitions are used.

6.5.1 Testing the Approach for Mining of Unordered Induced Subtrees

The UNI3 algorithm is evaluated by comparing it with the HybridTreeMiner (HBT) (Chi, Yang & Muntz 2004b) and RootedTreeMiner (RTM) (Chi, Yang & Muntz 2004a), which is Chi's implementation of the Unot algorithm (Asai et al. 2003). Throughout the discussion, each algorithm will be referred to using its abbreviation as indicated in the brackets. The existing algorithms for the mining of unordered induced subtrees take only the transaction-based support into account, while the UNI3 algorithm can also handle the occurrence match support. To identify when transaction-based support is used, our algorithm is preceded by 'T-' (e.g. T-UNI3).

In Section 6.4.1, two preconditions to avoid expensive CF ordering were indicated. In the notation, a '- Lx' symbol is appended at the end, where x corresponds to the precondition explained earlier (e.g. UNI3-L1). Furthermore, the full (k-1) pruning option can be disabled in the UNI3 algorithm to indicate the difference implied. For occurrence match support, full (k-1) pruning is necessary to avoid the possibility of generating pseudo-frequent subtrees. However, when transaction-based support is used, the downward closure lemma automatically holds, and full (k-1) pruning is not required since the set of detected frequent subtrees will remain the same. Furthermore, a reduction in run-time is expected since, when generating all (k-1)-subtrees from a potentially frequent k-subtree, all (k-1)-subtrees need to be ordered into their canonical form so that their frequency is correctly verified. When no full (k-1) pruning is performed, the proposed algorithm is denoted as UNI3(NP). The minimum support σ is denoted as (sxx), where xx is the minimum frequency. In all the experiments except for one at the end of the section, the transaction-based support is used since the algorithms being compared can use only this support definition. The artificial tree databases were obtained by using the TreeGenerator (Zaki 2005a) software. Experiments were run on 3Ghz (Intel-CPU), 2Gb RAM, Mandrake 10.2 Linux machine and compilations were performed using GNU g++ (3.4.3) with the –O3 parameter.

6.5.1.1 Time Performance Test

For this test, the real-world CSLogs dataset (Zaki 2005a) and a synthetic dataset characterized by deep tree structures consisting of 20000 transactions were used. The T-UNI3-L2 implementation was used as that is the most optimized version of the proposed algorithm. As can be seen in Fig. 6.11, for both datasets the T-UNI3-L2 algorithm enjoys the best time performance. The table containing the recorded values of the experiment is also included below the figure to enable a more detailed comparison. Additionally, by not performing full (k-1) pruning (T-UNI3-L2(NP)) there was an additional performance gain because CF checking does not need to be performed for all the (k-1)-subtrees of a potentially frequent k-subtree.

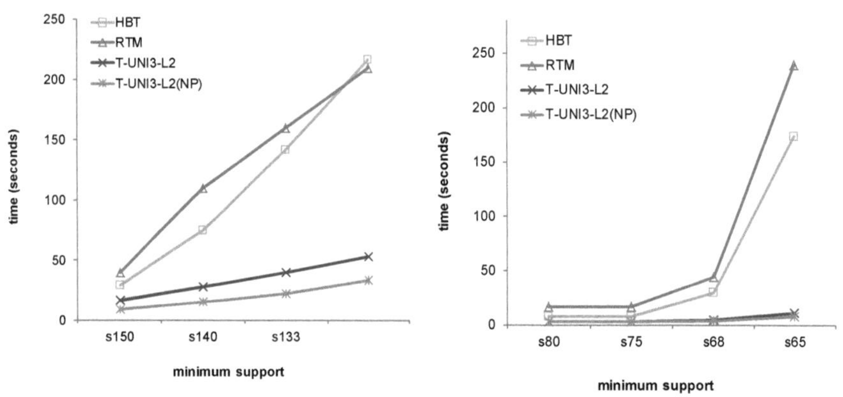

	T-UNI3-L2	T-UNI3-L2(NP)	HBT	RTM
s150	16.5115	8.89101	29.4412	39.5449
s140	27.6826	15.1947	74.671	109.493
s133	39.4045	22.3395	141.46	160.159
s130	52.6674	33.511	216.428	209.786

(a) CSLogs dataset

	T-UNI3-L2	T-UNI3-L2(NP)	HBT	RTM
s80	3.06816	2.84535	8.14945	16.6042
s75	3.06617	2.83918	8.18062	16.9161
s68	4.70423	3.8748	29.3586	44.6127
s65	10.8486	7.91074	173.741	239.532

(b) Deep tree dataset

Fig. 6.11 Time performance test

6.5.1.2 Scalability Test

For this experiment, a synthetic dataset consisting of 10,000 items was generated. The average depth and fan-out is equal to 40. The number of transactions generated

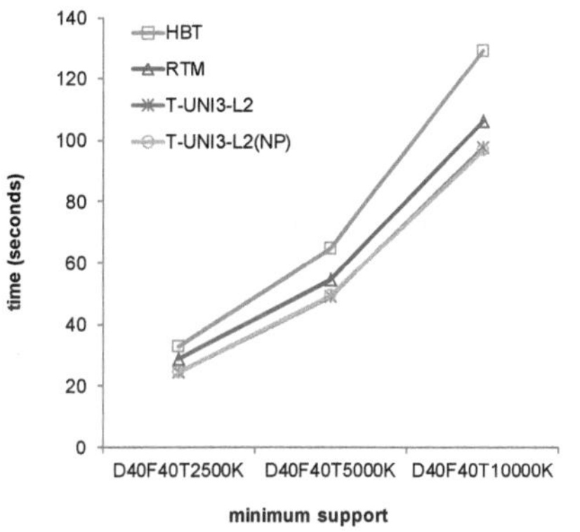

	s162 D40D40T2500K	s325 D40D40T5000K	s650 D40D40T10000K
HBT	32.5421	64.7807	129.175
RTM	28.8603	54.3724	106.483
T-UNI3-L2	24.286	48.8068	97.5331
T-UNI3-L2(NP)	24.5916	49.5139	96.934

Fig. 6.12 Scalability test

was 2.5M, 5M, 10M, for the different datasets, and the respective support threshold was 162, 325 and 650. From Fig. 6.12, one can see that all the tested algorithms are well scalable for the different dataset sizes used. The time performance is comparable among the algorithms with T-UNI3-L2 performing slightly better than others. When no pruning is performed (T-UNI3-L2(NP)), there is an additional performance gain. In general, performing full (k-1) pruning is expensive whenever numerous candidate subtrees are generated, since canonical transformations will be performed more frequently. However, for transaction-based support definition, full (k-1) pruning can be omitted since the downward closure lemma automatically holds, and becomes a problem only when repetitions of subtrees within transactions need to be counted, as is the case for occurrence-match or hybrid support.

6.5.1.3 Variations of UNI3 Test

This experiment compares the performance of the UNI3 variations. The aim is to demonstrate some of the important implementation issues that need to be taken into account when developing unordered tree mining algorithms. The L0 version does not implement any of the preconditions for avoiding the ordering of subtrees into their canonical form (CF). L1 implements the first precondition, and L2 enforces both preconditions explained in Section 6.4.1. Looking at the graph from Fig. 6.13, the implementation of both preconditions for CF ordering exceptions results in best time performance. This is because the CF ordering and checking is quite expensive and the avoidance of any of these operations will result in a time gain. Additionally, performing no full (k-1) pruning (NP) results in further time performance gain since, again, less CF ordering is needed among the (k-1)-subtrees of a potentially frequent k-subtree.

From Fig. 6.13(b), one can see why preconditions 1 and 2 affect the run-time performance of UNI3 variations. When no preconditions were applied, the number of canonical transformations performed by T-UNI3-L0 is 6,069,225. This number is further reduced when no full (k-1) pruning is performed, and consequently the run-time execution is sped up. By applying precondition 1, T-UNI3-L1, the number of canonical transformations performed is reduced to 2,231,049. In addition, if no full (k-1)-pruning is performed, T-UNI3-L1(NP), the number of canonical transformations performed, is reduced by about 0.3 times. Applying both preconditions will further increase the efficiency, as is evident from the performance of T-UNI3-L2 in

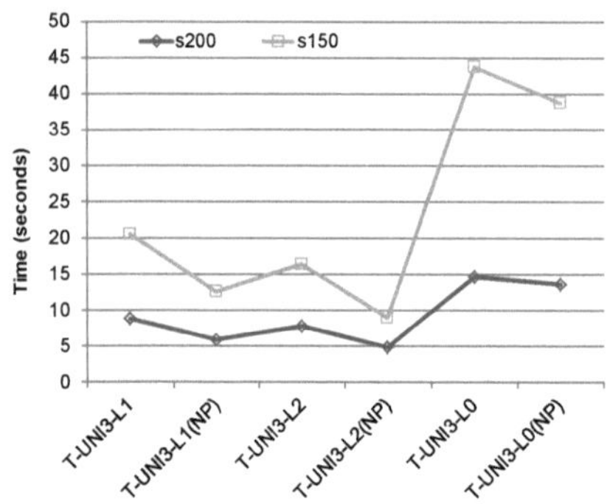

(a) Time Performance

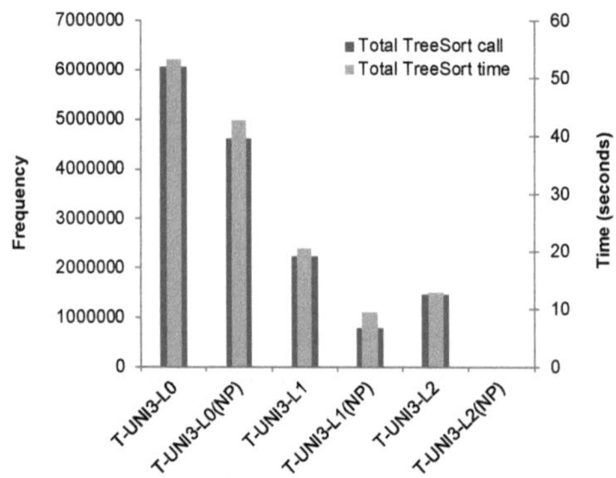

	T-UNI3-L0	T-UNI3-L0(NP)	T-UNI3-L1	T-UNI3-L1(NP)	T-UNI3-L2	T-UNI3-L2(NP)
TreeSort Call	6,069,225	4,613,622	2,231,049	775,446	1,455,603	0
TreeSort Time	53.2912	42.7983	20.5651	9.40213	12.9224	0

(b) Tree sorting statistics

Fig. 6.13 Testing UNI3 variations

Fig. 6.13. Finally, whenever no full (*k*-1) pruning is performed, T-UNI3-L2(NP), the optimum performance is achieved over all other variations of UNI3. There are no canonical transformations performed while the numbers of frequent unordered induced subtrees returned remain the same.

6.5.1.4 Occurrence-Match Support Test

The purpose of this experiment is to show the performance of the UNI3 algorithm when occurrence-match support is used. Since, to our knowledge, there are no current algorithms for mining induced unordered subtrees using occurrence-match support, we have included the performance of our algorithm for mining ordered induced subtrees for extra comparison. The dataset was artificially created using the TreeGen with |Tr|:1M, |N|:10,000, and |D|:40, |F|:40. Fig. 6.14 shows that our algorithm is well scalable when occurrence-match support is used. Furthermore, our T-UNI3-L2 algorithm enjoys the best time performance for this dataset when compared with HBT and RTM when transaction-based support is used.

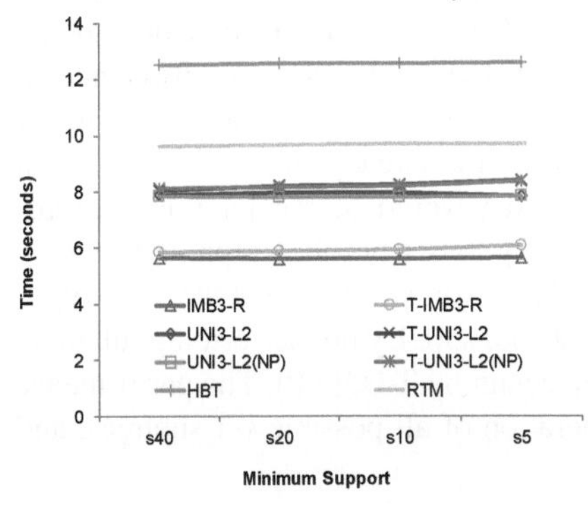

Fig. 6.14 Occurrence-match support test

6.5.2 Testing the Approach for Mining of Unordered Embedded Subtrees

The U3 algorithm is evaluated by comparing it with the SLEUTH (Zaki 2005b) algorithm which is the first algorithm that mines unordered embedded subtrees in a complete manner. Since the approach described in (Chehreghani et al. 2007) extracts maximal patterns rather than the complete set of frequent patterns, it is not suitable for comparison. For transaction-based support, the developed algorithm is preceded by 'T-' (e.g. T-U3). When no full (*k*-1) pruning is performed, (NP) is added at the end (e.g. U3(NP)). The differences caused by implementing the preconditions for avoiding unnecessary tree sorting were demonstrated in the previous section. In these sets of experiments, the U3 algorithm will always have both pre-conditions in place since that is the most optimal version of the algorithm implementation. TreeGenerator (Zaki 2005a) is used to obtain the artificial tree databases. The minimum support σ is denoted as (sxx), where xx is the minimum frequency threshold. Experiments were run on 3Ghz (Intel-CPU), 2Gb RAM, Mandrake 10.2 Linux machine.

6.5.2.1 Time Performance Test

For this test, we have used a reduced version (54%) of the real-world CSLogs dataset (Zaki 2005a) because when the full dataset is used all of the compared algorithms fail to return any results for a reasonable support threshold. This is a quite large dataset (59,691 transactions), and during the candidate enumeration process, many subtree occurrences have to be stored and processed which can cause some memory problems. To handle such situations, one could make use of distributed parallel processing so that the tree database is split, and the load balanced over a number of autonomous processing units. Merging of results would then take place to obtain a complete solution to the task. The dataset was progressively reduced by randomly removing some transactions and at 32,241 transactions, some interesting results appeared. As can be seen in Fig. 6.15, the T-U3 algorithm enjoys better time performance when compared with SLEUTH which has some performance issues for decreasing support value. We refrained from running it further for lower support thresholds (s100 and s150) since there was already such a large jump in performance from s205 to s200. The time taken would be too long and recording the result would increase the scale of the y-axis from the graph and thereby decrease the clarity of the displayed results. However, we show that the T-U3 algorithm still performed reasonably well for s100 and s50. By not performing full (k-1) pruning, there was an additional performance gain by T-U3(NP). The performance gain resulted firstly from avoiding the generation of all possible k-1 subtrees, and

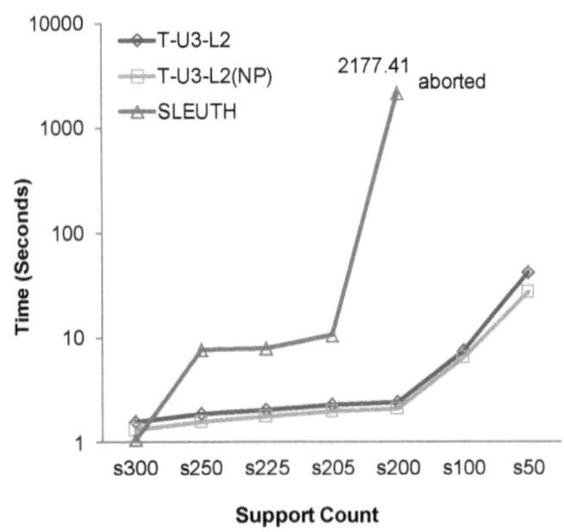

Fig. 6.15 Time performance test on CSLogs data

secondly because canonical form transformations do not need to be performed for all the (k-1)-subtrees of a potentially frequent k-subtree. Saying this however, it is worth noting that if occurrence match support were used, the time decrease by not performing full (k-1) pruning may not always occur. It is quite possible that there will be some trade-off with the enumeration of additional candidate and frequent subtrees (i.e. pseudo-frequent subtrees).

6.5.2.2 Scalability Test

For this experiment, a synthetic dataset was generated consisting of 10,000 items, with an average depth and fan-out of 40. The number of transactions was varied from 100K, 500K, 1M, for the different dataset sizes, and the respective transaction-based support thresholds were 25, 125 and 250. From Fig. 6.16, one can see that all the tested algorithms are well scalable for the different dataset sizes. The time performance is comparable among the algorithms with T-U3 performing slightly better than others.

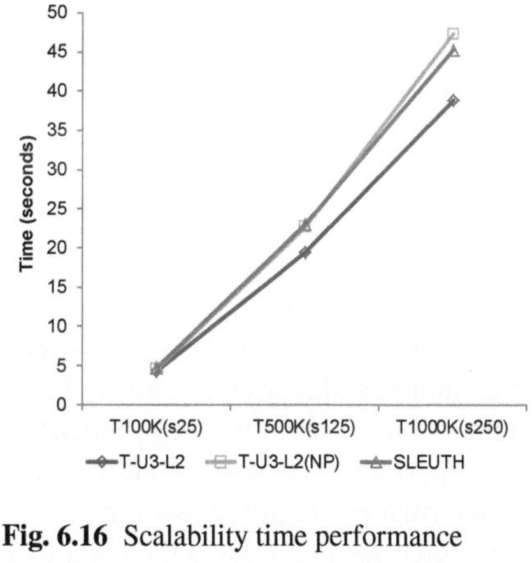

Fig. 6.16 Scalability time performance

6.5.2.3 U3 Variations Test

In this experiment, we compare the performance of the U3 variations. The dataset used for this experiment is generated by the TreeGen with |D|:40, |F|:40, |N|:10,000 and $|T_r|$:1M. We have shown earlier that the application of preconditions reduces the number of unnecessary canonical transformations. In this section, we are testing a variation of U3. L1 implements the *RMP* optimization and preconditions 1 & 2. L2 is the U3 version that implements the *RMP* and *RL* optimization. Please note that the *RMP* and *RL* optimization was discussed in detail in Chapter 4 and was experimentally tested in Chapter 5 when the TMG approach was used for the mining of ordered subtrees. The graph in Fig. 6.17 shows that the implementation of *RL* optimization gives a slightly better performance. First, it takes less time to generate

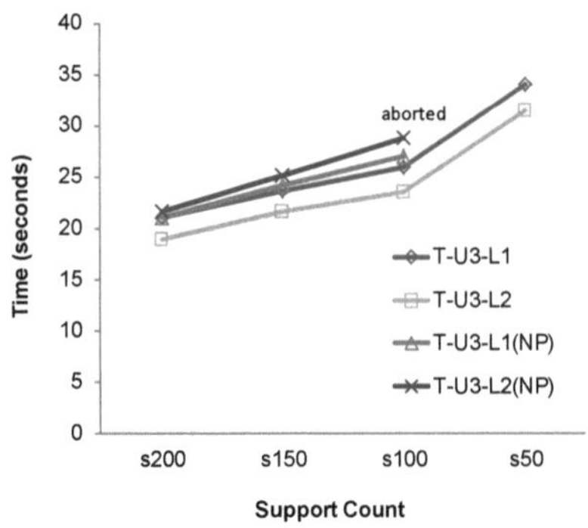

Fig. 6.17 Testing U3 variations

the *RL* than the *EL*. In addition, with the *RL* we perform less indirect calls to arrays, since the nodes coordinates stored in the *RL* are already the true coordinates of the nodes. With this dataset, the U3 variants that do not perform full (k-1) pruning perform slower and they generate many pseudo-frequent subtrees. The NP variants were aborted after the minimum support level was set to below 100 as the memory usage reached the capacity of the test machine. This is due to the large number of pseudo-frequent subtrees generated, when no pruning is performed.

6.5.2.4 Tree Database Canonical Transformation Test

Performing the canonical transformation is expensive. Thus, we try to avoid the canonical transformation process whenever possible, without affecting the quality of the result. By performing canonical transformation on the underlying tree structure of the original database, it is expected that more candidates will already be in their canonical form, thereby reducing the number of transformations required in the subsequent candidate generation process. This experiment was designed to confirm this by comparing the number of canonical transformations between two versions of U3: one that performs T_{db} canonical transformations (T_{db} (+CT)) and the other that does not perform canonical transformation on T_{db} (T_{db}(-CT)). CSLogs (Zaki 2005a) dataset was used.

Fig. 6.18 displays the difference in the number of canonical transformations performed at each k step (i.e. enumeration of subtrees with k nodes, at k = 2 no canonical form ordering is necessary). Transaction-based support definition was used and Fig. 6.18 reports results for support = 100. It can be seen that by performing canonical transformation on T_{db}, the numbers of canonical transformations performed during the candidate generation processes are reduced greatly. This is because most of the generated candidates are already in their canonical form. Using this approach, the computation of multiple unnecessary canonical transformations is avoided, which would have been necessary if the canonical transformation of the tree database was not performed. Consequently, the run-time performance is also further optimized by avoiding unnecessary canonical transformations during

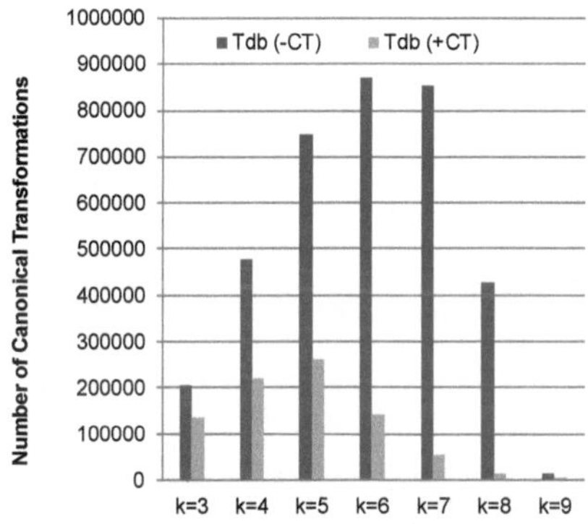

Fig. 6.18 Comparison of number of canonical transformations performed

candidate generation process as evident in Fig. 6.19. Performing the canonical transformation on the tree database is critical to overall performance, especially when mining embedded subtrees. The larger the numbers of canonical transformations to be performed, the more expensive is the overall processing. T-U3-L2(-), a version of U3 that performs no T_{db} canonical transformation clearly performs no better than does T-U3-L2(+) which performs T_{db} canonical transformation.

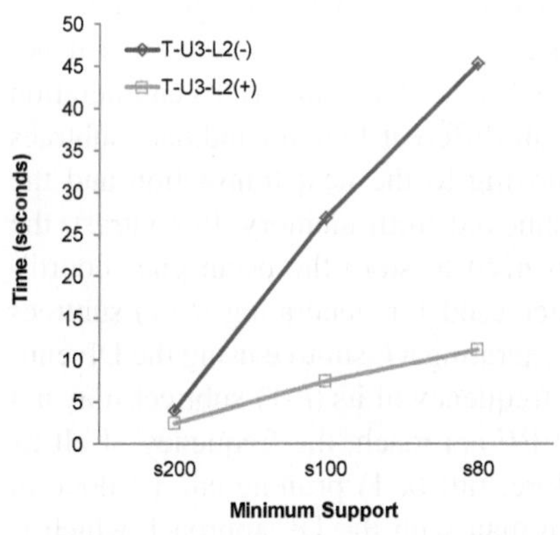

Fig. 6.19 Comparison of time performance on T_{db} canonical transformation

6.5.2.5 Occurrence Match Support Testing

These experiments were performed in order to check the performance of algorithms when the occurrence-match (weighted) support definition was used. We used an artificial data set where the underlying tree structure has an average depth and fan-out of 40, and consists of 1M transactions. As can be seen from Fig. 6.20(a), the U3

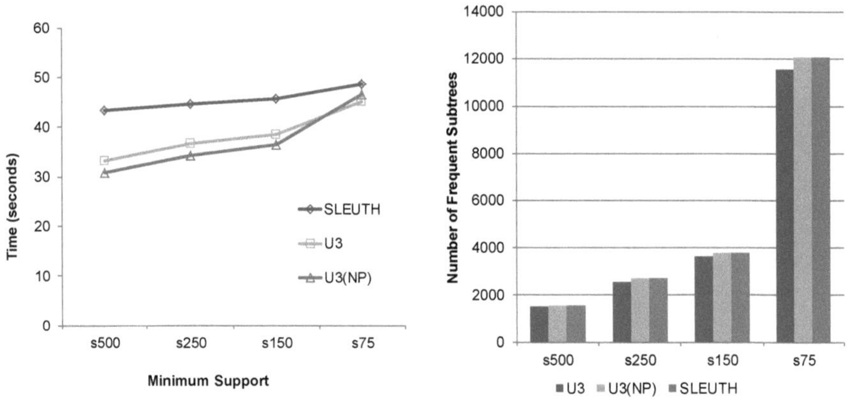

(a) Time performance (b) Number of frequent subtrees reported

Fig. 6.20 OS test

algorithm has a better time performance when compared with the SLEUTH algorithm. Whenever the minimum support is set to 75, we can see that both U3 variants start to degrade. By analyzing the process, we notice that both U3 variants get close to reaching the system memory capacity at which point they start to use the machine's virtual memory. Whenever a program uses virtual memory, it swaps data back and forth between the main memory and the secondary storage, which can significantly slow down any kind of processing. The SLEUTH on the other hand, conserves the system memory better as it utilizes the depth-first (DF) enumeration approach. A DF enumeration will generate all different length candidate subtrees from each transaction completely before moving to the next transaction and the information about that transaction can be removed from memory. In contrast, the breadth-first (BF) enumeration strategy will need to store the occurrence coordinates of generated k-subtrees which are later used for generating (k+1)-subtrees from the same transaction. However, when generating a k-subtree using the DF enumeration method, information regarding the frequency of its (k-1)-subtrees may not be available at that time; whereas, with the BF approach, the frequency of all its (k-1)-subtrees has been determined. Therefore, full (k-1) pruning can be done in a more complete way with the BF approach than with the DF approach which is forced to employ an opportunistic pruning (Zaki 2005a) strategy that prunes only infrequent subtrees in an opportunistic way. This may result in the generation of pseudo-frequent subtrees as is illustrated in Fig. 6.20(b). The SLEUTH algorithm and the no-pruning version of the U3 algorithm, U3(NP), generate additional frequent subtrees in comparison with the U3 approach that performs full (k-1)-pruning. Besides the reduction in performance of the U3 algorithm for lower support thresholds, in many cases as shown by other experiments, BF enumeration employed by the U3 can be very effective and more efficient whenever the memory space is not an issue.

6.5.2.6 C_k and F_k Generation

We have developed a simple dataset that represents the tree displayed at the top of Fig. 6.21. The aim is to show how the number of candidate and frequent subtrees enumerated by different approaches can vary even for such a simple dataset. As can be seen from the chart of Fig. 6.21, SLEUTH enumerates many more candidate subtrees due to the employed join approach which enumerates additional invalid subtrees. They are invalid in the sense that they do not conform to the underlying model of the tree structure. In contrast, the TMG approach is guided by the tree model and generates only valid candidate subtrees.

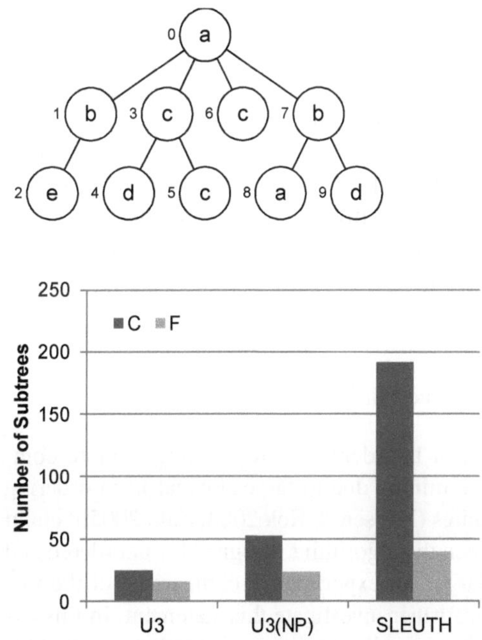

Fig. 6.21 Number of subtrees reported for the example tree at $\sigma = 2$

Because of the opportunistic pruning approach employed in SLEUTH, many more subtrees are considered as frequent in comparison with U3 (Fig. 6.21). As mentioned earlier in the book, we refer to these additional subtrees as pseudo-frequent subtrees. An example would be the subtree with encoding 'a b / c d' since it occurs twice in the tree (*oc*:0134 and *oc*:0734), whereas its infrequent (k-1)-subtree with encoding 'a c d' occurs only once (*oc*:034). Furthermore, to our surprise, the number of candidate and frequent subtrees enumerated differs between U3(NP) and SLEUTH. When we checked the extracted frequent subtrees, we noticed that SLEUTH considers certain subtrees as frequent (e.g. 'a b / b', 'a d / d'), while in our approach these are not considered as frequent. SLEUTH then further extends these subtrees during the candidate enumeration process which explains the large number of candidate and frequent subtrees enumerated by SLEUTH. The difference observed here is due to the difference in the U3 and SLEUTH algorithms, on what is considered as distinct occurrences of the subtree. To explain this difference, consider the subtree 'a b / b' with *oc*: '0 1 7'. When SLEUTH searches for the distinct occurrences of subtree 'a b / b' it finds *oc*: '0 1 7' and *oc*: '0 7 1', and considers these as distinct occurrences giving the subtree the support of 2. In the U3 algorithm, on the other hand, the subtree candidates are enumerated systematically as encountered in the tree database. In other words, the U3 algorithm works the other way, where we traverse the database and form candidates from all occurrences while ordering the corresponding encodings in canonical form to group all the different occurrences of the same unordered subtree entity as one. Hence, since subtree 'a b / b' is found at only one set of coordinates oc: '0 1 7' (regardless of the order), this candidate subtree will be generated only once, as U3 will not use the same set of occurrence coordinates twice during the candidate 3-subtree generation. In other words, our strategy is motivated by the fact that the 3 items of subtree 'a b / b' were associated together only once at occurrence oc:'0 1 7', since the order does not matter. To summarize, the SLUETH and U3 algorithms were developed with slightly different applications in mind, and the different way of counting the occurrences may be useful for different applications. In the U3 algorithm, a unique set of occurrences is considered only once, while SLEUTH also considers the variation of order in that set of occurrences.

6.5.2.7 Unordered vs. Ordered Subtree Mining

At the implementation level, the task of unordered subtree mining is more complex in comparison to ordered subtree mining due to the canonical form ordering that needs to be performed. Some studies (Nijssen & Kok 2003; Zaki 2005b) claim that the performance difference between the algorithms designed for unordered and ordered subtree mining is not very large. An experimental comparison of the two different tasks has been performed to further investigate this statement. In this experiment,we compare the developed framework for mining of unordered embedded subtrees with the developed framework for mining of ordered embedded subtrees (MB3) (Tan et al. 2005). The CSLogs dataset (Zaki 2005a) was used in this experiment. The minimum support σ is denoted as (sxx), where xx is the minimum frequency threshold. The transaction-based support definition was used and for the unordered embedded subtree mining algorithm, the notation used is T-U3; whereas, T-MB3 denotes the ordered embedded subtree mining algorithm. The T-U3(NP) corresponds to the U3 algorithm where the full (k-1) pruning option has been disabled. Since for the transaction-based support definition full (k-1) pruning does not need to be performed (as explained in Chapter 2), the resulting frequent subtree will be the same, but there is a reduction in time by not performing full (k-1) pruning.

One would expect that on the same dataset, the number of frequent unordered subtrees would be less than the number of frequent ordered subtrees, since many

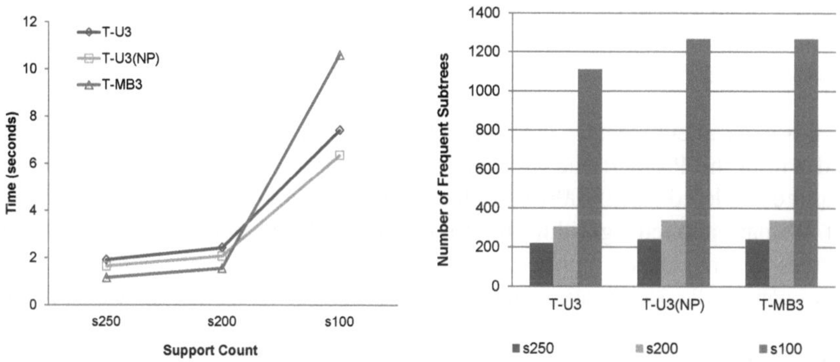

(a) Time performance ordered vs unordered (b) Number of frequent subtrees reported

Fig. 6.22 Ordered vs. Unordered test

ordered subtrees form the automorphism group of a single unordered subtree (Section 6.2). However, the results in Fig. 6.22(b) indicate that this is not necessarily the case and this can be explained as follows. When mining ordered subtrees, the order of sibling nodes becomes an additional criterion for grouping candidate subtrees. This makes the likelihood of two subtrees being grouped as the same candidate smaller for ordered subtree mining, and hence there is a reduced chance of a subtree being frequent. However, this does not mean that the run-time performance should always be faster for ordered subtree mining. As more unique candidates are generated, there is extra overhead added when performing frequency counting since more subtrees are hashed into the hash table. Each hashing operation costs processing time and the support threshold will be another factor determining the number of subtrees that need to be hashed at each candidate *k*-subtree generation. This explains the results from Fig. 6.22(a) since at lower support the time performance of T-MB3 starts to degrade.

To summarize, one could not draw a firm conclusion regarding the complexity difference between ordered and unordered subtree mining. It is highly dependent on the characteristics of the dataset used, since the frequency distribution of subtree patterns (attributes and values) throughout the dataset determines how many ordered and unordered frequent subtrees exist for a given support.

6.5.3 Additional Tests

In the previous section we have performed a number of experiments to evaluate existing approaches under different conditions with data of various complexity and structural characteristics. In this section we present some complementary experiments performed using more recently obtained real world data. For the remaining tests presented in this chapter, a different machine was used to conduct the experiments. They were run on Intel Xeon E5345 at 2.33 GHz with 8GB RAM and 4MB Cache Open SUSE 10.2.

6.5.3.1 Real World Data

Process Logs

For this experiment we have used publicly available process log dataset previously used within the ProM framework for process mining[1] (Dongen et al. 2005). The dataset considered ('a42f900R.xml) describes 1000 process instances in XML format. We have applied algorithms for mining both induced and embedded unordered subtrees on this dataset. Furthermore for the embedded case, the compared algorithms can handle occurrence match support, and hence we have included the results for both transactional (T-U3, T-SLEUTH) and occurrence match (U3, SLEUTH) support. The time performance is shown in Fig. 6.23(a). As one can see, for embedded subtree case, the U3 and T-U3 algorithms have better performance with respect to SLEUTH and T-SLEUTH, respectively. For induced subtrees, the T-UNI3 algorithm performs better when compared to HBT and RTM.

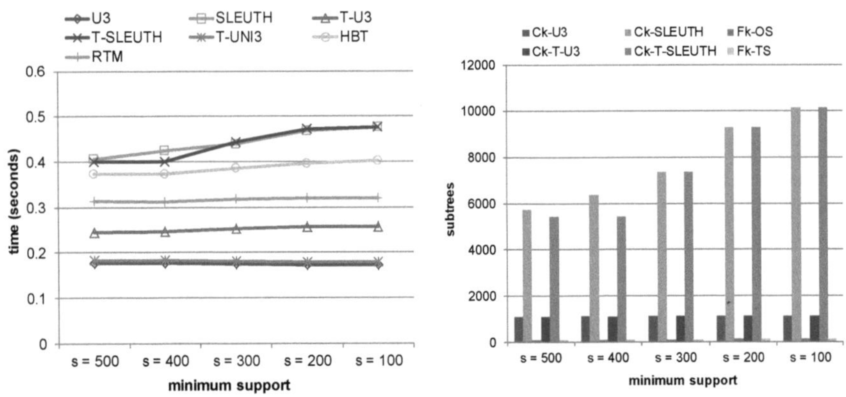

(a) Time performance

(b) Difference in number of subtrees enumerated

Fig. 6.23 Process log data

One interesting observation is the difference in time performance between SLEUTH and T-SLEUTH, and U3 and T-U3. When used for occurrence match support the SLEUTH algorithm takes slightly longer than T-SLEUTH, while in our implementations the opposite is true, as T-U3 takes longer than U3. This indicates another difference at the implementation level between the two approaches in the way that the frequency of the subtrees is counted. The reason for T-U3 taking longer is because of the extra logic (and space for transactional identifiers see Section 6.3.4) used for determining the transaction-based support. This can be confirmed by Fig. 6.23(b), where we show the number of candidate subtrees generated by the U3 (i.e Ck-U3, Ck-T-U3) and SLEUTH (i.e. Ck-SLEUTH, Ck-T-SLEUTH)

[1] See http://prom.win.tue.nl/tools/prom/

algorithms and the frequent subtrees (which are the same in both approaches) for occurrence-match (i.e. Fk-OS) and transaction-based (i.e. Fk-TS) support. We did not include the information regarding frequent induced subtrees as they are the same as for the embedded subtree case. For this dataset there was only a slight variation in the number of frequent subtrees obtained using occurrence-match or transaction-based support, which occurred at s = 500 (extra 3 in Fk-OS) and s = 400 (extra 9 in Fk-OS). It is also at these support thresholds that T-SLEUTH is faster because less candidate subtrees are generated as can be seen in Fig. 6.23(b). In our approach, the T-U3 is consistently slower than U3 because of the extra logic and space for storing transactional identifiers. However, the increase in time due to the additional candidate subtrees that had to be generated by U3, is not visible because of the TMG based enumeration approach rather than join.

University Courses

For this experiment we have used the publicly available XML data where each instance contains information regarding the available university courses (a total of 48640 transactions/instances)[2]. Using either support definitions would yield the same frequent subtrees as there are no repetitions of items within a transaction. Please note that we could not run the HBT and RTM algorithms on this data due to the format of the data not being suitable for those algorithms, thereby causing run time exceptions. In Fig. 6.24, we show the time performance and the number of frequent subtrees enumerated for the U3 and SLEUTH algorithms when transaction-based support is used. Please note that this dataset is characterized by rather shallow and wide trees, and as such algorithms utilizing depth-first enumeration approach are expected to perform better in comparison to algorithms utilizing a breadth-first approach (see Section 4.1). We have experimentally demonstrated this aspect in Chapter 5. This difference can be seen in Fig. 6.24 where for s = 5, the SLEUTH algorithm performs slightly better (by approximately 0.6 seconds). However, as the support is lowered, the U3 algorithm starts to perform better, because at such lower support thresholds the join approach for candidate enumeration utilised in SLEUTH will generate many more candidates in comparison to the TMG candidate enumeration approach utilized in U3. This can be seen at the bottom of Fig. 6.24, where we have shown the actual number of frequent subtrees (Fk) for each of the support thresholds and the number of candidate subtrees generated by the U3 (i.e. Ck-U3) and SLEUTH (i.e. Ck-SLEUTH) algorithms.

6.5.3.2 Hybrid Support Testing

This section provides an experiment to indicate the run time and the number of frequent subtrees extracted when the hybrid support definition is used within the proposed framework. With hybrid support being a newly proposed support definition, there are currently no other algorithms that the run time can be compared

[2] See http://library.cs.utwente.nl/xquery/docs/uni_5.xml

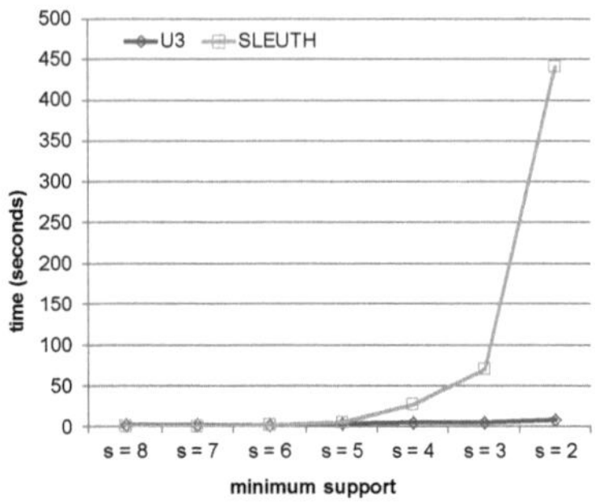

	s = 8	s = 7	s = 6	s = 5	s = 4	s = 3	s = 2
Fk	3557	4405	5795	13127	26633	40413	56950
Ck-U3	131412	131550	131737	203194	407935	635947	839383
Ck-SLEUTH	11540111	16968319	28540072	110822608	442943235	1029504968	1956387983

Fig. 6.24 University courses data

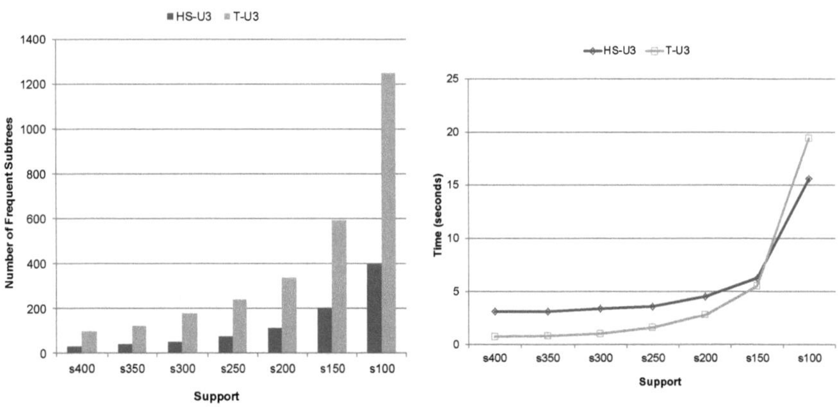

(a) Number of frequent subtrees (b) Time performance

Fig. 6.25 Hybrid support test

with. Since hybrid support can be seen as a specialization of the transaction-based support, the number of frequent subtree generated will be compared when hybrid support definition is used and when transactional based support is used. The U3 algorithm will be used to extract all frequent unordered embedded subtrees, and when hybrid support is used it is denoted as HS-U3, while for transaction-based support T-U3 notation is used as usual.

For this experiment, the CSLOGS dataset consisting of 32,421 transactions was used. Fig. 6.25(a) shows the number of frequent subtrees detected by the HS-U3 and T-U3 algorithms. The support thresholds displayed on the x-axis of the graph correspond to the transaction-based support used for the T-U3 algorithm. For the

HS-U3 algorithm the transactional part of the hybrid support remained the same and in addition to this the occurrence part was set to 2 for all the support values. Hence, for a given transaction support x, the HS-U3 algorithm will detect all the subtrees that occur at least 2 times in x transactions. The T-U3 algorithm will just check for the existence of a subtree in x number of transactions and, as can be seen from Fig. 6.25(a), the number of frequent subtrees detected by the T-U3 algorithm is larger, especially for lower support thresholds. Fig. 6.25(b) shows the time taken to complete the task by the HS-U3 and T-U3 algorithms. The HS-U3 algorithm usually takes longer to complete the task, and this is due to the additional logic required for determining whether a subtree is frequent. The T-U3 algorithm needs to check only for the existence of an item in the transaction, while the HS-U3 algorithm needs to take the number of occurrences in each unique transaction into account. The only time that T-U3 algorithm takes longer is for s100 (see Fig. 6.25(b)) and this is because many more frequent subtrees are enumerated for that support threshold (see Fig. 6.25(a)).

6.6 Conclusion

This chapter has presented a framework for the mining of unordered subtrees using the tree model guided (TMG) candidate subtree enumeration strategy. TMG is guided by the underlying tree structure of the document so that all the generated candidate subtrees are valid by conforming to this structure. The encoding of each enumerated subtree is ordered into the canonical form (if necessary) so that the tree isomorphism problem for unordered subtrees is appropriately handled. The TMG candidate generation process is effectively executed through the use of a suitable representation of the structural aspects of the tree database, i.e. the *recursive list* (RL). For efficient counting of subtree occurrences with respect to existing support definitions, we use the *vertical occurrence list (VOL)*. *VOL* stores a representation of a k-subtree encoding together with its occurrence information so that candidate (k+1)-subtrees can be generated by combining the occurrence information with information kept in the *RL*. The use of a general TMG framework in combination with the representative structures, allows us with some adjustments to mine unordered induced or embedded subtrees, using either transaction-based, occurrence match (weighted) or hybrid support.

The effectiveness of the developed framework is demonstrated through its high performance and scalability when experimentally compared with the current state-of-the-art algorithms. The main reason for the efficiency of the approach is the TMG candidate generation that uses the underlying structure of the tree database to generate only those candidates that conform to the tree model. The integration of two exceptions for avoiding expensive canonical form ordering was experimentally demonstrated to result in an improvement in time performance. Furthermore, the effective use of the *RL* structure has resulted in a more effective use of the available memory which often also reduces the time taken.

The experimental comparison of the approach with the algorithms for the mining of unordered induced subtrees has demonstrated the scalability and the efficiency of the proposed framework. The proposed algorithm produces better time performance than the current state-of-the-art algorithms. Furthermore, in the developed framework, any of the support definitions can be used, while the compared approaches can handle only the simpler transaction-based support.

To address the problem of mining unordered embedded subtrees, it was compared with the SLEUTH algorithm which is the first and currently state-of-the-art algorithm that tackles this problem in a complete manner. A number of experiments have indicated that the developed framework has a better time performance when compared with the SLEUTH algorithm. The SLEUTH algorithm cannot ensure that no pseudo-frequent subtrees will be generated when occurrence match support is used, as it does not employ full (k-1) pruning but rather adopts an opportunistic approach. When there are numerous pseudo-frequent subtrees, the overall processing cost outweighs the cost of performing full pruning. On the other hand, we notice that in some instances, the TMG approach has some limitations. One limitation is that, since it uses the horizontal enumeration approach, the memory usage tends to be larger than for vertical approaches. This can lead to poor performance on very large datasets where the minimum support is set to a very small value (i.e. enormous number of candidates), because during the candidate enumeration phase, it will start using the virtual memory. Performing complex computations in virtual memory is very slow. To overcome the limitation of the horizontal enumeration approach, one might consider a parallel architecture where the processing is split over a number of autonomous processing units.

Another performance degrading factor of the proposed framework is the expensive canonical transformations approach, which is basically a recursive sorting operation on labeled trees. Whenever many of candidate subtrees from the database consist of many nodes, this can become very expensive. Our approach for mining unordered induced subtrees performs well whenever we can minimize the number of canonical transformations performed during the candidate enumeration process. In most cases, our pre-conditions are very effective in preventing the canonical transformation process being performed on subtrees that are already in the canonical form. However, in the embedded subtrees, it is likely that the pre-conditions may not be sufficient to detect all subtree candidates already in their canonical form. In most cases, the pre-conditions appear effective in avoiding unnecessary transformation in our algorithm and hence, the task is completed in an efficient manner. However, on more difficult datasets characterized by complex tree structures, that can be many levels deep and can have many sibling nodes, the performance of our algorithm degrades accordingly. When we analyze the operations of the algorithm, we notice that the most expensive cases come from the non-tail extension, while our preconditions for no ordering are there only for tail extensions. This bottleneck can be improved by defining extra pre-conditions for avoiding canonical form ordering that will hold for the non-tail extensions.

One would expect the mining of unordered subtrees to be slightly more expensive than mining ordered subtrees because of the extra canonical transformations processing. We have found that this can differ from one case to another. The guiding principle is that the larger the number of unique subtrees generated, the higher the processing cost. From a given dataset, the numbers of generated unique automorphism groups $\subseteq$ ordered subtrees. We enumerate automorphism groups when we are mining unordered subtrees and we enumerate unique ordered subtrees when we are mining ordered subtrees. In cases where the size of the generated unique automorphism groups is much smaller than that of the generated ordered subtrees, we can expect that the run-time performance of algorithms that mine unordered subtree will be faster to a certain extent, despite the extra canonical transformation processing required.

References

1. Agrawal, R., Srikant, R.: Fast Algorithms for Mining Association Rules. In: Proceedings of the 20th International Conference on Very Large Data Bases (VLDB), Santiago de Chile, Chile, Septemebr 12-15, pp. 487-499 (1994)
2. Asai, T., Arimura, H., Uno, T., Nakano, S.-i.: Discovering Frequent Substructures in Large Unordered Trees. In: Grieser, G., Tanaka, Y., Yamamoto, A. (eds.) DS 2003. LNCS (LNAI), vol. 2843, pp. 47–61. Springer, Heidelberg (2003)
3. Chehreghani, M.H., Rahgozar, M., Lucas, C., and Chehreghani, M.H, Mining Maximal Embedded Unordered Tree Patterns. Paper presented at the Proceedings of the, IEEE Symposium on Computational Intelligence and Data Mining (CIDM), Honolulu, Hawaii, April 1-5 (2007)
4. Chi, Y., Yang, Y., Muntz, R.R.: Canonical forms for labeled trees and their applications in frequent subtree mining. Knowledge and Information Systems 8(2), 203–234 (2004a)
5. Chi, Y., Yang, Y., Muntz, R.R.: HybridTreeMiner: An Efficient Algorithm for Mining Frequent Rooted Trees and Free Trees Using Canonical Forms. Paper presented at the Proceedings of the 16th International Conference on Scientific and Statistical Database Management (SSDBM 2004), Santorini Island, Greece, June 21-23 (2004b)
6. Hadzic, F., Tan, H., Dillon, T.S.: UNI3 - Efficient Algorithm for Mining Unordered Induced Subtrees using TMG Candidate Generation. In: Proceedings of IEEE Symposium on Computational Intelligence and Data Mining (CIDM), Honolulu, Hawaii, USA, April 1-5, pp. 568–575. IEEE, Los Alamitos (2007)
7. Hadzic, F., Tan, H., Dillon, T.S.: U3 - mining unordered embedded subtrees using TMG candidate generation. In: Proceedings of the IEEE / WIC / ACM International Conference on Web Intelligence, Sydney, Australia, December 9-12, pp. 285–292 (2008)
8. Nijssen, S., Kok, J.N.: Efficient discovery of frequent unordered trees. In: Proceedings of the 1st International Workshop on Mining Graphs, Trees, and Sequences, Dubrovnik, Croatia (2003)
9. Tan, H., Hadzic, F., Feng, L., Chang, E.: MB3-Miner: mining eMBedded subTREEs using tree model guided candidate generation. In: Proceedings of the 1st International Workshop on Mining Complex Data in conjunction with ICDM 2005, Houston, Texas, USA, November 27-30, pp. 103–110 (2005)

10. Tan, H., Dillon, T.S., Hadzic, F., Chang, E., Feng, L.: IMB3-Miner: Mining Induced/Embedded Subtrees by Constraining the Level of Embedding. In: Ng, W.-K., Kitsuregawa, M., Li, J., Chang, K. (eds.) PAKDD 2006. LNCS (LNAI), vol. 3918, pp. 450–461. Springer, Heidelberg (2006)
11. Zaki, M.J.: Efficiently Mining Frequent Trees in a Forest: Algorithms and Applications. IEEE Transactions on Knowledge and Data Engineering 17(8), 1021–1035 (2005a)
12. Zaki, M.J.: Efficiently Mining Frequent Embedded Unordered Trees. Fundamenta Informaticae 66(1), 33–52 (2005b)

Chapter 7
Mining Distance-Constrained Embedded Subtrees

7.1 Introduction

For certain applications, the distance between the nodes in a hierarchical structure could be considered important and two embedded subtrees with different distance relationships among the nodes need to be considered as separate entities. The embedded subtrees extracted using the traditional definition are incapable of being further distinguished based upon the node distance within that subtree. In this chapter, we describe the extension of the general TMG framework, to enable the mining of distance-constrained embedded subtrees, (Hadzic 2008; Tan 2008). In such subtrees, the distances of the nodes relative to the root of the subtree need to be taken into account during the candidate enumeration phase. The distances of nodes relative to the root (node depth) of a particular subtree will need to be stored and used as an additional equality criterion for grouping the enumerated candidate subtrees. In Chapter 9, we will illustrate scenarios and applications where the mining of distance-constrained embedded subtrees would be preferable to mining of traditional embedded subtrees, since the extracted subtree patterns will be more informative. We also highlight the importance of distance-constrained subtree mining in the context of web log mining, where the web logs are represented in tree-structured form. In what follows, we will discuss the importance of distance-constrained embedded subtrees from a more general perspective and relate it to some previous work on extracting tree-structured queries.

As seen in the discussion in the book so far, there are a few options with respect to the choice of the tree mining parameters that can be used in the current frequent subtree mining framework. They all have their particular applications for which they will prove to be more appropriate than the other parameters. Once the set of all frequent subtrees has been extracted, one can perform a number of queries on the pattern set in order to detect specific information that is of interest to the user. For example, when posing queries on a tree database, it may be the case that some portion of the query tree is not known and/or is not considered interesting or important. This motivated the AtreeGrep system to perform approximate tree searching (Shasha et al. 2002) which allows for the queries that could contain 'don't

F. Hadzic, H. Tan, and T.S. Dillon: Mining of Data with Complex Structures, SCI 333, pp. 175–190.
springerlink.com

care' symbols to represent the irrelevant part of the query. They have been split into variable length don't cares (VLDCs) and fixed length don't cares (FLDCs). For a VLDC, the portion of the irrelevant part of the query tree can be substituted by 0 or more concepts in the data tree; whereas in FLDC, the irrelevant part is substituted by only a single concept. With the traditional subtree type definitions, the posing of such queries cannot be done directly or without any post-processing on the extracted frequent subtree set. In this sense, the distance-constrained embedded subtrees are related to queries containing VLDCs with the difference being that in the distance-constrained embedded subtrees, the distance of nodes relative to the root node is indicated. Queries with VLDCs are in a way equivalent to embedded subtrees since the irrelevant part of the query is replaced by the ancestor-descendant relationship between relevant parts of the query. However, there is no indication of the length of the irrelevant part and hence, if desired, one can use the distance constraint on embedding subtrees for additional information about the length of the irrelevant portion of the query. It is worth mentioning that the level of embedding mentioned in Chapters 4-6 could also be related in the sense that if we want to limit the allowed variable length in a query with VLDC, we can simply use the maximum level of embedding constraint. In fact, for queries with FLDCs where we are allowed only one level of the irrelevant portion, the maximum level of embedding constraint would be set to 2 so that we allow only ancestor-descendant relationships of depth 2. However, since the distance-constrained embedded subtrees indicate this information, they could be used for answering queries containing both VLDCs and FLDCs.

A similar remark could be made about the introduction of the wildcard symbol ('?') (Wang & Liu, 2000) in subtree patterns to match any label. By showing the distances of nodes relative to the root, the presence of a wildcard symbol in the subtree pattern can be worked out. For example, given two nodes a and b in a subtree where a is ancestor of b, and where the distance from a to the root is equal to x and distance from b to the root is equal to y, then if $(x - y) > 1$ it is known that there is a wildcard symbol between a and b. In fact, the expression $(x - y) - 1$ would indicate the number of wildcard symbols, or for VLDCs, the length of the tree pattern considered as don't care.

Speaking from a more general perspective, a distance-constrained embedded subtree is more informative than an embedded subtree and in some applications this additional information is considered important. While an induced subtree preserves the parent-child relationships from the original tree, in an embedded subtree the parent-child relationships are allowed to be ancestor-descendant relationships in the original tree. By mining embedded subtrees, one can detect commonly occurring relationships between data objects in spite of the difference in the level where the relationship in the document occurred. Certain concepts may be represented in a more specific/general way in certain documents. This specific information can often be in the form of additional child nodes of the concept, and hence, two general and related concepts may be separated by a number of levels in the document tree. If the user is interested only in the relationship between these two general concepts, such a relationship could be directly found in an embedded subtree set, while if induced subtrees were extracted, the information irrelevant to the user may be present in the patterns of interest. While mining of embedded subtrees is a generalization over induced subtrees, one limitation is that the context information may be lost in some patterns. For example, when analyzing a biomedical database containing patient records of potential illness-causing factors, one would be interested in the common set of data object properties that have frequently been associated with a particular disease. By allowing ancestor-descendant relationships, it may be possible to lose some information about the context in which the particular disease characteristic occurred. This is mainly due to the fact that some attributes of the dataset may have a similar set of values, therefore it is necessary to indicate which value belonged to which particular attribute. There appears to be a trade-off here and in this case allowing ancestor-descendant relationships can result in unnecessary and misleading information, but in other cases, it proves useful as it detects common patterns in spite of the difference in granularity of the information presented. A difficulty with embedded subtrees appears then to be that there is too much freedom allowed with respect to the difference in the distances between the nodes. All occurrences of one particular relationship are considered the same and valid even if the distance between the related data objects is so different making it possible that it occurred in a different context and has a different intended meaning. One way to avoid this characteristic is to further distinguish the embedded subtrees based upon the distance between the nodes, as is achieved by mining of distance-constrained embedded subtrees.

This notion of distance-constrained embedded tree mining will have some important applications in biological sequences, web information systems and conceptual model analysis. It is important to note here that by no means, is any claim being made that mining distance-constrained embedded subtrees should replace the mining of embedded subtrees. Mining of embedded subtrees without any constraint has still many important applications as will be discussed in Chapter 9.

In this study, we describe the necessary adjustments to the TMG framework to extract all ordered/unordered embedded subtrees with node distance information. It adds more granularity to the problem as the occurrences of the embedded subtrees with different distances among the nodes are now considered as different candidates (hence the name distance-constrained). Overall, the mining of distance-constrained subtrees is more expensive in terms of space and time required than when mining any other subtree types. The chapter is organized as follows.

In Section 7.2, we provide a formal definition of an distance-constrained embedded subtree, and formulate the problems addressed in this chapter. An example illustrating the scenario where imposing the distance-constraint on embedded subtrees will allow one to answer more specialized queries, is provided in Section 7.3. Section 7.4 describes the necessary adjustments to the general framework to mine ordered/unordered distance-constrained embedded subtrees. At the current stage, there are no other algorithms developed for this problem and hence, in Section 7.5 we will provide a few experiments that demonstrate mainly the scalability and highlight the differences in the number of subtree types detected. The conclusion of the chapter is given in Section 7.6.

7.2 Distance-Constrained Embedded Subtrees

As mentioned in the introduction, in order to allow for more specialized queries, an extension to the definition of subtree types to be extracted may be needed. One of the extensions made is to the current embedded subtree definition so that the distances of nodes relative to the root node in the original tree are taken into account. The updated definition of an embedded subtree can be stated as follows for unordered and ordered embedded subtrees:

Given a tree $S = (vs_o, V_S, L_S, E_S)$ and tree $T = (vt_o, V_T, L_T, E_T)$, S is an ***unordered distance-constrained embedded subtree*** of T, denoted as $S \prec_{ude} T$ *iff* (1) $V_S \subseteq V_T$; (2) $L_S \subseteq L_T$, and $L_S(v) = L_T(v)$; (3) if $(v_1, v_2) \in E_S$ then $parent(v_2) = v_1$ in S and v_1 is ancestor of v_2 in T; and (4) $\forall v \in V_S$ there is an integer stored indicating the level of embedding between v and the root node of S in the original tree T.

Given a tree $S = (vs_o, V_S, L_S, E_S)$ and tree $T = (vt_o, V_T, L_T, E_T)$, S is an ***ordered distance-constrained embedded subtree*** of T, denoted as $S \prec_{ode} T$ *iff* (1) $V_S \subseteq V_T$; (2) $L_S \subseteq L_T$, and $L_S(v) = L_T(v)$; (3) if $(v_1, v_2) \in E_S$ then $parent(v_2) = v_1$ in S and v_1 is ancestor of v_2 in T; (4) $\forall v \in V_S$ there is an integer stored indicating the level of embedding between v and the root node of S in the original tree T; and (5) the left-to-right ordering among the sibling nodes in T is preserved in S.

With these definitions in place, we now define the two problems addressed in this chapter:

1. Extract a set of subtrees *ST* from T_{db} where *ST* contains the complete and non-redundant set of subtrees (*S*) from T for which the following conditions hold: (a) $S \prec_{ude} T$ (i.e. S is an unordered distance-constrained embedded subtree of T) and (b) $\sigma(S) \geq \sigma$.
2. Extract a set of subtrees *ST* from T_{db} where *ST* contains the complete and non-redundant set of subtrees (*S*) from T for which the following conditions hold: (a) $S \prec_{ode} T$ (i.e. S is an ordered distance-constrained embedded subtree of T) and (b) $\sigma(S) \geq \sigma$.

Within these problems, any support definitions can be considered since, the use of *vertical occurrence list* structure (*VOL*) (Chapter 4 to 6) allows one to determine the support of a subtree for any support definition. The *VOL* and the frequency determination process remain unchanged in the adjustment to incorporate distance constraint; hence, with the same framework, all support definitions can be considered.

The main implication of mining embedded subtrees using this constraint is that the distance between the nodes will be used as an additional candidate grouping criterion. Since many embedded subtrees can form one candidate, this adds more granularity since each of those subtrees is now grouped as a different candidate when the distance among the nodes is different. Hence, mining distance-constrained subtrees is more expensive in terms of space and time required. As a simple illustration of the difference that this additional distance constraint will impose on the task of mining embedded subtrees, consider the example tree shown in Fig. 7.1. If the traditional

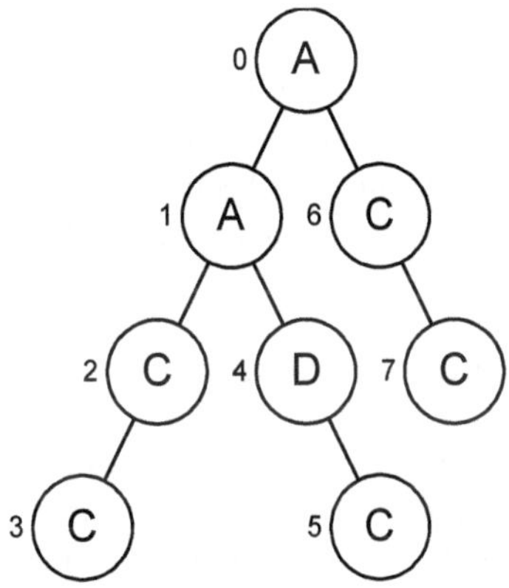

Fig. 7.1 Example tree with labeled nodes ordered in pre-order traversal

mining technique for embedded subtrees is used, a subtree 'A C' by occurrence-match support definition would have support equal to 8. On the other hand, if the distance equality constraint is added, we would need to divide this candidate into three candidates depending on the varying distance between the nodes. Hence, the three 'A C' subtree candidates would have varying distances of 1, 2 and 3 and the support of 2, 4 and 2 respectively.

For subtrees with more nodes, the stored distance for each node will correspond to its distance from the root of that particular subtree. To illustrate the difference in mining of different subtree types, please consider the example tree in Fig. 7.2 where the label of each node is shown with its pre-order position on the left. If ordered or unordered induced subtrees are mined, *st* occurs only once in *T* with occurrence coordinates *oc*:[0, 1, 5, 6, 9], while for ordered embedded subtrees it also occurs at *oc*:[0, 1, 5, 8, 9] since the ancestor-descendant relationships between nodes 'c' (*oc*:5) and 'd' (*oc*:8) are allowed. With unordered embedded subtrees, the order can also be exchanged and hence *st* also occurs at *oc*:[0, 1, 5, 8, 7]. If unordered distance-constrained subtrees are mined, each of the three occurrences of *st* will be considered as a separate subtree depending on the distance of the nodes to the root

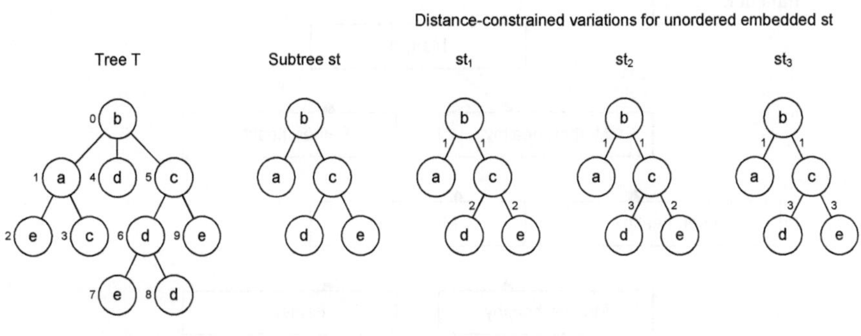

Fig. 7.2 Example tree *T* and subtree *st* with its distance-constrained variants

of the subtree as detected in T (i.e. st_1, st_2 and st_3 in Fig. 7.2.). The numbers next to the link indicate the distance between the nodes connected by that link in T. Hence, st_1 is a representative of the subtree with *oc*:[0, 1, 5, 6, 9], st_2 of *oc*:[0, 1, 5, 8, 9] and st_3 of *oc*:[0, 1, 5, 8, 7]. Valid ordered distance-constrained embedded subtrees are st_1 and st_2 since in st_3 the nodes 'd' and 'e' occur in different order than in T.

7.3 Motivation for Integrating the Distance Constraint

In the introduction we explained the limitation of embedded subtrees and how the distance information that distance-constrained embedded subtrees provide can be useful in certain applications. In this section, we will use an example to further illustrate this aspect. In Fig. 7.3 we display two example trees which indicate a part of the ancestor family tree from two ill patients (the examples were extracted from an image of a disease family tree obtained from (Access Excellence 2007)). Such information is used for linkage analysis of an illness by performing gene testing which can provide information about someone with a disease-related gene mutation.

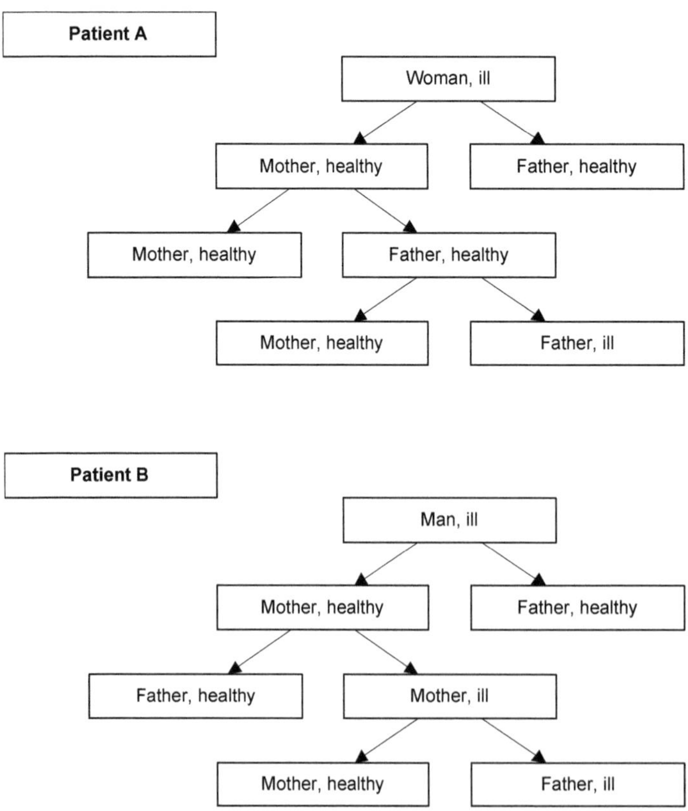

Fig. 7.3 Example representation of two ill patients (A and B) with common ancestors

When looking for a disease gene, scientists often start by studying DNA samples from family members over several generations who have a number of relatives who have developed an illness (Access Excellence 2007).

For example, a scientist may want to discover how many ill relatives an ill patient has had and to discover the number of generations that separates them. Using the traditional embedded subtree definition, we can extract information only about the number of ill relatives, but cannot obtain the information about the number of generations that separate the patient from the relatives that have a common disease. This is because the traditional embedded subtree definition does not have this kind of expressive capability. In contrast, by utilizing the distance-constrained embedded subtrees, we can find out exactly how many generations separate them, by inspecting the distance information stored between the nodes. From Fig. 7.3, patient A has only one diseased ancestor and it is her great-grandfather, while patient B has two diseased ancestors: a grandmother and great-grandfather.

Even though the figure does not provide such an example, it is worth noting that it could well be the case that an ill patient will have two ancestors of the same gender with the illness. In this case, the traditional embedded subtree definition would group these subtree occurrences as one candidate and indicate wrongly that there is only one ancestor with a disease. On the other hand, by mining distance-constrained embedded subtrees, both occurrences will be considered as separate entities due to the difference in the distance to the root node which is used as an additional candidate grouping criterion.

7.4 Extending TMG Framework to Extract Distance-Constrained Embedded Subtrees

In this section, we describe the necessary adjustments to the general TMG framework in order to enable the mining of distance-constrained embedded subtrees (DCES). The core process of candidate enumeration remains largely unchanged and hence we will not be repeating the general steps for ordered and unordered mining using the TMG framework that was explained in Chapters 6 and 7, respectively. To correctly enumerate DCES, at the implementation level this means that for counting purposes, when each subtree is represented uniquely by its string encoding, the distance of nodes should be taken into account. Therefore, the only necessary adjustment is in the way candidate subtrees are represented using the string encoding which is explained next.

7.4.1 Tree Representation

When the string encoding of a subtree is to be determined, the general idea of listing the node labels as encountered in the pre-order traversal of a tree remains, with the following adjustment. For each node, the distance to the root is worked out from the node levels stored in the *recursive list* (*RL*). These node levels correspond to the levels of the nodes in the original tree to be mined. The root of the subtree is assigned the depth of 0 and all other nodes are assigned the difference between their depth and the depth of the root of the subtree. The stored distance information corresponds now to the depths of nodes within the newly encoded subtree. A string is more easily manipulated when there is a uniform block size. Since each node now has an additional integer stored to indicate its distance to the root node, the block size is larger than the one occupied by a backtrack symbol ('/') in the encoding. For this reason, the string encoding was further modified to store a number next to each backtrack '/' symbol indicating the number of backtracks in the subtree, as opposed to storing each of those backtracks as a separate symbol.

To summarize, a string encoding for distance-constrained embedded subtrees can be generated by adding node labels in a pre-order traversal of a tree and appending an integer number denoting the distance to the root node, and appending a backtrack symbol ('/') whenever we backtrack from a child node to its parent node with an integer number denoting the number of backtrackings that occur between the current child node to the next node. To illustrate this modified string encoding, please refer again to Fig. 7.2. The following are the encodings for the trees presented in that figure: $\varphi(T)$:'b0 a1 e2 /1 c2 /2 d1 /1 c1 d2 e3 /1 d3 /2 e2 /2', $\varphi(st_1)$:'b0 a1 /1 c1 d2 /1 e2 /2', and $\varphi(st_2)$: ''b0 a1 /1 c1 d3 /1 e2 /2'. The backtrack symbols can be omitted after the last node, e.g. $\varphi(st_3)$: ''b0 a1 /1 c1 d3 /1 e3'.

7.4.2 Mining Ordered Distance-Constrained Subtrees

To enumerate all the ordered distance-constrained embedded subtrees, the process remains exactly the same as described in Chapters 5 and 6, except that all subtrees during the candidate enumeration phase are represented as described in Section 7.4.1. Since the representation is unique for every ordered distance-constrained embedded subtree, all the occurrences will be counted appropriately by hashing the unique string encoding.

7.4.3 Mining Unordered Distance-Constrained Subtrees

The process for the enumeration of all the unordered distance-constrained embedded subtrees remains exactly the same as described in Chapter 6, except that all subtrees during the candidate enumeration phase are represented as described in the previous section. Since the representation is unique for every unordered distance-constrained embedded subtree, all the occurrences will be counted appropriately by hashing the unique string encoding. However, another adjustment that needs to be made to the general framework is the ordering of unordered distance-constrained subtrees into their canonical form. In Chapter 6, we explained the depth-first canonical form (DFCF) ordering scheme used within our framework. The rules for canonical form (CF) ordering, as described in the previous chapter, perform this correctly when there is no distance information associated with each label. When there is an integer associated with each label, this needs to be taken into account during the ordering process. Hence, in addition to the ordering rules described in Chapter 6, we add another condition in cases where the node labels being compared are the same. In this case, the distance information associated with each node will be considered and the nodes with smaller distances to the root of the tree are placed to the left. This is necessary in order to count the occurrences correctly. In a standard unordered subtree, when all the conditions are checked, the nodes with the same labels and either no children or the same descendant, are just left as they are found. That is, they are considered as the same subtrees and hence it is not necessary to order them. However, when there is distance information associated with each node, it needs to be used as an additional check. This is because, if they are not ordered according to a rule, it is possible that the string encoding of the same entity will be counted as different entities. To illustrate this aspect, please consider Fig. 7.2 again. Let us say that we are dealing with 3-subtrees and are hashing the encodings of distance-constrained instances of embedded subtree with encoding b e / e. This subtree occurs three times in *T*, i.e. *oc*:[0,2,7], *oc*:[0,2,9] and *oc*:[0,7,9] and the respective distance-constrained subtree encodings for those subtrees are: b0 e2 / e3, b0 e2 / e2 and b0 e3 / e2. If we did not take the distance information into account, then it would be considered that these subtrees do not need to be sorted since the sibling nodes have the same label. However, they all have different encodings and when hashed would all be considered as separate candidates. Given that the tree *T* represents a single transaction in a database, for occurrence-match support set to two, none of these subtrees would be frequent. In fact, subtrees b0 e2 / e3 and b0 e3 / e2, are both members of the automorphism group of an unordered distance-constrained embedded subtree. With the additional condition, subtree b0 e3 / e2 with *oc*:[0,7,9] would have the order of sibling nodes exchanged. Since the same string encoding is then hashed on both instances, the subtree would correctly be determined to be the same entity as candidate subtree b0 e2 / e3 with *oc*:[0,2,7] and its support correctly set to 2. Compared with the processes required when mining other subtree types, the mining of unordered distance-constrained embedded subtrees can be considered as the most complex task in the TMG framework.

7.5 Experimental Results and Discussion

This section presents a number of experiments performed for evaluating the described framework for the mining of distance-constrained embedded subtrees. The approach for the mining of ordered distance-constrained embedded subtrees is referred to as RAZOR (Tan et al. 2006), while the approach for unordered distance-constrained embedded subtrees is referred to as u3Razor (Hadzic, Tan & Dillon 2008a). To the best of our knowledge, there are no other algorithms available that mine embedded subtrees with the additional distance constraint. These experiments are designed to test the scalability and compare the difference in the number of frequent subtrees enumerated.

7.5.1 Ordered Distance-Constrained Embedded Subtrees

In this section, we present some of the tests performed using the RAZOR algorithm. Firstly, we show the scalability of the approach followed by the comparisons with the MB3 (ordered embedded) and iMB3 Miner (ordered induced) algorithms (Chapter 5) on the grounds of the number of frequent subtrees generated for varying support and level of embedding thresholds. When addressing the problem of mining frequent subtrees, most of the time and space complexity comes from the candidate enumeration and counting phase. The distance-constrained subtrees are much larger in number than are induced or embedded subtrees and, in general, an algorithm for this task would require more space and run-time. Note that the occurrence-match support definition is used in all the experiments. The minimum support σ is denoted as (sxx), where xx is the minimum frequency. Experiments were run on 3Ghz (Intel-CPU), 2Gb RAM, Mandrake 10.2 Linux machine and compilation was performed using GNU g++ (3.4.3) with the –O3 parameter.

7.5.1.1 Scalability Test

The purpose of this experiment is to test whether the algorithm is scalable with respect to the increasing number of transactions present in a database. We use the 100Ks, 500Ks, and 1Ms artificial databases with minimum supports 50, 250, and 500 respectively. The result in Fig. 7.4 shows that the time to complete the operation scales linearly with the increase in transaction size.

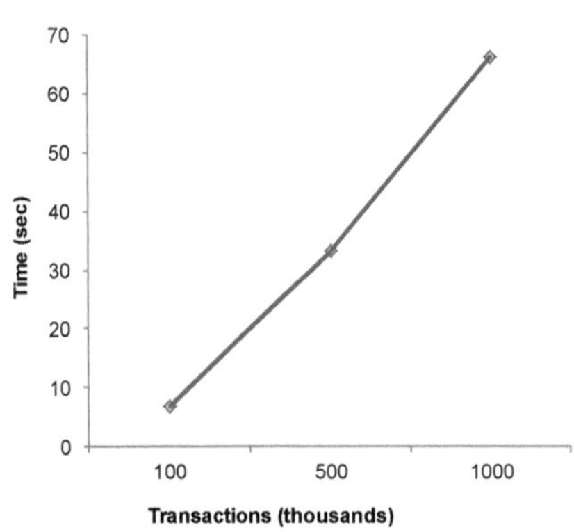

Fig. 7.4 Scalability test

7.5.1.2 Frequent Subtrees with Different Support

For this experiment, we used a reduction of the CSLogs dataset. If the occurrence-match support definition is used, the transaction number would need to be reduced

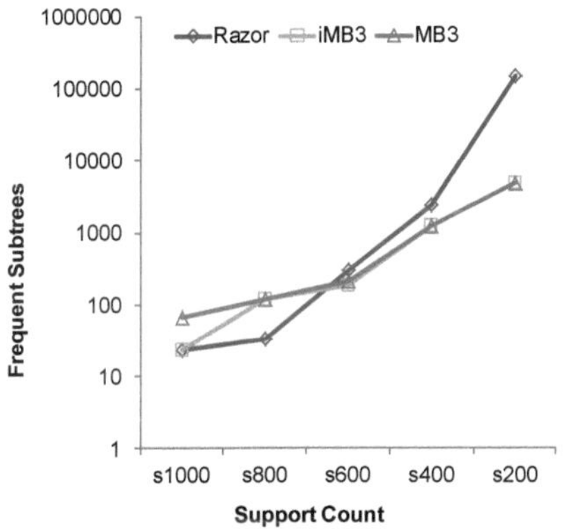

Fig. 7.5 Number of frequent subtrees detected for varying support thresholds

so that the tested algorithms could return the set of all frequent subtrees. We have randomly reduced the number of transactions from 52,291 to 32.421. The comparison of the number of frequent subtrees generated among the MB3, iMB3 and RAZOR algorithms, for varying support thresholds is presented in Fig. 7.5.

iMB3 and RAZOR were set with the maximum level of embedding (δ) constraint (Chapter 4) equal to 5. Consequently, the number of detected frequent subtrees differs slightly between the iMB3 and MB3 algorithms. In Fig. 7.5, we can see that the number of frequent subtrees detected by the RAZOR algorithm increases significantly when the support is lowered. This is due to the additional distance relationship used to uniquely identify candidate subtrees. As shown earlier, an embedded subtree may be split into multiple candidate subtrees when the distance is taken into account. As the support decreases, many more subtrees will be frequent and hence, many more variations of those subtrees with respect to the distance among the nodes will also be frequent. This explains such a large jump in the number of frequent subtrees detected by the RAZOR algorithm, when the support threshold is lowered sufficiently.

7.5.1.3 Varying the Level of Embedding

The purpose of this experiment is to compare the number of frequent subtrees detected by the algorithms when the maximum level of embedding (δ) is varied. We use the Deeptree dataset whose characteristics were explained in Table 5.1 from Chapter 5. The minimum support threshold was set to 100. The MB3 algorithm does not restrict the level of embedding and hence, the number of frequent subtrees detected remains the same for all cases. As the largest level of embedding of the dataset tree is 17, the MB3 miner detects the largest number of frequent subtrees.

Variation in the number of frequent subtrees detected by the RAZOR and the iMB3 algorithm can be observed in Fig. 7.6. At δ:1, the same number of subtrees

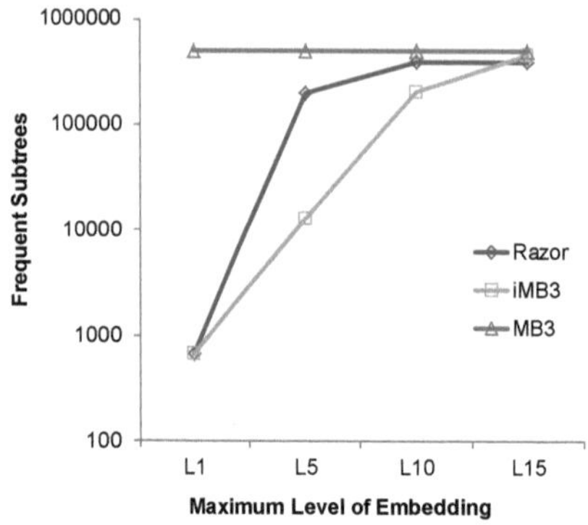

Fig. 7.6 Varying the level of embedding

is detected since no extra candidates can be derived when mining induced subtrees (i.e. the distance between the nodes is always equal to 1).

When the δ threshold is increased, the RAZOR algorithm detects more frequent subtrees for the reasons explained in the previous section. However, when the δ threshold was increased to 15, the iMB3 algorithm detected more subtrees as frequent. When such a high level of embedding is allowed, many previously infrequent embedded subtrees will become frequent as there is more chance for their recurrence. On the other hand, the RAZOR algorithm may further divide each of those subtrees based upon the distance of nodes relative to the root, and the frequency of the new candidate subtrees may not reach the support threshold. This explains why iMB3 has detected a larger number of frequent subtrees at δ:15, in comparison with the RAZOR algorithm. As expected, it can also be seen that as the δ threshold is increased, the number of frequent subtrees detected by iMB3 algorithm gets closer to the number detected by the MB3 algorithm where no δ restriction applies.

7.5.2 Unordered Distance-Constrained Embedded Subtrees

This section presents a number of experiments performed for evaluating the presented framework for the mining of unordered distance-constrained subtrees, referred to as the u3Razor (Hadzic, Tan & Dillon 2008a). The first experiment is to show the scalability of the approach with the additional distance constraint imposed on the embedded subtrees to be mined. The second experiment shows the difference between the number of frequent subtrees detected among the u3Razor, UNI3 (Hadzic, Tan & Dillon 2007) and U3 (Hadzic, Tan & Dillon 2008b) algorithms. UNI3 mines unordered induced, and U3 mines unordered embedded subtrees. The experiments were run on Intel Xeon E5345 at 2.33 GHz with 8 cores, 8 GB RAM and 4MB Cashe Open SUSE 10.2.

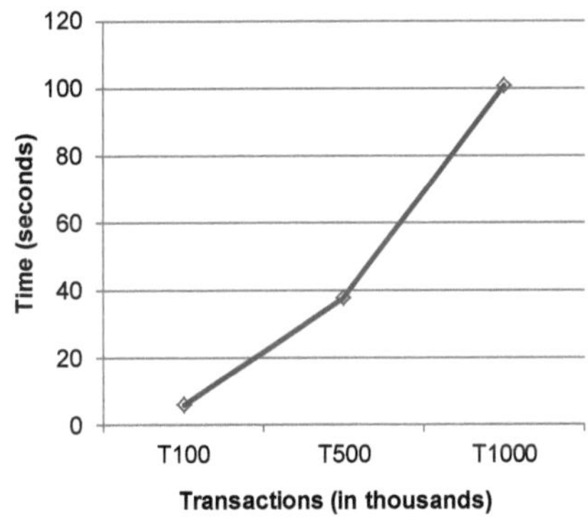

Fig. 7.7 Scalability – time performance / number of transactions

7.5.2.1 Scalability

The purpose of this experiment is to test whether the proposed u3Razor algorithm is scalable with respect to the increasing number of transactions present in a database. An artificial database was created, where the size of the transactions for each test was varied from 100,000, 500,000 to 1 million with minimum support 50, 250, and 500 respectively. For this test, the occurrence-match support definition was used. The result displayed in Fig. 7.7 shows that the time to complete the operation scales approximately linearly with the increase in transaction size.

7.5.2.2 Number of Frequent Subtrees Detected

For this experiment, the CSLogs data set consisting of 32421 transactions was used and transaction-based support definition was used. The comparison of the number of frequent subtrees generated among the T-U3Razor, T-U3 and T-UNI3 algorithms, for varying support thresholds is presented in Fig. 7.8. The number of frequent subtrees detected by the T-U3Razor and T-UNI3 algorithms is equal for support thresholds 1000, 800 and 600. At these supports, the U3 algorithm detects additional subtrees as frequent. These results imply that for all those support thresholds, all the additional embedded subtrees detected occur with a different distance among the nodes in the original tree. Otherwise, a number of them would have been detected by the U3Razor algorithm, where the distances have to be the same. It should be noted here that there may be some embedded subtrees that occur with the same distance among the nodes, but the number of occurrences of such subtrees is not sufficient to be considered as frequent by the U3Razor algorithm for the given support. Since there are no embedded levels among the nodes in induced subtrees (i.e. the level of embedding among all the nodes is equal to 1), both U3Razor and UNI3 detect the same frequent subtrees in this scenario. For lower support thresholds, more subtrees will be considered as frequent. The difference between the number

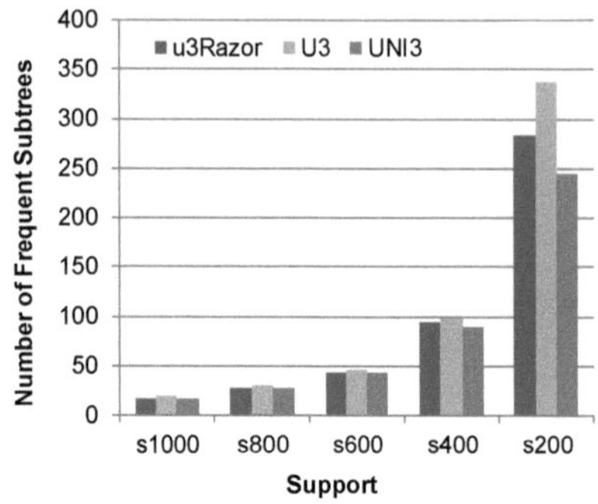

	u3Razor	U3	UNI3
s1000	17	19	17
s800	27	30	27
s600	42	46	42
s400	94	99	90
s200	283	337	244

Fig. 7.8 Number of frequent subtrees detected for varying support thresholds

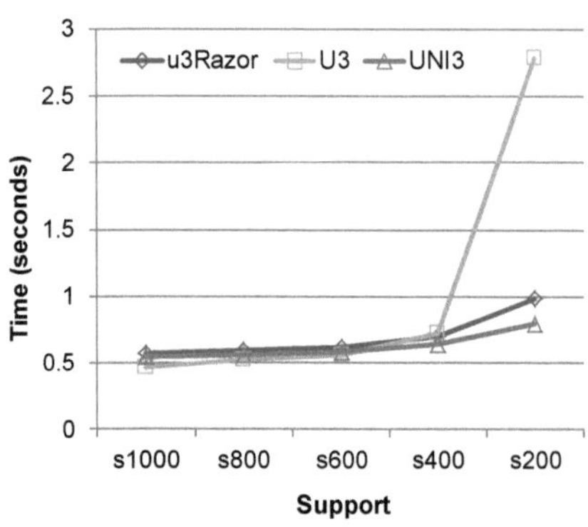

Fig. 7.9 Time performance

detected by U3Razor and UNI3 in Fig. 7.8 indicates that there are some sufficient occurrences of embedded subtrees where the distance among the nodes is the same in the original tree. There are 4 such embedded subtrees for s400 and 39 for s200.

Fig. 7.9 shows the time taken by the algorithms to complete this task. For most support thresholds, the u3Razor algorithm takes slightly longer, due to the fact that the level information needs to be stored for all the nodes of the enumerated subtrees.

At s200, the U3 algorithm takes the longest which is explained by the additional 54 subtrees that it considers as frequent in comparison with U3Razor (see Fig. 7.8).

7.6 Conclusion

In this chapter, we have extended the traditional definition of embedded subtrees in order to take into account the distance amongst the nodes. As the traditional embedding definition allows too much freedom with respect to the frequent subtrees extracted, we felt that an extra grouping criterion was required. We have discussed the motivation and some important applications for mining of distance-constrained embedded subtrees. The necessary adjustments to the general TMG framework to incorporate distance information during candidate enumeration were discussed. We have presented an encoding strategy to efficiently enumerate candidate subtrees with the additional grouping criterion and the efficient TMG approach to candidate enumeration was preserved in yet another subset of the tree mining problem.

A number of experiments were used to demonstrate the scalability of the resulting algorithms and to highlight the difference in the number of frequent subtrees extracted by different approaches.

From the experiments, we learn that enforcing the distance constraint might add extra complexity to the task of mining embedded subtrees as more candidate subtrees will need to be enumerated and counted. Nevertheless, this may differ from one case to another. We note that the cost of enumeration is the same as for the traditional embedded subtrees. The frequency counting process, on the other hand, might be more expensive due to the larger number of candidate subtrees that would be generated by imposing the distance constraint. However, while the number of unique candidate subtrees to be enumerated will increase because of the added granularity, this constraint can be used to filter out certain embedded subtrees based on the distance criterion. In addition, this constraint can be used to further distinguish what would be otherwise just an embedded subtree into its distance-coded variants. The comparison of the results with the algorithms mining traditional subtree types, indicate the potential for more specific data analysis from tree-structured documents by considering the distance between the nodes.

References

1. Access Excellence The National Health Museum. Understanding gene testing: What does a predictive gene tell you (2007), `http://www.accessexcellence.org/AE/AEPC/NIH/gene14.html`
2. Hadzic, F., Tan, H., Dillon, T.S.: UNI3 - Efficient Algorithm for Mining Unordered Induced Subtrees using TMG Candidate Generation. In: Proceedings of IEEE Symposium on Computational Intelligence and Data Mining (CIDM), Honolulu, Hawaii, USA, April 1-5, pp. 568–575. IEEE, Los Alamitos (2007)
3. Hadzic, F.: Advances in knowledge learning methodologies and their applications. Curtin University of Technology, Perth (2008)

4. Hadzic, F., Tan, H., Dillon, T.S.: Mining Unordered Distance-constrained Embedded Subtrees. In: Boulicaut, J.-F., Berthold, M.R., Horváth, T. (eds.) DS 2008. LNCS (LNAI), vol. 5255, pp. 272–283. Springer, Heidelberg (2008a)
5. Hadzic, F., Tan, H., Dillon, T.S.: U3 - mining unordered embedded subtrees using TMG candidate generation. In: Proceedings of the IEEE / WIC / ACM International Conference on Web Intelligence, Sydney, Australia, December 9-12, pp. 285–292 (2008b)
6. Shasha, D., Wang, J.T.L., Shan, H., Zhang, K.: ATreeGrep: Approximate Searching in Unordered Trees. Paper Presented at the Proceedings of the 14th International Conference on Scientific and Statistical Database Management, Edinburgh, Scotland, UK, July 24-26 (2002)
7. Tan, H., Dillon, T.S., Hadzic, F., Chang, E.: Razor: mining distance-constrained embedded subtrees. In: Proceedings of the Workshop on Ontology Mining and Knowledge Discovery from Semistructured documents (MSD) in conjunction with ICDM 2006, Hong Kong (2006)
8. Tan, H.: Tree Model Guided (TMG) enumeration as the basis for mining frequent patterns from XML documents. University of Technology Sydney, Sydney (2008)
9. Wang, K., Liu, H.: Discovering Structural Association of Semistructured Data. IEEE Transactions on Knowledge and Data Engineering 12(3), 353–371 (2000)

Chapter 8
Mining Maximal and Closed Frequent Subtrees

8.1 Introduction

In general, for frequent pattern mining problems, the candidate enumeration process exhaustively enumerates all possible combinations of itemsets that are a subset of a given database. This process is known to be very expensive since, in many circumstances, the number of candidates to enumerate is quite large, and also the frequent patterns present in real-world data can be fairly long (Bayardo 1998). Efficient techniques attacking different issues and problems relevant to the enumeration problem are therefore highly sought after. In addition to the enumeration problem, another important problem of frequent pattern mining is to efficiently count and prune away any itemsets discovered to be infrequent. Due to the large number of candidates that can be generated from the vast amount of data, an efficient and scalable counting approach is critically important. Another problem when extracting all frequent subtrees from a complex tree database, is that the number of patterns presented to the user can be very large, thereby making the results hard to analyze and gain insights from.

Mining of maximal and closed patterns is one of the directions taken in the association rule mining research to alleviate the complexity of enumerating all frequent patterns (Han & Kamber 2006; Taouil et al. 2000; Uno et al. 2003; Wang, Han & Pei 2003; Zaki & Hsiao 2002). A maximal pattern is a frequent pattern of which no proper superset is frequent, and a closed pattern is a pattern that has no superset with the same support. Discovering all maximal patterns, however, does not come with a complete support count of each frequent itemset and thus cannot completely satisfy the requirements for association rule mining (Bayardo 1998; Chi et al. 2004; Pei, Han & Mao 2000). On the other hand, discovering closed itemsets instead of the complete set of frequent itemsets, may be more desirable, since all frequent patterns and their support information, can be obtained from a set of closed patterns. The trade-off comes with the extra step of decoding all the sub-patterns from a set of closed patterns. With these characteristics and benefits, some work in frequent subtree mining has also focused on extracting maximal and closed subtrees. It is also often the case that not all the patterns found in a frequent pattern set are of

F. Hadzic, H. Tan, and T.S. Dillon: Mining of Data with Complex Structures, SCI 333, pp. 191–199.
springerlink.com

interest to the user or useful for the application at hand, as they can contain irrelevant attributes or be redundant in the sense that the association they represent has already been captured by another pattern.

The algorithms described in earlier chapters generally differ in the candidate enumeration and counting strategy, tree representation and representative structures. However, a common factor is that all candidate subtrees are first enumerated and then counted by employing a bottom-up lattice search. Even though the existing algorithms are well scalable for large datasets, a difficulty that remains is that, if the instances in the tree database are characterized by complex tree structures, the approach will fail due to the enormous number of candidate subtrees that need to be enumerated. This was demonstrated in Chapter 5 where we presented the mathematical analysis to estimate the worst case performance for the TMG candidate enumeration strategy. The analysis is valid for any other horizontal enumeration approach, as it would need to enumerate at least as many candidates, if not more. Hence, similar to the itemsets mining problem, due to the combinatorial explosion, the number of frequent patterns that need to be enumerated grows exponentially with the database size and/or complexity of the underlying database structure. As was the case in association rule mining from relational data, to alleviate the complexity problem and reduce the number of candidate subtrees that need to be enumerated, some work has shifted toward the mining of maximal and closed subtrees (Chi et al. 2004, Chehreghani et al. 2007). As mentioned earlier, a closed subtree is a subtree for which none of its proper supertrees has the same support, while for a maximal subtree, no supertree exists that is frequent. No information is lost since the complete set of frequent subtrees can be obtained from closed subtrees (including support information) and maximal subtrees. As an example, please consider Fig. 8.1, where a tree database (T_{db}) is shown consisting of three transactions, together with the set of frequent maximal and closed subtrees when transaction-based support is set to 2. We can see that there is an additional frequent closed subtree with support equal to 3. This subtree is a subset of the larger subtree displayed on the left, and this is the reason why it is not a maximal subtree, but is a closed subtree because of the support that is different from that of its supertree.

By mining closed and maximal subtrees, larger and more complex databases can be handled. The reason is that we do not need to generate all the frequent subtrees, and a majority of candidates generated during the process can be pruned when they are a subset of a closed/maximal subtree and satisfy the respective support condition. Hence, the mining of maximal/closed subtrees is one of the important topics in the general area of frequent subtree mining.

The rest of the chapter is organized as follows. In Section 8.2, we provide formal definitions of the maximal and closed subtrees. We explain some existing methods for maximal and closed subtree mining in Section 8.3. A popular algorithm for the mining of closed/maximal subtrees is explained in more detail in section 8.4. Section 8.5 concludes this chapter.

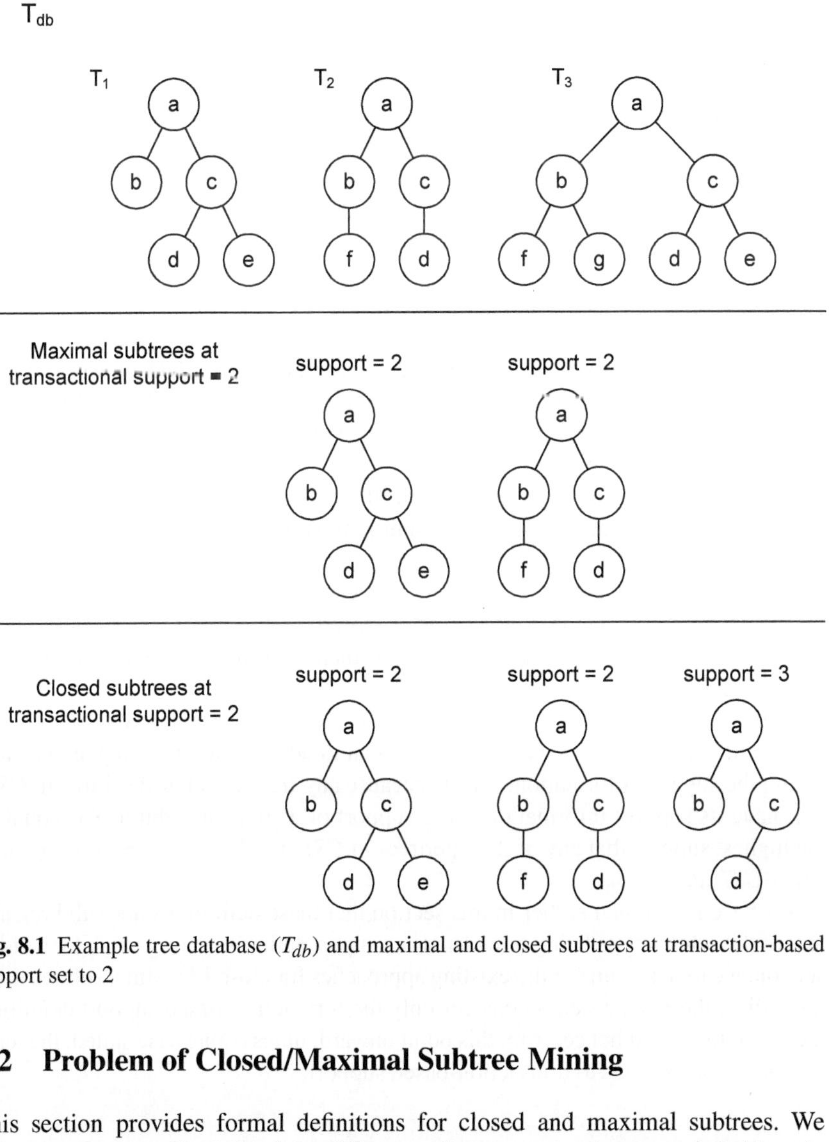

Fig. 8.1 Example tree database (T_{db}) and maximal and closed subtrees at transaction-based support set to 2

8.2 Problem of Closed/Maximal Subtree Mining

This section provides formal definitions for closed and maximal subtrees. We have already formally defined the different subtree types and support definitions in Chapter 2. Those subtree definitions also apply to closed and maximal subtrees, and hence we can have induced/embedded ordered/unordered closed/maximal subtrees. One can also add constraints to such subtrees (such as the maximum level of embedding or distance constraint from Chapter 4), but in this chapter we will narrow our discussion to those subtree types for which the work in closed/maximal subtree mining has been done. The aspect that makes a subtree closed or maximal is dependent on the properties of a subtree with respect to the other frequent subtrees from a tree database with respect to the given support threshold. The properties considered among the frequent subtrees are those of subset/superset relationships. We next provide some formal definitions, and the term subtree used here can correspond to any of the different subtree types defined in Chapter 2.

A subtree S is a proper superset of another (sub)tree S', denoted as $S' \subset S$, iff (1) S' is a subtree of S (see Chapter 2), and (2) size(S) > size(S').

Given a set of frequent subtrees, ST from a tree database T_{db}, a subtree $S \in ST$ is a frequent **maximal subtree** iff there are no proper supersets of S in ST. For a given support, the set of all frequent subtrees can be obtained from the set of maximal frequent subtrees, but the support information will not be available.

Given a set of frequent subtrees, ST from T_{db}, where a support of a subtree S is indicated as $sup(S)$, a subtree $S \in ST$ is a frequent **closed subtree** iff, there is no other subtree S' in ST, where S' is a proper superset of S and $sup(S) = sup(S')$.

For a given support, the set of all frequent subtrees can be obtained from the set of closed frequent subtrees, and in contrast to maximal subtrees, all the support information can also be obtained.

The following summarizes the problems that this chapter is concerned with:

Given a tree database T_{db} and minimum support threshold s, extract all closed/maximal subtrees that occur at least s times in T_{db}.

Given a tree database T_{db} with minimum support threshold (s), where a set of frequent subtrees is denoted as ST, set of closed subtrees as CST, and the set of maximal subtrees as MST, for the specified s, then the following relationships hold: $MST \subseteq CST \subseteq ST$.

One can derive ST from MST because any frequent subtree is a subset of one or more maximal subtrees. Given CST,one can in addition to obtaining all ST, also obtain the support information. This is because any frequent closed subtree in CST, will have its support information and a support of a frequent subtree S is equal to the highest support that any of its supertrees in CST has, i.e. $sup(S) = \max(sup(S_C))$ $\forall S_C \in CST$ and $S \subset S_C$.

We have mentioned earlier in this section that these definitions are valid regardless of the subtree type being mined and/or support definition being used. In the next section, we look at some of the existing approaches for closed/maximal subtree mining. All of these approaches consider only the transaction-based support definition (see Chapter 2) and hence, from this point onward, unless otherwise stated, the term 'support' corresponds to transaction-based support.

8.3 Methods for Mining Closed/Maximal Subtrees

Compared with the available frequent closed/maximal itemset mining algorithms, the number of available frequent closed/maximal subtree mining algorithms is much less. This is to be expected as tree mining itself is a fairly new area which is much more complex than the traditional itsemset/sequence mining problem. In this section, we briefly overview some of the existing work in this area and in the next section we provide further details for a chosen approach.

The first algorithm that appeared in the context of closed/maximal subtree mining was Pathjoin (Xiao, Yao, Li & Dunham 2003), developed to extract all maximal ordered induced subtrees. The algorithm works by first extracting all maximal frequent paths and then joining the paths to form the set of frequent subtrees. The frequent subtrees that have a superset in the set (i.e. are not maximal) are pruned. Pathjoin uses a compact data structure, FST-Forest, which is a compressed representation of the tree and is used for finding the maximal paths. CMTreeMiner algorithm (Chi et al. 2004) is an algorithm developed for the mining of both closed and maximal ordered induced subtrees without first discovering all frequent subtrees. It traverses an enumeration tree that systematically enumerates all subtrees. During this process, an enumeration DAG (directed acyclic graph) is used to prune all the branches of the enumeration tree that cannot correspond to closed and/or maximal frequent subtrees. In (Ozaki & Ohkawa 2006), the differences in the search strategies are discussed with respect to the mining of closed ordered subtrees. Two algorithms are presented: one which uses a breadth-first enumeration strategy, and the other which uses depth-first/breadth-first enumeration strategies. These two versions correspond to adjustments to the AMIOT (Hido & Kawano 2005) and TreeMiner (Zaki 2005) algorithms so that closed subtrees are obtained by applying pruning strategies. SCCMETreeMiner algorithm (Ji & Zhu 2008) mines frequent closed and maximal ordered embedded subtrees using length-decreasing support constraint. It uses the right-most path extension enumeration strategy and a projection technique to construct the candidate subtree set. The branches of the enumeration tree that do not correspond to closed and/or maximal subtrees under the length-decreasing constraint are pruned. The length-decreasing constraint is motivated by noting that in applications, small subtrees with high support are considered interesting, while the larger subtrees may be considered interesting even when their support is relatively small. Hence, the length-decreasing support constraint reduces the support threshold as the length of the tree increases. SEAMSON (Scalable and Efficient Algorithm for Maximal frequent Subtree extractiON) algorithm (Paik, Nam, Hwang & Kim 2008) discovers frequent maximal ordered embedded subtrees by using their newly proposed *L-dictionary* structure which is essentially a compressed representation of the database. *L-dictionary* is created by storing the label, pre-order position, parent positions, and tree (transaction) indexes, for each node in the database tree in form of lists. The frequency of each label is worked out from the number of lists, and lists from frequent labels are joined together through the parent node information and tree index in order to arrive at maximal subtrees. Exit-B algorithm (Paik, Nam, Won & Kim 2008) extract frequent maximal ordered embedded subtrees, and the novel idea is to represent trees with bit sequences, store them in specialized data structures, and extract all maximal subtrees. The main purpose is to increase time performance and space efficiency by avoiding disadvantages caused by storing string labels. The *TDU* approach (Chehreghani et al. 2007) is a top-down approach for extracting all frequent maximal embedded unordered subtrees. It starts by computing the set of all frequent nodes by incrementing their count in a 1-dimensional array and an *intermediate representation tree* is constructed. Each tree that is infrequent is fragmented into its *k-1*-subtrees until a frequent subtree is reached. This subtree is then checked for a subset/superset relationship with other frequent subtrees, to eliminate any subtrees of an already frequent maximal subtree.

In the next section, we describe the CMTreeMiner algorithm in more detail, and use the simple database from Fig. 8.1 to illustrate the process.

8.4 CMTreeMiner Algorithm

The CMTreeMiner algorithm (Chi et al. 2004) can mine both closed and maximal subtrees from databases of rooted labeled ordered and unordered trees. In this description, we will focus on its implementation for mining of databases of rooted labeled ordered trees.

The CMTreeMiner algorithm uses a data structure called the *enumeration DAG* (directed acyclic graph) to represent the subtree/supertree relationships on the set of all frequent subtrees. Each level in this data structure consists of the frequent subtrees, the size of which (number of nodes) is equal to the level number. The neighboring levels are then connected by directed edges which emanate from the subtree to the supertree to represent a subtree/supertree relationship between two frequent subtrees in the neighboring levels. The enumeration DAG for all the frequent subtrees of the example database from Fig. 8.1 is shown in Fig. 8.2. Please note that in this example, the frequent subtrees correspond to frequent ordered induced subtrees when the transaction-based support threshold is set to 2. The purpose of the enumeration DAG is to determine whether a given frequent subtree is maximal or closed and to prune away trees that do not correspond to frequent closed or maximal subtrees. The notion of a *blanket* of a frequent subtree *st* is used which is defined as the set of supertress of *st* that are frequent and contain only one additional node in comparison to *st*. Hence, for a frequent subtree *st* at level l_x, in the enumeration DAG of all frequent subtrees, the blanket consists of the frequent subtrees in level l_{x+1} that are reachable through the directed edges emanating from subtree *st*. As an example, consider the subtree at level 2 with string encoding (see Chapter 5) 'A B'. Its blanket consists of its supertrees 'A B / C' and 'A B F' at level 3 as it is connected to them in the enumeration DAG. Let us denote the additional node of a supertree from the blanket of a frequent subtree *st* as *n'/st*. Hence when *st* = 'A B' then the *n'/st* of 'A B / C' is equal to node 'C'. The maximal frequent subtree *st* is then defined as the subtree in enumeration DAG whose blanket is empty, and a closed frequent subtree is defined as the one where, for each supertree in its blanket, the support of *n'/st* is smaller than the support of the subtree *st*. Using these definitions, the frequent closed and maximal subtrees can be easily determined by traversing the enumeration DAG.

In order to avoid the potentially expensive traversal of the complete enumeration DAG, several pruning strategies are adopted. The motivation behind the pruning strategies is that, under certain conditions, the branches of the enumeration DAG can be pruned since it can be guaranteed that they do not contain any frequent closed or maximal subtrees. The authors use the notion of *left-blanket* and *right-blanket*

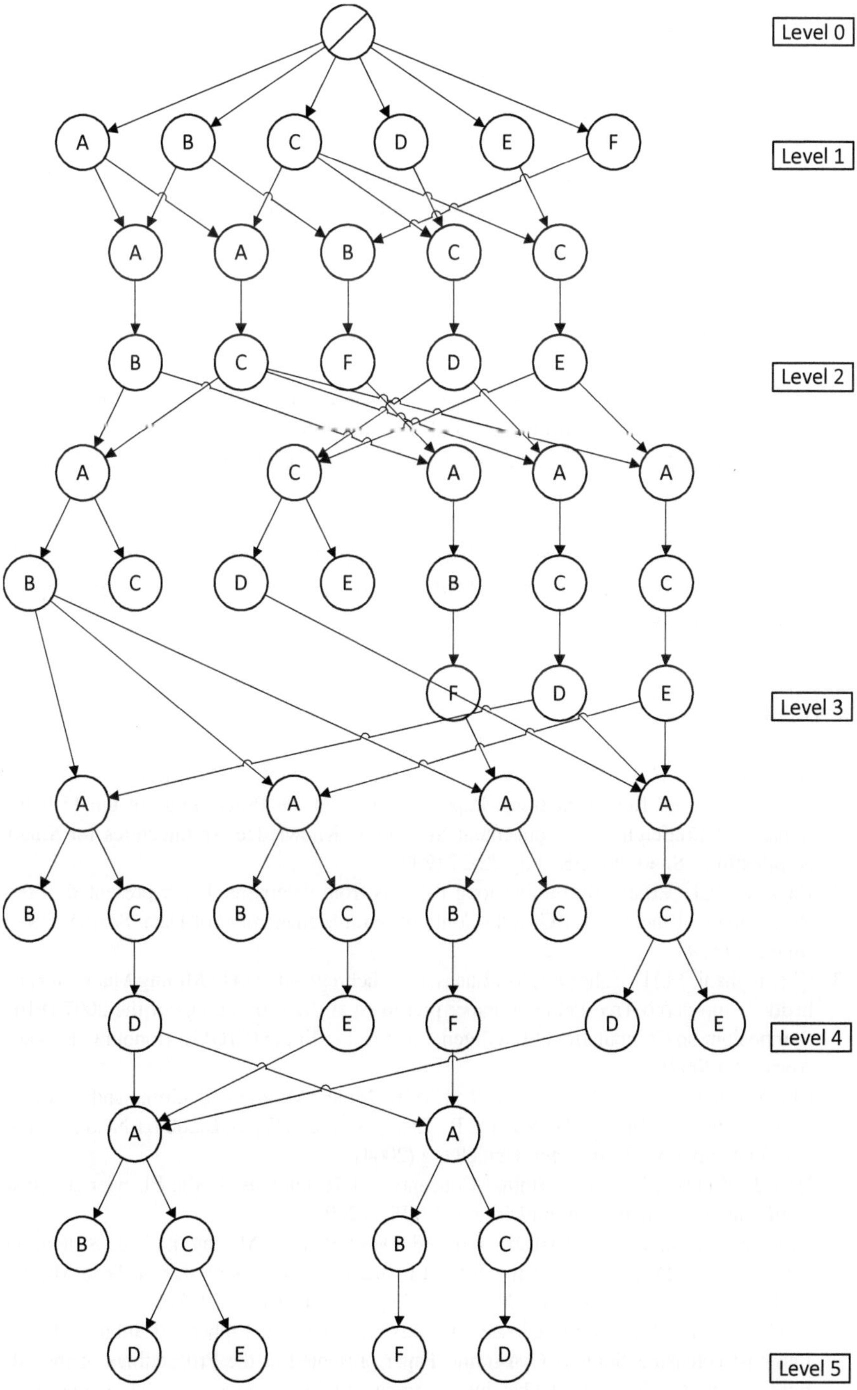

Fig. 8.2 Enumeration DAG of T_{db} from Fig. 8.1 (transaction-based support = 2)

pruning and consider the information regarding the repetitions of the subtrees in the transactions in order to effectively prune branches of the enumeration DAG guaranteed to contain no frequent maximal or closed subtrees. For more detail about the pruning strategies and the algorithm as a whole including the pseudo code, we refer the interested reader to (Chi et al. 2004).

While the CMTreeMiner algorithm mines both closed and maximal subtrees with a slight modification of the algorithm, the results will contain either closed or maximal subtrees as was presented in (Chi et al. 2004).

8.5 Conclusion

This chapter has looked at the problem of frequent closed/maximal subtree mining. We have discussed the importance of this research direction in the field of tree mining. A number of available algorithms were described for these tasks. As can be seen, in comparison to frequent subtree mining algorithms, frequent closed/maximal subtree mining algorithms are much fewer. Hence, at this stage, the research in this area is yet to mature, as there is still no general framework according to which all subtree types can be mined, and which can incorporate all existing support definitions and constraints.

References

1. Balcazar, J.L., Bifet, A., Lozano, A.: Mining Frequent Closed Unordered Subtrees Through Natural Representations. Paper presented at the Proceedings of the 15th International Conference on Conceptual Structures: Knowledge Architectures for Smart Applications, Sheffield, UK, July 22-27 (2007)
2. Bayardo, R.J.: Efficiently mining long patterns from databases. Paper presented at the Proceedings of the ACM SIGMOD Conference on Management of Data, Seattle, USA, June 2-4 (1998)
3. Chehreghani, M.H., Rahgozar, M., Lucas, C., Chehreghani, M.H.: Mining Maximal Embedded Unordered Tree Patterns. Paper presented at the Proceedings of the 2007 IEEE Symposium on Computational Intelligence and Data Mining (CIDM), Honolulu, Hawaii, April 1-5 (2007)
4. Chi, Y., Yang, Y., Xia, Y., Muntz, R.R.: CMTreeMiner: Mining both closed and maximal frequent subtrees. In: Dai, H., Srikant, R., Zhang, C. (eds.) PAKDD 2004. LNCS (LNAI), vol. 3056, pp. 63–73. Springer, Heidelberg (2004)
5. Han, J., Kamber, M.: Data Mining: Concepts and Techniques. 2 edn. Elsevier, Morgan Kaufmann Publishers, San Francisco, CA, USA (2006)
6. Hido, S., Kawano, H.: AMIOT: Induced Ordered Tree Mining in Tree-Structured Databases. In: Proceedings of the 5th IEEE International Conference on Data Mining (ICDM), Houston, Texas, USA, November 27-30, pp. 170-177 (2005)
7. Ji, G.L., Zhu, Y.W.: Mining Closed and Maximal Frequent Embedded Subtrees Using Length-Decreasing Support Constraint. Paper presented at the Proceedings of the 7th International Conference on Machine Learning and Cybernetics, Kunning, China, July 12-15 (2008)

8. Ozaki, T., Okhawa, T.: Efficient Mining of Closed Induced Ordered Subtrees in Tree-structured Databases. Paper presented at the Proceedings of the 6th IEEE International Conference on Data Mining - Workshops, Hong Kong, December 18-22 (2006)
9. Paik, J., Kim, U.M.: A Simple Yet Effective Approach for Maximal Frequent Subtree Extraction from a Collection of XML Documents. Paper presented at the Web Information Systems - WISE 2006 Workshops, Wuhan, China, October 23-26 (2006)
10. Paik, J., Nam, J., Hwang, J., Kim, U.M.: Mining Maximal Frequent Subtrees with List-Based Pattern-Growth Method. In: Zhang, Y., Yu, G., Hwang, J., Xu, G. (eds.) APWeb 2008. LNCS, vol. 4976, pp. 93–98. Springer, Heidelberg (2008)
11. Pei, J., Han, J., Mao, R.: CLOSET: An Efficient Algorithm for Mining Frequent Closed Itemsets. Paper presented at the Proceedings of the ACM SIGMOD Workshop on Research Issues in Data Mining and Knowledge Discovery, Dallas, Texas, USA, May 14 (2000)
12. Taouil, R., Pasquier, N., Bastide, Y., Lakhal, L.: Mining Bases for Association Rules using Closed Sets. Paper presented at the Proceedings of the 16th International Conference on Data Engineering (ICDE 2000), San Diego, CA, USA, February 28 - March 3 (2000)
13. Uno, T., Asai, T., Uchida, Y., Arimura, H.: LCM: An Efficient Algorithm for Enumerating Frequent Closed Item Sets. In: Paper presented at the Proceedings of the ICDM 2003 Workshop on Frequent Itemset Mining Implementations (FIMI 2003), Melbourne, Florida, USA (2003)
14. Wang, J., Han, J., Pei, J.: CLOSET+: Searching for the Best Strategies for Mining Frequent Closed Itemsets. Paper presented at the Proceedings of the 9th ACM SIGKDD International Conference on Knowledge Discovery and Data Mining, New York, NY, USA, August 24-27 (2003)
15. Zaki, M.J., Hsiao, C.-J.: CHARM: An Efficient Algorithm for Closed Itemsets Mining. Paper presented at the Proceedings of the 2nd SIAM International Conference on Data Mining, Arlington, VA, USA, April 11-13 (2002)
16. Zaki, M.J.: Efficiently Mining Frequent Trees in a Forest: Algorithms and Applications. IEEE Transactions on Knowledge and Data Engineering 17(8), 1021–1035 (2005)

Chapter 9
Tree Mining Applications

9.1 Introduction

The aim of this chapter is to discuss the applications of tree mining algorithms with respect to the general knowledge analysis task, as well as some specific applications. The implications of using different tree mining parameters (i.e. subtree types, support definitions, constraints) are discussed, and illustrative scenarios are used to indicate useful application areas for different parameters. This kind of overview will be particularly useful for those not so familiar with the area of tree mining as it can reveal useful applications within their domain of interest. It gives guidance as to which type of tree mining will be most useful for their particular application.

Tree mining has already been shown to be useful in areas such as analysis of phylogenetical databases (Shasha, Wang & Zhang 2004), protein structures (Hadzic et al. 2006), and it was mentioned that processing tree-structured data is important for applications in vision, natural language processing, molecular biology, programming compilation etc. (Wang, Zhang, Jeong & Shasha 1994). In this chapter, we will take a number of applications and discuss them in more detail. These involve applications of tree mining algorithms to the analysis of health information, mining of web logs, matching of heterogeneous knowledge structures and protein structure analysis.

The rest of the chapter is organized as follows. In Section 9.2, different types of knowledge representations are overviewed. The types suitable for analysis by applying tree mining algorithms are indicated together with any pre-processing that needs to take place. A general discussion of the importance of different support definitions and the important applications for each of the different subtree types and their constrained extensions is provided in Section 9.3. Some guidelines for applying the tree mining algorithms for the analysis of health information are given in Section 9.4. The way that the Web log mining problem can be recast as the tree mining problem is discussed in 9.5 and a number of experiments are provided using synthetic and real-world data. Section 9.6 is concerned with the knowledge matching task, where we discuss a way in which the tree mining algorithms can aid in

F. Hadzic, H. Tan, and T.S. Dillon: Mining of Data with Complex Structures, SCI 333, pp. 201–247.
springerlink.com

matching heterogeneous knowledge structures, and experiments on real-world data are provided to support the discussion. Section 9.5 gives some guidelines for applying the tree mining algorithms to the mining of protein structures. The chapter is concluded in Section 9.7.

9.2 Types of Knowledge Representations Considered

The purpose of this section is to discuss the types of knowledge representations for which the tree mining algorithms are very applicable. Furthermore, since the tree mining algorithms assume the database to be in a certain format, any kind of pre-processing that needs to be done prior to the application is discussed. The parallelism between XML documents and tree structures was already discussed in Chapter 2, where we have seen how XML documents can be effectively modeled as a rooted ordered and labeled tree, without the loss of any information. Generally speaking, XML documents are widely used to represent any domain-specific information that can be hierarchically represented. Hence, in this scenario the type of knowledge referred to is quite general and it corresponds to any domain aspect that is being captured by the tree-structured document. For example, XML schemas often indicate the structural organization of some domain-specific information. In general, this group of knowledge representations includes any document where the underlying structure can be modeled as a tree. Decision trees and concept hierarchies are a popular means of representing domain knowledge and they are discussed separately in what follows.

Decision trees are simple to construct and can be easily processed by both humans and machines. As the name implies, they are hierarchical by nature and tree mining algorithms can prove useful for applications in general tasks such as knowledge comparison, integration and analysis. A decision tree is a directed graph consisting of labeled nodes and directed edges. The nodes correspond to the attribute of some domain which is posed as a question, while the edges emanating from the nodes are mutually exclusive values for that attribute. The leaf nodes represent the value of the attribute whose value prediction is enabled by the represented knowledge. The main difference between the decision tree representation and the tree database expected as input by a tree mining algorithm, is the fact that the edges in a decision tree are labeled with corresponding attribute value tests. The tree mining algorithms do not consider a tree database with labeled edges and hence, a pre-processing stage is required. Each of the labeled edges would become a node itself with the label equal to the edge label, its parent node being the node from which the edge is emanating and its child node being the node to which the edge is connecting.

A concept hierarchy is another type of knowledge representation where the application of tree mining can aid with tasks related to general knowledge management. It is suitable for describing inheritance relationships between the concepts of the domain, where concept is an abstraction for representing a collection of domain attributes. The edges in a concept hierarchy are not necessarily labeled and they represent the 'is-a' relationships between the concepts. Hence, a concept hierarchy can also be represented as a rooted ordered labeled tree where the edges indicate the 'is-a' relationship between the interconnected nodes (i.e. the concept node to which the edge leads is a specialization of the concept node from which the edge emanates). As such, the amount of pre-processing is minimal and it may occur for optimization purposes (e.g. mapping string labels to integers).

Production rules are one of the most common knowledge representations in AI. A production rule consists of condition-action pairs, where the condition part indicates a number of premises (attribute values or constraints) that need to be met in order for the action to occur. The action part usually corresponds to the value of the target attribute for which the knowledge is represented. In the cases where hierarchical relationships exist between the target attribute values, it may be possible to represent the set of production rules in the form of a decision tree or concept hierarchy, in which case it may be possible to apply the tree mining algorithms for analyzing the presented knowledge. However, in this chapter, the production rules will not be considered as suitable knowledge representation to be analyzed by the tree mining algorithms, since pre-processing is required and in many cases the rules may not be hierarchically related to each other.

Another popular knowledge representation is the semantic net which is a directed graph consisting of a set of nodes (attribute objects or concepts) and a set of edges describing the associated semantics. The relationships in the semantic net are not necessarily limited to the inheritance relationships. The underlying structure of a semantic net can contain cycles and hence, in this chapter, a semantic net is not considered as a suitable knowledge representation where analysis can be performed by the application of tree mining.

Frames are structures for capturing a collection of interrelated knowledge about a stereotyped concept, attribute, system state or an event (Dillon & Tan 1993). There can be 'is-a' relationships and multiple 'is-a' relationships between the concepts or classes in a frame. The 'is-a' relationship is the type of inheritance where a sub-concept (subclass) inherits the properties of its superclasses. As is the case in a concept hierarchy, the superclass represents the generalizations of the subclasses, while the subclass of a given class represents specialization of the classes situated higher in the hierarchy (Dillon & Tan 1993). When the relationships are limited to these simple inheritance relationships, the knowledge represented by a frame could be modeled as a rooted ordered labeled tree, and hence, tree mining could be applied for general analysis. However, it is possible to have multiple 'is-a' relationships between the classes or concepts in a frame, which has also been referred to as multiple inheritance relationships (Dillon & Tan 1993). In this case, the structure of the frame is a lattice of classes, where each class may have one or more immediate superclasses. Therefore, in a multiple inheritance relationship, a class can inherit properties from two or more classes which themselves are not hierarchically related. With the existence of multiple inheritance relationships, the underlying structure of the frame is a graph; for this reason, frames are not considered as suitable knowledge representations to be analyzed by the application of tree mining.

A general note for semantic nets and frames is that when the relationships are limited to inheritance relationships (i.e. the underlying structure is an acyclic graph), the tree mining algorithms could be applicable for knowledge analysis. This is because the structure in which the concepts are represented can be modeled as an ordered labeled tree. Furthermore, if the number of cycles in the graph structure is not so large, then it may well be possible to transform the knowledge representation into a tree structure without a great loss of semantic information, and tree mining could be applied to analyze this subset of knowledge that can be represented as a tree. However, this transformation and application is beyond the scope of this book, and hence the focus is narrowed to the knowledge representations where all information can be modeled as a tree, without any significant pre-processing.

This section has considered some different types of knowledge representations and has indicated those where the application of tree mining can aid in tasks related to general knowledge management. In general, one could state that the tree mining algorithms are very applicable to any documents where the underlying structure used to represent the information can be modeled as a rooted ordered labeled tree (see Chapter 2). In this chapter, the knowledge representations considered are XML documents and decision trees. In Section 9.7, we will discuss more specifically how the information found in Web logs can be effectively represented using tree structures in order to allow the extraction of more informative patterns for the particular application at hand.

9.3 Application for General Knowledge Analysis Tasks

In each of the chapters describing the different subtree mining problems, we have provided a brief discussion of why the mining of those particular subtree types is important. In this section, we may replicate some of the discussion, as we are discussing the implications behind all tree mining parameters, including support definitions and constraints.

9.3.1 Implications of Using Different Support Definitions

This section describes the motivation for the use of existing support definitions within the current tree mining framework. The support definitions are discussed in order of increasing complexity.

9.3.1.1 Transaction-Based Support

As mentioned earlier in the book, this support definition just checks for the existence of items in the transaction and considers the repetition of items as irrelevant. In traditional frequent itemset mining from relational data, checking whether an item exists within a transaction is sufficient to determine the traditional support definition. Hence, using transaction-based support would appear to be the obvious choice when moving from frequent pattern mining from relational data to frequent pattern mining from tree-structured data. When converting the information in relational data to be presented in a tree-structured form, it is common for an instance in the relational document to be described by one transaction (independent subtree) in a tree-structured document. This has made transaction-based support the focus of many tree mining works and, from the available support definitions, it is the one that is most often considered.

9.3.1.2 Occurrence Match Support or Weighted Support

Applications using occurrence match support (OC) consider the repetition of items in a transaction to be important and the subtree occurrences in the database as a whole are counted. To illustrate the importance of occurrence match support, the dataset describing a protein ontology instance store for Human Prion Proteins

```
...
<ATOMSequence>
    <ProteinOntologyID>PO0000000009</ProteinOntologyID>
    <_ATOM_Chain>A</_ATOM_Chain>
    <_ATOM_Residue>GLN</_ATOM_Residue>
    <AtomID>1497</AtomID>
    <Atom>CD</Atom>
    <ATOMResSeqNum>217</ATOMResSeqNum>
    <X>-2.127</X>
    <Y>2.685</Y>
    <Z>6.088</Z>
    <Occupancy>1</Occupancy>
    <TempratureFactor>0</TempratureFactor>
    <Element>C</Element>
</ATOMSequence>
<ATOMSequence>
    <ProteinOntologyID>PO0000000009</ProteinOntologyID>
    <_ATOM_Chain>A</_ATOM_Chain>
    <_ATOM_Residue>GLN</_ATOM_Residue>
    <AtomID>1498</AtomID>
    <Atom>OE1</Atom>
    <ATOMResSeqNum>217</ATOMResSeqNum>
    <X>-1.623</X>
    <Y>3.754</Y>
    <Z>6.433</Z>
    <Occupancy>1</Occupancy>
    <TempratureFactor>0</TempratureFactor>
    <Element>O</Element>
</ATOMSequence>
...
```

Fig. 9.1 Snapshot of the representation of Human Prion Protein dataset in XML format in XML format (Sidhu et al. 2005) will be considered again. The partial XML representation of protein data is displayed in Fig. 9.1. The original dataset describes a protein ontology instance store for Human Prion Proteins in XML format (Sidhu et al. 2005). Protein Ontology (PO) provides a unified vocabulary for capturing declarative knowledge about the protein domain and classifies that knowledge to allow reasoning. Using the PO format, ATOMSequence labels can be compared easily across PO datasets for distinct protein families to determine the sequence and structural similarity among them. Structured ATOMSequence labels, with repetition of Chain, Residue and Atom details can be used to compare a new unknown protein sequence and structure with existing proteins in the PO dataset, which helps users with drug discovery and design. In this case, the repetition in the structure of the protein is of considerable importance.

Another scenario where occurrence match support may be important is when performing specialized queries on a tree-structured database. As an example, consider a library-based application where author information is stored separately in each transaction, as is illustrated in Fig. 9.2. A user may be interested in finding out information about the total number of books that were published in a certain year (X) by a certain publisher (Y). To satisfy this query, the repetition of book-year(X)-publisher(Y) relations within a transaction will need to be considered. The answer is given by the total number of occurrences of relation book-year(X)-publisher(Y) in the whole database. The total number of occurrences of the query subtree A first shown on the right of Fig. 9.2 is of interest. In these scenarios, the repetition of items within a transaction is considered important and the knowledge of the number of repetitions provides useful information.

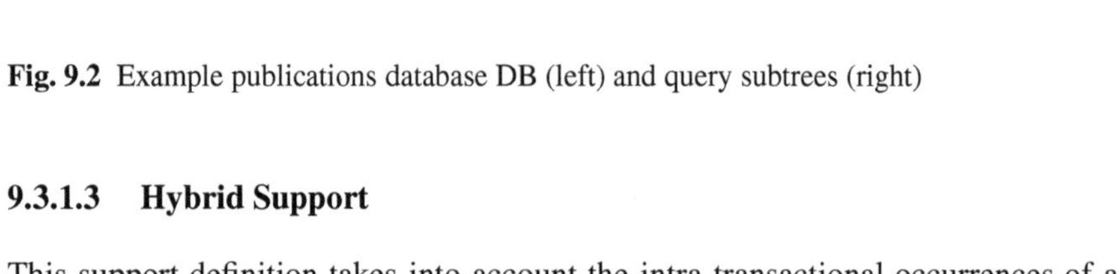

Fig. 9.2 Example publications database DB (left) and query subtrees (right)

9.3.1.3 Hybrid Support

This support definition takes into account the intra-transactional occurrences of a subtree. It can be seen as a specialization of the transaction-based support so that a subtree is considered to be supported by a transaction only if it occurs as many times as the user-supplied occurrence support. Hence, given a hybrid support of $x|y$, a subtree will be considered as frequent if it occurs in x transactions and it occurs at least y times in each of those transactions. As was already discussed in Chapter 2, for certain applications, the number of times that a subtree occurs within a transaction is important. Such applications could occur in many web information systems, where the user's goal is to extract specific information by posing specialized queries. Please consider Fig. 9.2 again. If the user is interested in finding n number of authors that have published at least x books with publisher Y, then such specific patterns could not be found in the extracted pattern set if either occurrence-match or transaction-based support was used. This is because the repetition of author-book-publisher(Y) relation within a transaction will need to be considered. On the other hand, if the hybrid support definition is set to $n|x$, then all the extracted patterns would have occurred at least x times in n transactions. The user can then search for its query of book-publisher(Y) relation in the extracted set of patterns. Hence, using the hybrid support, additional constraints can be imposed for a pattern to be considered as frequent, and this in turn will allow for more specific queries to be answered by searching through the extracted pattern set.

9.3.2 Implications for Mining Different Subtree Types

This section discusses some of the applications of the commonly mined subtree types within the current tree mining framework. The subtree types are discussed in the order of increasing complexity.

9.3.2.1 Ordered Induced Subtrees

This type of subtree is the simplest among the ones considered in this chapter, and the main implication is that the parent-child relationships must remain the same as in the original tree. As such, they have been extensively used and are the most common type considered by the tree mining algorithms. A reason for this would be that when transforming from relational data into tree structured data, the derived tree structure itself has only a few levels, and hence, all the meaningful information is effectively represented by an induced subtree.

As an example, consider again the tree representation of a publication database DB from Fig. 9.2. If all the frequent ordered induced subtrees are extracted, then all the possible queries could be answered with respect to the minimum support threshold used. The example query subtree on the right in Fig. 9.2 is of the induced subtree type, and in fact all particular queries could be answered by searching the extracted pattern set that contains induced subtrees. Hence, an algorithm for mining induced subtrees would be sufficient for this application.

Another desired aspect of an induced subtree for some applications, is that all the knowledge concepts will remain within the context in which they have occurred in the original tree database. An example could be taken from an application in the health domain where patient records are stored as separate transactions in an XML document. Each record contains information about various causal factors of a disease. The aim is to extract the patterns where particular patient information has to remain in the context in which it occurred. In other words, every pattern extracted should be meaningful in the sense that, from the pattern itself, it is known what feature (causal characteristics) a particular value is describing. In these cases, if embedded subtrees were mined (i.e. ancestor-descendant relationships are allowed), some information could be lost about the context in which the particular disease characteristic occurred. This is mainly due to the fact that some features of the dataset may have a similar set of values, and it is necessary to indicate which value belonged to which particular feature.

The popularity of mining induced subtrees could be partly because they are the most easily detected subtree type by a human observer and it is natural to think of subtrees in this sense. However, when certain applications have indicated the need for extending the parent-child relationship between the nodes to ancestor-descendant, the focus shifted toward the mining of embedded subtrees. This allows one to detect information embedded deeply within the tree structure and the importance of this is discussed next.

9.3.2.2 Ordered Embedded Subtrees

When structurally rich information is represented using XML, it is quite common for the document to be organized into several levels and each transaction can be many levels deep. In these cases, many query trees posed on an XML document may be of the embedded subtree type so that ancestor-descendant relationships embedded deeper in the tree can be detected. Furthermore, certain concepts may be represented in a more specific/general way in certain documents. In a tree-structured document, this specific information can be in the form of additional child nodes of the concept, and hence, two general and related concepts may be separated by a number of levels. If the user is interested only in detecting the relationship between these two general concepts, such a relationship could be directly found in an embedded set of subtrees, while if induced subtrees were extracted, the relationship between these concepts may not be found.

When trying to find the common structures among knowledge representations, if embedded subtrees are mined, we are allowing structures to be considered as similar even if they occur at different levels in the tree. In an embedded subtree, the relationship is not limited to parent-child, and hence by allowing ancestor-descendant relationships, this enables the extraction of more sub-structures where the levels of embeddings between the nodes are not limited to one and can be different. For example, if in knowledge representation 'A', the level of embedding between the nodes 'a' and 'b' representing some domain concepts is much larger than the level of embedding between the nodes representing the same concepts in knowledge representation 'B', then 'A' stores more specific knowledge about the concept represented by node 'a'. Since it is common for knowledge representations to differ in the amount of specific knowledge stored, the mining of embedded subtrees is more suitable for general knowledge comparison.

To illustrate this aspect, an application of tree mining to matching of knowledge structures of decision tree type will be presented (Hadzic & Dillon 2007). Later in the chapter, more experiments and discussion will be provided in regards to matching of document structures in general.

The representations used in our example from Fig. 9.3 correspond to the knowledge models used for classification purposes, and hence, the extra specific knowledge corresponds to the additional number of attribute constraints used for further separation of class values. The knowledge models were obtained using data mining tools on the publicly available datasets describing the 1984 United States Congressional Voting Records Database (Blake, Keogh & Merz, 1998). We used different subsets of data and the feature subset selection approach presented in (Hadzic & Dillon 2006) was performed in order to mimic a real world scenario when different organizations collect their own sets of data and find different features to be relevant. Each of the knowledge models is represented as a separate subtree (transaction) within the tree database, and transaction-based support is used to extract the largest embedded subtree that occurs in each transaction. If we have k different models, then a subtree will be considered frequent only if it occurs in all k models. Hence, the transaction-based support was used with the threshold of 3.

Please note that, since in this domain we are dealing with a binary classification problem, and the knowledge model is represented by a binary decision tree, we filter out any of the subtree patterns where any of the nodes has a degree larger than 2. These patterns would be meaningless as there would be three class values distinguished by two attribute constraints. They do not indicate substructures that imply true shared knowledge since their classificatory purpose has been lost.

The largest frequent ordered embedded subtree (K_S) would display the largest common structure between the *KMs* and is displayed last in Fig. 9.3. By comparing K_S with the knowledge models, we can see that it was necessary to mine embedded subtrees in order to detect the common knowledge structure. If induced subtrees were mined, only the root node and its value nodes would be considered frequent. KM_2 most closely matches the shared structure K_S from Fig. 9.3. KM_2 differs from the K_S only in the sense that it has one extra attribute after the last node which splits the class values further into 'D' and 'R', rather than generalizing it to 'R'. The knowledge model KM_1 stores two additional attributes after the first attribute node, while KM_3 stores an additional three attributes at the last node. One can see that KM_2 is more general than KM_3, but KM_2 is only more general than KM_1 between the first and second attribute while it is more specific at the last node.

Knowledge discovery tasks in general can be hard and time consuming, and hence sharing the already developed knowledge representations is desirable. This particularly occurs when a number of organizations from the same domain would like to have a knowledge basis on which they can integrate their organizations' specific knowledge. Having a general knowledge model would save time and costs associated with having to acquire general knowledge about the domain at hand. Another related area where the merging of tree structured knowledge models will be useful, is in a classification ensemble which is a multi-classifier system where each classifier is developed for the same domain problem (Zhang, Street & Burer 2005; Stolfo et al. 1997; Prodromidis, Chan & Stolfo, 2000). Hence, if the classifiers are of the decision tree type, the shared knowledge structure would indicate the general knowledge of the domain. The new classifier is then expected to have better

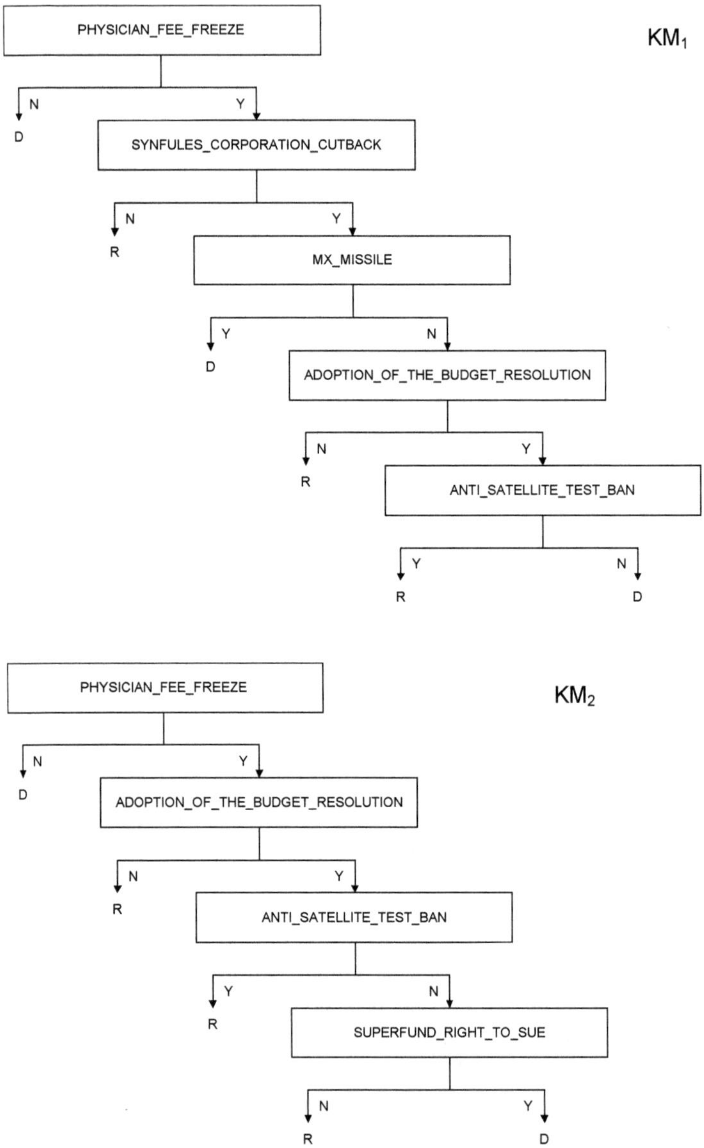

Fig. 9.3 Different knowledge models (KM_1, KM_2, KM_3) and the shared knowledge structure (common embedded subtree K_S) (notation N-No, Y-Yes, R-republican, D-democrat)

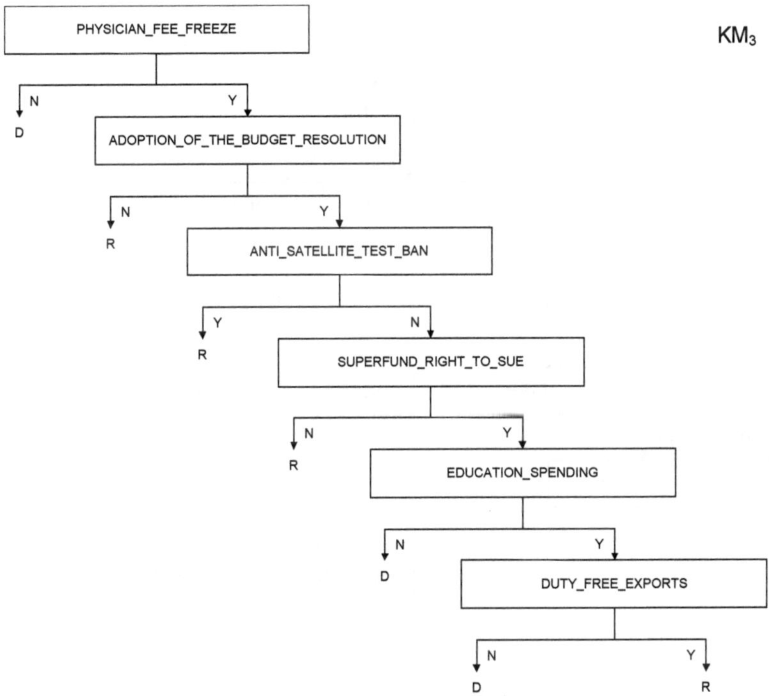

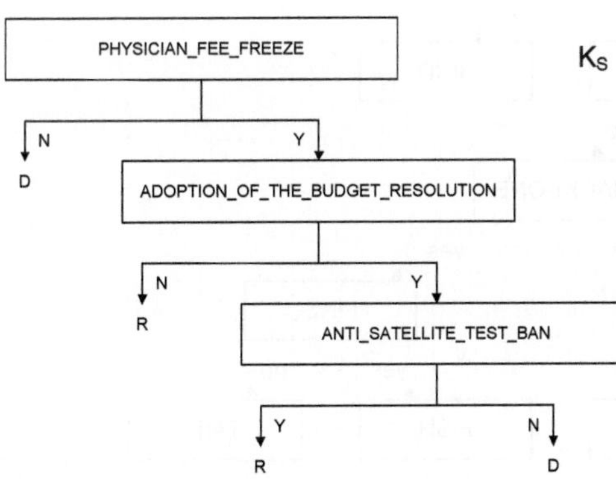

Fig. 9.3 *(continued)*

generalization capability and hence, be more accurate in classifying future unseen data objects. Merging of knowledge structures has been of interest for a long time and many useful applications can be found in e-commerce, enterprise application integration and the general management of scientific knowledge.

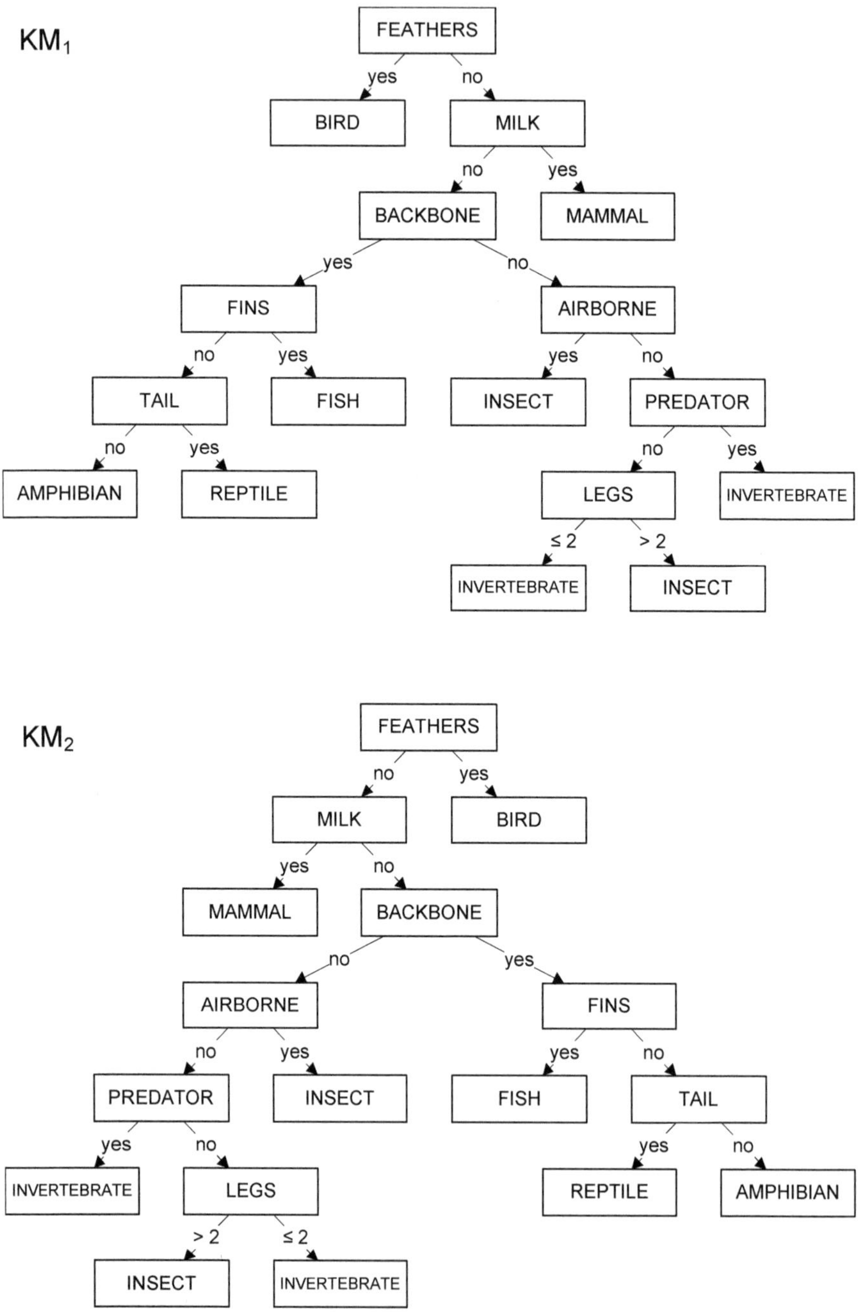

Fig. 9.4 Two knowledge models (KM_1 and KM_2) different in sibling node order but representing same information

As already mentioned in Chapter 5, mining of ordered subtrees is useful for queries performed on a single database where the sibling node order is already known and hence, the order restriction can be placed on the subtree. In the example presented in this section, the largest ordered embedded subtree was sought, and it happened to be that the common subtree was ordered in the same order in both knowledge models. However, this assumption cannot be made since the information content would remain the same even when the order among sibling nodes is exchanged in any way. When comparing conceptual knowledge models, different sibling node order usually does not make the structure match any less. Hence, the order of sibling nodes does not need to be preserved, in which case we are talking about unordered subtrees, which are discussed next.

9.3.2.3 Unordered Subtrees

In many cases, the order among the sibling-nodes is considered irrelevant to the task and is often not available. Mining of unordered subtrees becomes more suitable than mining of ordered subtrees, since a user can pose queries and does not have to worry about the order. All matching sub-structures will be returned, with the difference being that the order of sibling nodes is not used as an additional candidate grouping criterion. For many knowledge management related tasks, the unordered mining is starting to have better applications. The differences in the subtree types with respect to the knowledge matching task will be revisited again in Section 9.6.

For a simple illustration of the importance of unordered subtree mining when comparing knowledge models, consider the two example knowledge models represented in Fig. 9.4. These were learned from the publicly available 'zoo' dataset (Blake, Keogh & Merz, 1998). Even though at first sight they appear as different models, when the classification rule underlying the structure is examined, it implies exactly the same classification information. Hence, the order of sibling nodes did not play any role in the information content of the structure. Furthermore, since the user cannot be sure of the order of sibling-nodes, many queries posed in such domains would ignore the order. More discussion in regards to the importance of unordered subtree mining in context of Web log/content mining will be provided in Section 9.5.

9.3.3 Implications for Mining Constrained Embedded Subtrees

This section looks at some additional constraints that can be imposed on an embedded subtree. The focus is narrowed to the actual constraints where the general problem of frequent subtree mining is not modified, but the traditional definition of an embedded subtree has been changed according to the added constraint. The maximum level of embedding constraint is already a part of the general TMG framework since it is used to either mine induced or embedded subtrees. Therefore, only some possible application of it is indicated. The distance constraint imposed on embedded subtrees called for some adjustments in the general TMG framework. These adjustments are described together with some experiments indicating the differences between mining of embedded subtrees with and without the distance constraint. Each constraint is discussed with some application areas where its enforcement would be useful.

9.3.3.1 Maximum Level of Embedding Constraint

The maximum level of embedding constraint was discussed in Chapters 4 and 5 as a way of tackling the complexity of mining embedded subtrees. By restricting the level of embedding, the number of possible candidates can be reduced and less time and space is required to complete the task.

As was shown in the previous section, there is some difference between the mining of induced and embedded subtrees. In certain scenarios, allowing the extra embeddings can result in unnecessary and misleading information, but in other cases, it proves useful as it detects common structures in spite of the difference in concept granularity. In contrast to our examples presented in the previous section, the level of embeddings at which the knowledge structures can differ could in reality be very large. Hence, to make the large change from mining embedded subtrees where the allowed level of embedding is equal to the depth of the tree to induced subtrees where the level is limited to one, is a rather large jump and many useful patterns could be missed. In this sense, the maximum level of embedding constraint may prove useful as one could progressively decrease the level of embedding which can give clues about some general differences among the knowledge representations being compared.

9.3.3.2 Distance-Constrained Embedded Subtrees

This section starts by discussing the general implications of mining distance-constrained embedded subtrees. The discussion is general for the ordered or unordered contexts. The difference in application of ordered and unordered distance-constrained subtrees remains the same as for the cases where no distance constraint has been imposed. The definition of unordered/ordered distance-constrained embedded subtrees was given in Chapter 7, where the implications of mining distance constrained subtrees were discussed together with a motivating scenario. It was also mentioned in Chapter 7 that when posing queries, it can be the case that some portion of the query tree is not known and is not considered interesting or important. For this purpose, (Shasha et al. 2002) have introduced the idea of queries that can contain 'don't care' symbols to represent the irrelevant part of the query. These are known, as variable length 'don't cares' (VLDCs) and fixed length 'don't cares' (FLDCs). The distance constrained subtrees are related to queries containing FLDCs where the distance of nodes relative to the root node is indicated. The embedded subtrees are related to queries with VLDCs since the irrelevant part of the query is replaced by the ancestor-descendant relationship between relevant parts of the query. However, since the distance-constrained embedded subtrees contain distance related constraints, they could be used for answering queries containing both, VLDCs and FLDCs.

In Chapter 7 we have also noted the relationship of distance-constrained subtrees to the introduction of subtree patterns containing a wildcard symbol ('?') to match any label (Wang & Liu 2000). In a distance-constrained subtree, given two nodes *a* and *b* where *a* is ancestor of *b*, and where the distance from *a* to the root is equal to *x* and distance from *b* to the root is equal to *y*, then if $(x - y) > 1$ it is known that there is a wildcard symbol between *a* and *b*. The expression $(x - y) - 1$ can be used to work out the number of wildcard symbols, while if we are interested in VLDCs, the expression indicates length of the tree pattern considered as 'don't care'.

Mining of distance-constrained embedded subtrees will have some important applications in biological data analysis, web information systems and many knowledge management related tasks. While in this section we have provided some general discussion of the usefulness of the constraints, their potential use with respect to the Web log mining application will be discussed in Section 9.5. It is important to note here that by no means, is any claim made that mining distance-constrained embedded subtrees should replace the mining of embedded subtrees. Mining of embedded subtrees without any constraint has still many important applications as discussed earlier and as will become evident in later sections of this chapter.

```
<?xml version="1.0" ?>
<element name = "Type">
    <element name = "Cause">
    <element name="Genetic">
        <element name="Gene a" />
        <element name="Gene b" />
        <element name="Gene c" />
    </element>
    <element name="Environmental">
        <element name ="Climate" />
        <element name ="Drugs misuse" />
        <element name ="Family conditions" />
        <element name ="Economic conditions" />
        <element name ="Stress" />
    </element>
    </element>
</element>
```

Fig. 9.5 XML representation of patients' records

9.4 Mining of Healthcare Data

This section is concerned with some case studies showing how the tree mining algorithms can be successfully applied for the mining of health data (Hadzic et al. 2008). More specifically, the aim is to demonstrate the potential of the tree mining algorithms in deriving useful knowledge patterns in the mental health domain.

The section starts by providing an overview of some general issues that arise when wanting to analyze health information using tree mining. It then provides an experiment to illustrate the application. At this stage, real-world data describing the patient records in the mental health domain could not be obtained. The experiments were performed on a synthetically created XML dataset analogous to a common representation of patients' records from the mental health domain. An example of such an XML document is shown in Fig. 9.5.

9.4.1 *Mining of Patients' Records*

This section starts by providing some general data mining guidelines for the mining of patient records and then narrows its focus to the issues related to the application of tree mining to the problem.

In order to perform an effective application of tree mining to health data analysis, the following steps need to be carried out:

1. Data selection and cleaning
2. Data formatting
3. Tree mining
4. Pattern discovery
5. Knowledge testing and evaluation

9.4.1.1 Data Selection and Cleaning

Most of the databases also contain information that is not needed by the application; for instance, a database may also contain records related to diseases which fall outside the domain of interest. The irrelevant information, as well as noise and inconsistent data, should be removed from the data set to be mined. Extra care should be taken during this process since some data may appear to be noisy but in fact represents a true exception case.

9.4.1.2 Data Formatting

In order to simultaneously analyze all the data within this given data set, the following two conditions need to be fulfilled:

1. all the data needs to be put into the same format
2. the chosen format needs to be understandable by the data mining application

For example, if there is access only to relational type of data and a set of features can be grouped according to some criteria, XML can be used to represent the relationships between data objects in a more meaningful way. Hence, the relational data could be converted into XML format.

9.4.1.3 Tree Mining

When using tree mining algorithms, one has to consider the particular type of subtree that is most suitable for the application at hand. Patients' records are most likely to be stored in the same format and attribute values will be ordered the same away among the collected records.

The format and ordering of the data coming from one organization is expected to be the same. In respect to the current tree mining framework, this means that the mining of ordered subtrees will be most suitable. However, if the collected data originates from separate organizations and these data are found in different formats, then unordered trees would be more suitable. The reason is that the subtrees where the order of sibling nodes is different will still be grouped to the same candidate, and hence the common characteristics of a particular illness would still be found.

Another choice to be made is whether one should be mining induced or embedded subtrees. Since the aim is to extract the patterns where particular patient information has to remain within the context in which it occurred, the relationship of nodes in the extracted subtrees should be limited to parent-child relationships. By allowing ancestor-descendant relationships, some information could be lost about the context in which a particular disease characteristic occurred. This is mainly due to the fact that some features of the dataset may have a similar set of values, thereby making it necessary to indicate which value belonged to which particular feature. Taking these observations into account, the developed approach for mining of ordered induced subtrees will be used.

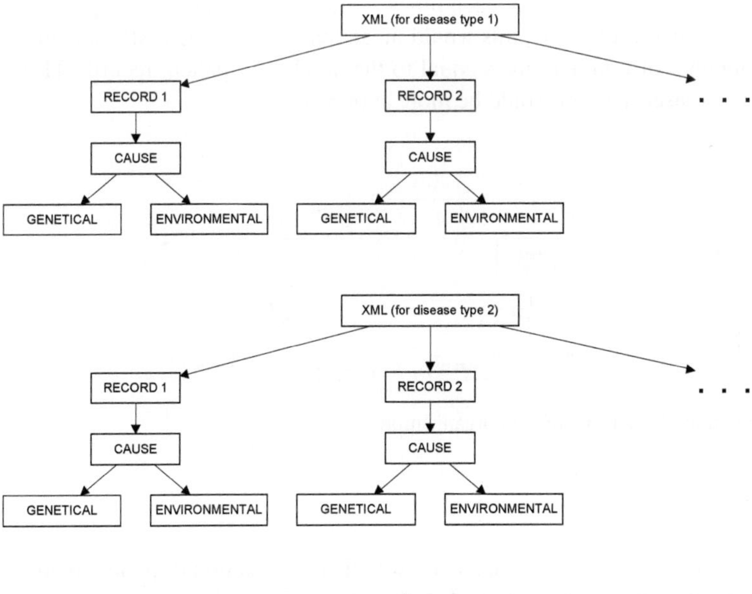

Fig. 9.6 Illustrating Scenario 1 of data organization

One further choice to be made is the type of support definition that should be used, which is dependent on how the data is organized. Consider the following three scenarios.

Scenario 1

Each patient record is stored as a separate subtree or transaction in the XML document and separate XML documents contain the records describing causal information for different mental illnesses. An illustrative example of this scenario for organizing the mental health data is displayed in Fig. 9.6. For ease of illustration, the specific illness-causing factors for each cause are not displayed, but one can assume that they will contain those displayed in Fig. 9.5. In this figure and in all subsequent figures, the three dots indicate that the previous structure (or document) can be repeated many times. For this scenario, both occurrence-match or transaction-based support can be used. This is because the aim is to find a frequently occurring pattern for each illness separately, and the records are stored separately for each illness. Hence, the total set of occurrences as well as the existence of a pattern in each transaction would yield the same result.

Scenario 2

The patient records for all types of mental illnesses are stored as separate subtrees or transactions in one XML document, as is illustrated in Fig. 9.7. It does not distinguish whether the records for each illness type are occurring together, as is the case in Fig. 9.7. In our example, this would mean that there is one XML document where the number of transactions is equal to the number of patient records. Here, the transaction-based support would be more appropriate.

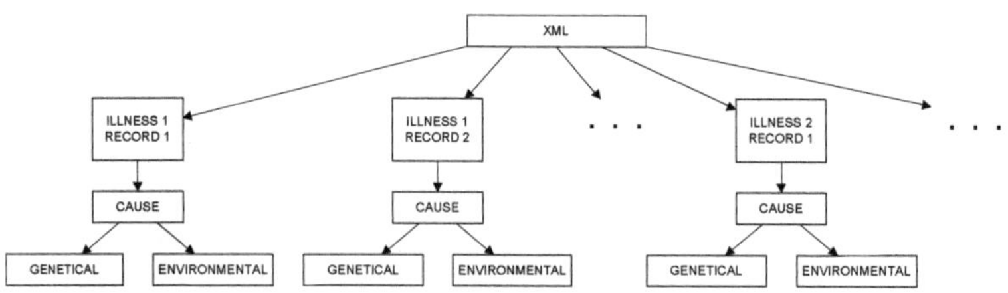

Fig. 9.7 Illustrating Scenario 2 of data organization

Scenario 3

The XML document is organized in such a way that a collection of patient records for one particular illness is contained in one transaction. This is illustrated in Fig. 9.8. In our example, this would mean that there is one XML document where

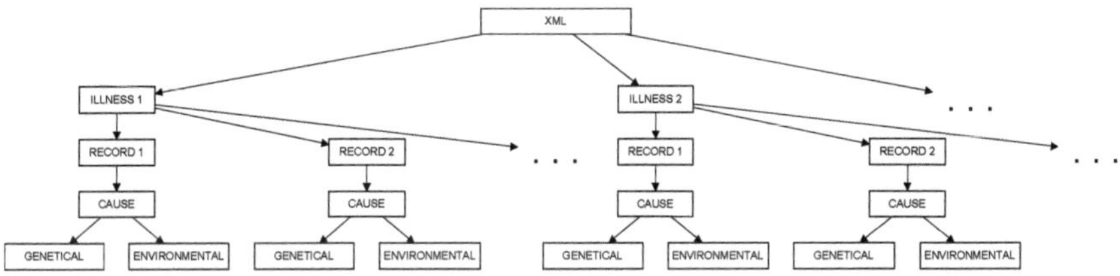

Fig. 9.8 Illustrating Scenario 3 of data organization

each transaction corresponds to a specific illness type. Each of those transactions would contain records of patients associated with that particular illness type. Hybrid support definition is most suitable in this case.

In order to find the characteristics specific to mental illness under examination, in scenarios 1 and 2 the minimum support threshold should be chosen to approximate the number of patient records (transactions) that the XML dataset contains about a particular illness. Due to noise often being present in data, the minimum support threshold can be set lower. However, it should not be set too low as there is a higher chance then for irrelevant factors to be picked up as important. For scenario 3, the number of mental illness types described would be used as the transactional part of the hybrid support, while the approximate number of patient records would be used as the requirement for occurrence of a subtree within each transaction. In this section, we are concerned with the second scenario.

Before data mining takes place, the data set can be split into two subsets, one for deriving the knowledge model ('internal data' from Fig. 9.9) and one for testing the derived knowledge model ('external data' from Fig. 9.9). External data can also come from another organization.

During the pattern discovery phase, precise combinations of genetic and environmental illness-causing factors associated with each type of the three mental disorders are identified. The results that establish the interdependence between genetic and environmental factors would make a breakthrough in the research, control and prevention of mental illnesses.

Testing and evaluation of the knowledge derived through the data mining applications is illustrated in Fig. 9.9. The 'internal data' corresponds to the subset of the data used to derive the hypothesis while the 'external data' corresponds to the subset of data 'unseen' by the tree mining algorithm. The 'external data' is used to verify the hypothesis so that it can become reliable enough to extend the current knowledge. In many cases, the choice of various data mining parameters can affect the nature and granularity of the obtained results. Where the hypothesis is not supported, the parameters will be adjusted and the previous steps alternated.

9.4.2 Experiment

The synthetically created XML document takes the form of the document displayed in Fig. 9.5 and it was made up of 30 transactions corresponding to patient records.

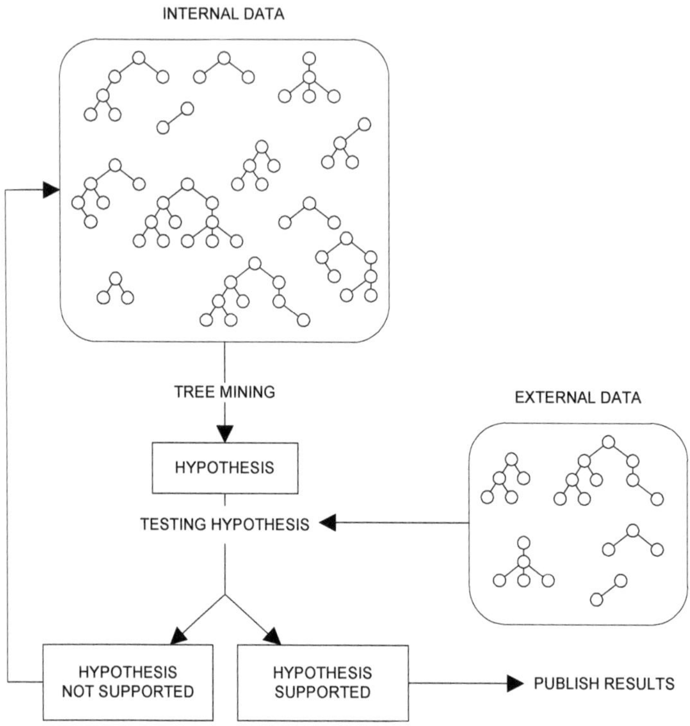

Fig. 9.9 Testing and evaluation of the derived knowledge

The document contains records for three different illness types: schizophrenia, depression, bipolar. The IMB3-Miner algorithm (Tan et al. 2006) was applied on the dataset using the transaction-based support definition. The minimum support threshold chosen was 7. A number of subtree patterns were detected as frequent. Only the largest subtree patterns (i.e. containing most nodes) were analyzed, since the smaller subtree patterns were subsets of the larger ones. The illness type was indicated, and this allows one to associate the causal patterns with the specific illness type.

Fig. 9.10 shows three different patterns, each associated with a specific mental illness type. As can be seen, the patterns are meaningful in the sense that the causes are associated with an illness. Note that this is only an illustrative example used to clarify the principle behind the data mining application, and by no means indicates the real causes. The real-world data would be of a similar nature, but is much more complex in regard to the information contained within the patient's records. This would pose no limitation to the application of the algorithm since, as demonstrated earlier in the book, it is well scalable for large and complex data.

The number of mentally ill people is increasing globally each year. This introduces major costs in economic and human terms to the individual communities and the nation in general, both in rural and urban areas. The newly derived knowledge could help in the prevention of mental illness and assist in the delivery of

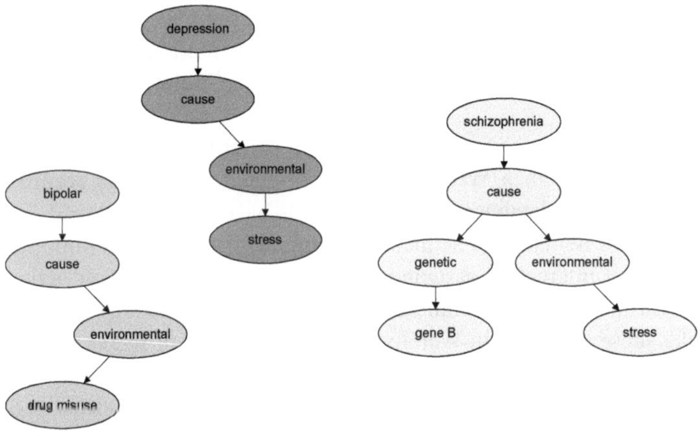

Fig. 9.10 Subtree patterns discovered from the artificial dataset

effective and efficient mental health services. Various community groups could greatly benefit from information systems based on data mining technology:

1. physicians would receive support in early diagnosis and treatment of mental illnesses;
2. patients and their care givers would receive support in dealing with, managing and treating illness;
3. accurate, reliable and up-to-date information would be available for the general public that will help understanding of mental illness. This will support management and prevention of mental illness;
4. medical researchers would receive support in advancing their research in identifying the disease-causing factors and effective patient treatments. This would reduce the possibility of redundant research (saving research time, effort and resources) and facilitate development of technologies for maintaining good health;
5. the cost of the mental health budget would be significantly reduced and the spending power of money spent on health and medical research enhanced by providing better information use.

9.5 Web Log Mining

Web mining is a general term that can correspond to either Web content mining i.e. extraction of information from the Web, or Web usage mining (also referred to as Web log mining) which is striving for user access and browsing patterns (Cooley, Mobasher, Srivastava 1997). To fully explore online business opportunities, it is important to track the utilization of a particular Web site and the service it provides. This can be done through the use of logging facilities during customer interaction with the site. The type of information commonly sought through the analysis of Web logs is: users' browsing patterns, analysis of users visiting the same Web pages, most frequent usage paths, paths occurring frequently inside user interactions with the site, detection of users who visit certain pages frequently, etc. (Cooley, Mobasher & Srivastava 1999).

This information can be extracted using the frequent pattern mining algorithms where the user interactions with a Web site are represented as separate records in transactional or sequential databases. Many methods have been developed for this purpose, and frequent itemset and sequence mining algorithms are commonly used to efficiently extract interesting associations, that reveal common user activities and interests (Eirinaki & Vazirgiannis 2003; Facca & Lanzi 2005; Pierrakos et al. 2003). Besides these advances in Web log mining, a number of works started to emerge where the Web log data is represented in a more complex form so that the navigational pattern and the overall structure of the Web site can be described in the data (Borges & Levene 1998; Buchner et al. 1999; Menesalvas et al. 2002). One example is the LOGML language proposed by Punin, Krishnamoorthy and (Zaki 2001a, 2001b) which allows for a more detailed and informative representation of Web logs, using an XML template. The set of requested Web pages and traversed hyperlinks in a Web log file are represented as a Web graph (tree) which is a subset of the Web site graph of the Web server being analyzed. In (Zaki 2005), an algorithm for mining ordered embedded subtrees was presented and it was demonstrated how it can be useful for extracting additional information from Web log data represented in tree-structured form. Generally speaking, tree mining algorithm methods can be very useful for determining common user activity and interests, by extracting frequently occurring informative sub-structures with respect to the user-supplied support. Hence, from an XML/LOGML document representing customer log activities, the extracted patterns will indicate customer electronic behaviour and this information can be useful for improving B2C and B2B transactions. Furthermore, by analyzing the visited paths in a Web site, the business can improve the design of Web pages and the structure of the Web site, as well as customize the look and the feel. In the context of tree mining, a tree-structured pattern will contain more descriptive information in comparison with an itemset or sequence pattern. This was demonstrated in (Zaki 2005) where it is shown that representing Web logs in a tree-structured form allows one to mine for substructures which are more informative, because they often reveal more information regarding the structure of the Web site and users' navigational patterns. To illustrate this here, we discuss a single user session obtained from the Web server log files of the Website for the CSSE journal. For clarity purposes, we omit the access details and consider only the actual pages visited within the site.

```
index.php/csse
index.php/csse/issue/archive
index.php/csse/search/authors
index.php/csse/search/titles
index.php/csse/information/readers
index.php/csse/information/authors
index.php/csse/information/librarians
```

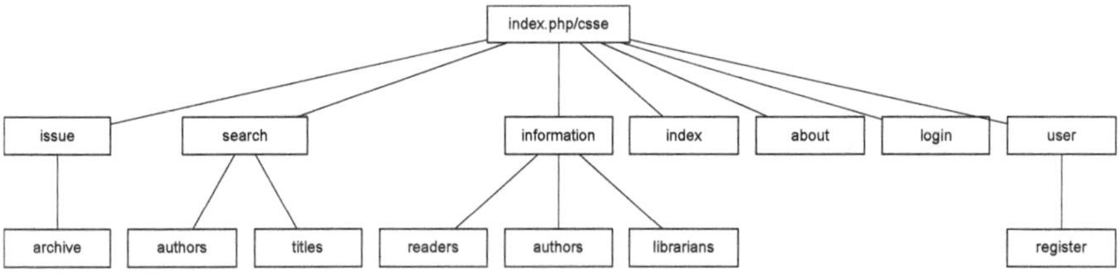

Fig. 9.11 Example log from CSSE journal website

```
index.php/csse/index
index.php/csse/about
index.php/csse/login
index.php/csse/user/register
```

Considering the 'index.php/csse' as the home page, the above sequence of logs can be represented in a tree-structured way as shown in Fig. 9.11. Please note that in the tree shown, the child node always contains the path of its parent node and hence we do not repeat the whole name of the Web page in each of the nodes, but rather the name of the specific page accessed. As can be seen, the corresponding tree structure is more informative as it captures the structure of the Web site, and navigational patterns over this Website. By organizing them in such a way, specific pages can be considered within the same context. An example of this is the three pages being grouped under the 'information' parent node in the tree. In this example, the internal nodes (i.e. nodes with children) in the tree reflect the structure of the tree, while the leaves in the tree correspond to the pages navigated to. This particular user session shows that after the user has found some journal specific information from the Website, involving the search for specific titles, and information regarding readership and authorship, the user has decided to register with the journal.

Web content in general is increasingly being represented in semi-structured or tree-structured form. The reason is that for many practical applications in domains such as Web mining, biology, chemistry, and network analysis, the expressional power of relational or structured data is not capable of effectively capturing the necessary relationships and semantics that need to be expressed in the domain. Semi-structured data sources are capable of dealing with 2-dimensional (2D) relationships among data entities that are manifested through structural relationships among attribute nodes. We have discussed in Chapter 2 how XML is effectively represented as a tree and hence mining of XML documents is usually recast as mining of rooted ordered labeled trees. We next discuss how Web log mining can be recast as a tree mining problem.

9.5.1 Transforming Web Usage Patterns to Trees

In Fig. 9.12, we provide a simplified example of a structure of a Web site used for selling some products and services. As mentioned earlier, the XML/LOGML

Fig. 9.12 Simplified example of a Website and Web usage patterns

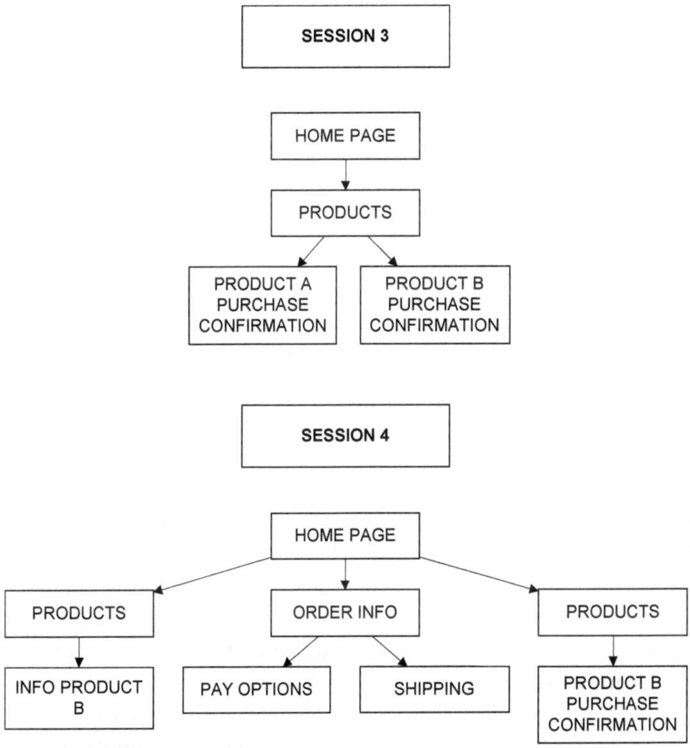

Fig. 9.12 *(continued)*

documents can be modeled as a rooted ordered labeled tree, and in the specific example of Fig. 9.12, each node in the tree corresponds to a Web page within the analyzed Web site. Four example interactions with the Website are illustrated as different sessions, and the sample tree illustrates the visited Web pages from left to right. In other words, the order of the Web pages visited in a session reflects the pre-order traversal of the tree. Hence, in session 3 the customer first visits the 'Products' page, from which product A is purchased, and then goes back to the products page to purchase product B. An assumption is made that on the 'Products' page of the Web site, there is an option to either read the information about a product or to directly purchase it, as was done in this particular case in session 3. Furthermore, representing the Web logs in this way will avoid the existence of cycles within the representation and makes the extraction of frequent patterns a more feasible prospect. This is the case in the LOGML (Punin Krishnamoorthy and Zaki 2001a), Web log representation language as the underlying structure is a tree and there are no cycles between the nodes of the document structure.

A user may be interested in determining the number of times that a certain set of Web pages was accessed from the home Web page. In this query, the order in which the Web pages were accessed is irrelevant and needs to be ignored when counting the occurrences of the particular Web pages being accessed. Hence, in such scenarios it is more suitable to use unordered subtree mining since, for a specified support, all the subtree patterns will be extracted that contain information regarding the same sets of pages visited, rather than only those where those particular Web pages were visited in the same order. For example, in Fig. 9.12, in both user sessions 1 and 2, the 'Products' and 'Order Info' Web pages were accessed, but in different order.

Within the current tree mining framework, the two most commonly mined subtrees are induced and embedded. An induced subtree preserves the parent-child relationships of each node in the original tree. In addition to this, an embedded subtree allows a parent in the subtree to be an ancestor in the original tree and hence, ancestor-descendant relationships are preserved over several levels. For example, the underlying tree of session 1 in Fig. 9.12, is an induced subtree of the original Web site tree, as each node that is a parent of a node in the subtree is also a parent of that same node in the original tree. On the other hand, the underlying tree of session 3 is an embedded subtree of the original Web site tree since the relationship between the 'Products' page and 'Purchase Confirmation' pages is not just parent-child but is ancestor-descendant since there is another node in the original site graph, i.e. 'Info Prod' between the 'Products' and "Purchase Confirmation' page. In other words, the parent of node 'Purchase Confirmation' is 'Products' in the session 3 tree, while it is an ancestor (grand-parent) node in the original site tree. With respect to Web log mining, induced subtrees are suitable when the user wants to have the information available of all the Web pages visited during an interaction with the Website, i.e. all the Web pages visited between any two nodes (Web pages). On the other hand, embedded subtrees are useful when the user is not interested in all the detailed page visits during an interaction with a Website, but is interested, for example, in the number of users who have accessed a particular Web page on the Web site, and have ended up buying a particular product. In the simplified example of Fig. 9.12, the purchase of a product x is represented as arriving at the 'Prod x Purchase Confirmation' page. In this scenario, the pages visited between the two nodes (Web pages) are irrelevant for the query and only if embedded subtrees are mined will this information be correctly revealed for a given user-specified support. From the example in Fig. 9.12, product B was purchased during sessions 2, 3 and 4, as indicated by the arrival at the 'Prod B Purchase Confirmation' page. For the query, the relevant part of the subtree is the link between the 'Home Page' and 'Prod B Purchase Confirmation' page. The pages visited between those pages are irrelevant, and only when embedded subtrees are mined will this relationship be detected to have occurred in three cases. If induced subtrees were extracted, some further analysis of the patterns would be required to detect the common relationship.

At other times, it may be necessary to limit the allowed number of pages visited between any two nodes. For example, the user will not be interested in the relationship if it took more than x pages to go from Web page a to Web page b. It may be suspected that it was just some random browsing of the Web site and any such type of user behaviour indicates inexperienced use of the Website. The user may be interested only in analyzing the usage by a frequent customer, i.e. someone who has used the Website many times. This is often the case when one wants to improve the layout of the pages for the selected group of customers and provide a specialized service to the frequent consumers of their service. Since induced subtrees allow only parent-child relationships between the nodes and embedded subtrees allow ancestor-descendant relationships, there is a need to limit the number of nodes that are allowed to exist between two ancestor-descendant nodes in the original tree database. The maximum level of embedding constraint (Chapter 4) can be used for this purpose. By imposing a limit on the level of embedding allowed in the extracted embedded subtrees, one can search for more specific patterns where the number of clicks needed to reach a particular page from another page is limited.

In yet another scenario, it may well be the case that the user wants to see the exact number of pages visited before one particular page is reached. For example, a user may be interested in finding the number of cases where a customer purchases a product after visiting x number of other pages within the site, i.e. determining the number of clicks the user made before he/she purchased a particular product. Similarly, a Web page that the user starts with will be the ancestor node in the subtree, whereas the Web page indicating the purchase of a product will be the descendant node of the subtree containing that information, as is the case in the example given in Fig. 9.12. Hence, the number of nodes between the ancestor and descendant nodes can indicate the number of other Web pages visited during the purchase activity within the Website. For example, session 3 may indicate a typical interaction with the Website of a frequent customer, since the product is purchased after a minimum number of clicks, and as opposed to sessions 2 and 4, no "Info Prod' or 'Order Info' pages were accessed. The distance (number of nodes) between the 'Products' page (node) and 'Prod B Purchase Confirmation' page (node) in session 2 is equal to two, while in session 3 it is equal to one. Hence, the usage pattern from the customer of session 3 indicates a more experienced user who knows the whole purchase process as no other information was checked and the Web site was used solely to purchase the products. In such scenarios, it may be useful to indicate the distance between the nodes in the subtree in order to distinguish the patterns more specifically. If embedded subtrees are mined, the relationship of the product being bought will be there but it does not have the expressive power of indicating the number of pages visited between the 'Home Page' and "Prod B Purchase Confirmation" page. In such application aims, it is more useful to mine embedded subtrees where the distance information between the nodes is indicated and used as an additional candidate grouping criterion. Furthermore, the distance information can be used to improve the layout of the pages and to test whether the new layout is more effective. For example, given old Web log data from the Web site using the traditional layout and new Web log data with the newly proposed layout, one could determine whether the new layout is better than the old one. This would be determined with respect to increased efficiency of a user visiting a Web page and then buying a product after a minimum number of clicks. For example, if there are relatively more patterns with a smaller distance between the two pages extracted from the new Web log data, then the layout of the Website has been improved as there are now more customers who buy a product or reach a desired Web page with fewer clicks (other page visits).

9.5.2 Experiments

In this section, a variety of experiments are presented to demonstrate the usefulness of unordered subtree mining when applied to the problem of Web mining (logs and content). In the developed framework (Chapter 6), induced subtrees are mined when the maximum level of embedding constraint is set to 1, and embedded when no such constraint is imposed. The former case is referred to as the UNI3 algorithm (Hadzic, Tan & Dillon 2007), while the latter is referred to as the U3 algorithm (Hadzic, Tan & Dillon 2008a) in the experiments provided in this section. The algorithm for the mining of unordered distance-constrained subtrees (as described in Chapter 7) is referred to as U3Razor (Hadzic, Tan & Dillon 2008b). The first experiment uses a synthetic file consisting of 100 transactions that is representative of the user interactions with the example Website provided in Fig. 9.12. This simple example will clearly illustrate some of the issues and main differences implied by mining different subtree types (and using different support definitions). The second experiment uses the publicly available real-world data (Zaki 2005). The underlying tree structure of this data is always ordered in the same manner (i.e. sibling nodes always appear in same order) and the types of subtrees that can be extracted are all of induced type. Hence, for this experiment, ordered subtree mining algorithms are more applicable and we demonstrate the type of patterns that can be extracted using our IMB3 algorithm (Tan et al. 2006) for mining of induced/embedded ordered subtrees (Chapter 5). To demonstrate the Web content mining aspect, we make use of the real-world movie database describing people involved in 'The Godfather' movie, their roles and previous 'filmography' history (http://www.imdb.com). The movie database as a whole was previously used in (Asai et al. 2003). Please note that in these experiments we do not compare our algorithms with the corresponding state-of-the-art algorithms as done in earlier chapters. Rather, the aim here is to show the applicability of unordered subtree mining algorithms to the problem of Web mining, with stronger focus on Web log mining that is represented in semi-structured form (e.g. LOGML). The experiments were run on Intel Xeon E5345 at 2.33 GHz with 8 cores, 8 GB RAM and 4MB Cashe Open SUSE 10.2.

9.5.2.1 Synthetic Web Log Data

This dataset consists of 100 transactions that correspond to different user interactions with the Web site used for selling some products and a few example transactions were provided in the 'sessions' of Fig. 9.12. Fig. 9.13(a) shows the number of frequent subtrees detected by the compared algorithms for given support thresholds, while Fig. 9.13(b) displays the time taken for the different algorithms to complete the task for the given support. As can be seen from Fig. 9.13(a), the U3 algorithm extracts the largest number of subtrees as frequent. This may be surprising considering that in the ordered tree mining algorithm, the candidate subtrees where the order among the sibling nodes is different but the set of nodes is the same, are considered as separate entities. However, for higher support thresholds, those separate entities will not have occurred often enough to be considered frequent, whereas in the unordered subtree mining scenario, they are grouped into one entity making that entity occur more often and hence, frequently. Results indicated that when the support threshold was set to 1 (i.e. all candidates are frequent), then the IMB3 algorithms had indeed many more candidates. The IMB3 algorithm always has a better time performance than its unordered counterpart U3 (Fig. 9.13(b)) because of the expensive canonical form orderings that need to be performed, as well as fewer subtrees being enumerated as frequent. The same observation can be made in regards to the time performance and number of subtrees considered as frequent when considering induced subtrees, and comparing results of ordered (IMB3-L1) and unordered (UNI3) case.

Let us now consider some patterns extracted by the algorithms and their usefulness with respect to Web log mining tasks. We will use the flat string representation to describe the patterns rather than the tree structures themselves. The main difference in terms of Web log analysis is that the ordered subtree patterns will retain the information about the order in which the Web pages were visited, while in an unordered subtree this information will be lost because they all grouped to one candidate that is a canonical form representation of all ordered candidates with different permutations among sibling nodes. However, if the analyst were interested in obtaining information related only to the number of sessions where a certain set of Web pages was visited, then the order of Web pages visited is irrelevant, and for a given support, the ordered subtree mining algorithm will miss these as it considers them as separate entities. For example, some interesting patterns found from the set extracted by the U3 algorithm are to do with the associations between products. At support = 20, the pattern found in the frequent pattern set, "Home_Page Prod_B_Purchase_Confirmation / Prod_C_Purchase_Confirmation (support = 24)" indicates that in 24 cases products B and C are bought together. Similarly, the pattern "Home_Page Prod_A_Purchase_ Confirmation / Prod_B_Purchase_Confirmation

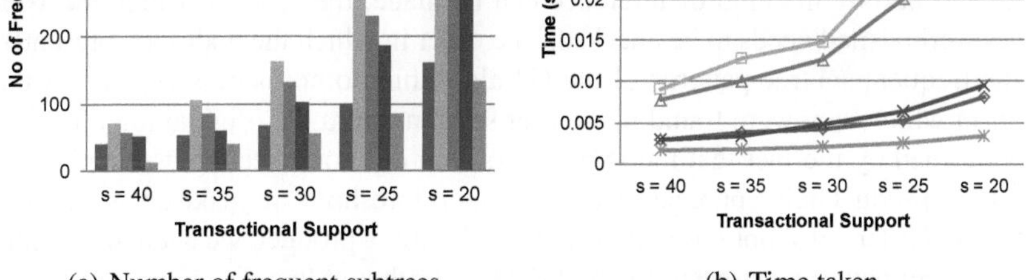

(a) Number of frequent subtrees (b) Time taken

Fig. 9.13 Synthetic Web Logs

(support = 32)" indicates that in 32 cases, products A and B are bought together. When looking through the patterns extracted by the IMB3 algorithm for support = 20, we could not find any associations between the products. In all the patterns, the pages indicating the purchase confirmations of different products never occurred together. This is due to the fact that those pages were not necessarily visited in the same order, and hence, the common association was not found since in ordered subtree mining they would be considered as separate entities. This highlights the importance of unordered tree mining when we want to find just the associations between Web pages (e.g. products) and do not care about the order in which they were visited. However, we are not claiming here that ordered subtree mining is not applicable to Web log analysis as, in many scenarios, the order may well be important. For example, the aim may be to detect a visit to a Web page after a specific number of other pages were visited in a given order. The choice is highly dependent on the way that the log data is organized in the file, and a combination of ordered and unordered subtree mining may need to be used to analyze the available information thoroughly.

To illustrate this aspect in this experiment, we next consider some frequent patterns extracted at support = 35. The largest pattern extracted by the IMB3 algorithm consists of 5 nodes "Home_Page Order_Info Pay_Options / Shipping / / Products (s = 40)". This pattern indicates only some browsing behaviour and does not contain information that implies any necessity to adjust business strategies or Web site design. The largest patterns extracted by the U3 algorithm consist of 6 nodes and these are: (1) "Home_Page Services / Products / Order_Info Pay_Options / Shipping (s = 35)"; (2) "Home_Page Products Info_Prod_B / / Order_Info Pay_Options / Shipping (s = 37)"; (3) "Home_Page Products Prod_B_Purchase_Confirmation / / Order_Info Pay_Options / Shipping (s = 35)". The first two patterns are essentially similar to the pattern extracted by the IMB3 algorithm and do not contain so much novel information. Pattern (3) may show some association between Product B and the 'Order_Info'. For example, looking at the remaining patterns (1) and (2), it appears that some users after seeing the 'Shipping' and 'Pay_Options' pages have not purchased any products, while the last patterns show that product 'B' was purchased. This may indicate that the current 'Shipping' and 'Pay_Options' options are not suitable for products 'A' and 'C' while they are suitable for product 'B'. Also, the organization may need to look at the Web page for shipping to check whether all the options are clearly explained or add new options in order to encourage more sales. However, before this kind of inference can be made, the order in which the Web pages were visited needs to be checked. The order in which the nodes are presented in the frequent subtree patterns of the U3 algorithm do not necessarily reflect the order in which they were found in the user sessions from the original log file. This is confirmed by the fact that the IMB3 algorithm did not detect an association involving a product being purchased at that support threshold. To make an association between 'Order-Info' options and their suitability for a product, we need to examine the frequent patterns extracted by the IMB3 algorithm at lower support thresholds. At support = 5 we could find two patterns potentially implying this kind of information, i.e. "Home_Page Services / Products Info_Prod_A / Info_Prod_B / Info_Prod_C

/ / Order_Info Pay_Options / Shipping / / Products Prod_A_Purchase_Confirmation (s = 6)", "Home_Page Services / Products Info_Prod_A / Info_Prod_B / Info_Prod_C / / Order_Info Pay_Options / Shipping / / Products Prod_B_Purchase_Confirmation (s = 7)".

There were other patterns involving purchases of products but the characteristics of these did not indicate a user browsing for all general information first before buying a product. Rather, the purchases were made directly or the 'Order_Info' options were viewed after the product was purchased. Since in this scenario we want to investigate any relationship between 'Order_Info' options and products being purchased, the order is important where the information about products is viewed first, then the order information, followed by a product being purchased. This kind of order of page visits would imply a new user or a user inexperienced with this Website so that he/she is not aware of all the services/products provided and the way that the business operates (shipping, pay options). An experienced user will usually not view all the products or services and will just purchase the products directly as he/she is already aware of all the general information of the Website. The above two patterns satisfy the order of page visits that indicate a new user of the Web site, since all information regarding products and pay options and shipping options is viewed first followed by a product being purchased. As seen in these two patterns, there were cases in which products 'A' (6 cases) or 'B' (7 cases) have been purchased after all the general information has been viewed. Hence, the initial inference indicated by unordered subtree mining is wrong as it appeared as if only product 'B' were being purchased. This is because the order was not taken into account, which itself indicates a new user of the Web site. Furthermore, the support of that pattern was much higher because all the occurrences of those sets of pages were counted without the order being taken as another grouping criterion. Hence, when it is possible to make inferences from the order in which the Web pages are visited, ordered subtree mining is more informative. On the other hand, unordered subtree mining is suitable when looking for a common set of Web pages occurring in sessions irrespective of the order in which they were visited.

To illustrate the difference between induced and embedded subtree mining, consider again the pattern indicating the associations between the different products detected by the U3 algorithm at s = 20. When exploring the patterns extracted by the UNI3 algorithm for that support, we could not find any patterns that associate two or more products together. However, the reason in this case is not because of the order, but rather is due to the different levels of embedding at which the associations occur. The patterns extracted by U3 (discussed above) start from the "Home_Page" node (root node) and hence will detect the product associations independent of the number of pages that are visited in between products being purchased. The UNI3 algorithm will count only those patterns that have the same pages in between. For example, if someone purchased a product after checking the product information (e.g. Product B from Session 2 in Fig. 9.12), and if someone purchased a product directly from the products pages (e.g. session 3 in Fig. 9.12), then those occurrences of the association will be considered as separate entities. Hence, if we are interested only in products being purchased, and the pages visited are irrelevant, then embedded mining will be more applicable as it will detect those deeply embedded relationships. This aspect can be further confirmed with the results obtained by the U3Razor algorithm as embedded patterns will be extracted and the distance between the nodes is used as the additional grouping criterion. Hence, we need to look at lower support thresholds to find those more specific patterns.

The U3Razor algorithm has detected the following variations of the pattern indicating the association between product B and C:

1. Home_Page Root Prod_B_Purchase_Confirmation 3 / Prod_C_Purchase_Confirmation 2 (s = 8);
2. Home_Page Root Prod_B_Purchase_Confirmation 3 / Prod_C_Purchase_Confirmation 3 (s = 14);
3. Home_Page Root Prod_B_Purchase_Confirmation 2 / Prod_C_Purchase_Confirmation 2 (s = 12);
4. Home_Page Root Prod_B_Purchase_Confirmation 2 / Prod_C_Purchase_Confirmation 3 (s = 7).

One would expect that the sum of the supports for each of these patterns will be equal to the support of the corresponding single pattern detected by U3. However, this is not the case in this example as 17 more occurrences were detected by the U3Razor algorithm. When we analyzed the results, we realized that in exactly 17 transactions, either the Prod_B_Purchase_Confirmation (9 transactions) or Prod_C_Purchase_Confirmation (8 transactions) pages are repeated. These repetitions occur with a different number of pages in between the home page and the corresponding product page which explains the 17 additional variations detected by the U3Razor algorithm. As mentioned earlier in this section, the distance constraint can also be useful when considering an adjustment to the Website. For example, from the just listed patterns one would like to increase the occurrence of pattern 3, as minimal clicks or Web page visits occurred before the product was purchased. Hence, by inspecting the increase/decrease of such patterns, one could infer whether a particular change to the Web site has resulted in the desired outcome (e.g. more experienced users/loyal customers, less browsing before a product is purchased).

Generally speaking, we have provided here some examples that illustrate the importance of being able to mine different subtree types. The mining of just one type may be suitable for specific queries and here we have highlighted some important differences and the type of information that can be inferred from each. Furthermore, by comparing the results of different subtree miners, one can obtain some insight into the general characteristics of the document/database. This kind of analysis would not easily be achieved if the Web logs were represented as itemsets or sequences, and some comparisons performed by (Zaki 2005) using different representations of CSLogs data confirmed this. In itemset mining, the navigational information would be ignored and only the unique Web pages visited would be listed.

As an example, in this experiment one would miss the information that can be inferred from a tree-structured pattern, when a node is revisited in the structure. For example, if a user first visits the "Prod_B" Web page, and then visits the 'Shipping' or 'Pay_Info" to then go back to "Prod_B" followed by "Prod_B_Purchase_Confirmation", one can infer that the order options are suitable for product B, in comparison with the products that were first viewed and then not purchased after order options were considered. This was the case earlier where the two largest frequent patterns detected by IMB3 algorithm at s = 5 indicated that 'order_info' options appear suitable for products 'A' and 'B', but possibly not for C as its purchase did not occur sufficiently enough in this kind of behaviour to indicate a new user. This information would not be arrived at if itemset mining had been applied.

If the available information were represented as sequences, then order would be imposed on the visited Web pages, and one could, to a certain degree, detect this kind of information. However, information regarding the structure of the Website, including the way it is traversed during a user session, would not be so easily obtainable. When sequential data is used to represent Web logs, a sequence consists of the longest forward links in user sessions and a new sequence is generated every time a user goes back to a page previously traversed (Zaki 2005). Hence, it is possible to lose some contextual information since the user may have gone back to a page to continue browsing the same aspect of the domain as was the case in the sequence before that. It could just be that the user has gone back to further check some information regarding this aspect. In this experiment, an example of general aspects would be the 'Order-Info' or 'Products' nodes which are parent nodes in the tree structure of the Web site. For example, if the user has first looked at information regarding product 'A' and then has gone back to view product 'B', then these activities would be stored as different sequences. In order to realize that both of these sequences are related, in the sense that the same general information is viewed, one would need to perform some analysis or search for the same sets of pages within different sequences. In a tree-structured representation of this activity, all the browsing behavior regarding a single aspect of the domain would be grouped under the general node representing this aspect. This is because a single user session is represented as a subtree of the original document tree and hence all the information reflecting the structure of the Web site is preserved.

Hence, a subtree pattern combines both the information contained in unordered items of an itemset, as well as the partial order information from sequences, into the tree structure that reflects the actual structure of the Website and the way it is navigated as a whole during a user session. As such, it can be more useful as it will contain combined information of the pages being associated together, user navigational behaviour and the structure of the Web site. By mining all this information collectively, one can gain further insights, not just related to customer behaviour but to the evaluation of the Web site design and/or business strategies. In this experiment, we showed how different types of subtree patterns can contain different information and how a number of inferences can be made from them, highlighting the importance of combining different frequent subtree miners (i.e. different types of subtrees or constraints) for a deeper insight. Even though the example used in this experiment is rather simple, the discussed differences and implications are likely to occur on a much larger scale in real-world datasets.

9.5.2.2 US1924 Data

The third experiment is performed on the publicly available server log data (US1924) described in LOGML (Punin, Krishnamoorthy & Zaki 2001a, 2001b) format. This dataset is organized in such a way that any Web page visited within the site will occur at the same position, and hence, the ordering among the sibling nodes is set in the tree. Furthermore, the underlying tree structure of the transactions has a depth of 1 and hence, all subtree types to be extracted are of induced type. With these characteristics, this particular dataset is sufficiently analyzed using an algorithm for the mining of ordered induced/embedded subtrees. In Fig. 9.14, we present the time performance and the number of frequent subtrees extracted using the IMB3 algorithm (Tan et al. 2006). Similar to the CSLogs dataset, this dataset is composed of users' access trees to the CS department Website at RPI. As an example of an interesting pattern, at s = 1000, it was found that all the frequent users' access tree ($k > 1$) contained the "…/madonna_lyrics/Homepage.html" page.

9.5.2.3 Web Content Mining

The dataset describes all the people (producers, directors, actors) involved in 'The Godfather' movie trilogy and provides 'filmography' (previous movies in which they were involved) information about each. The underlying structure of this dataset is a single tree (transaction) and hence, occurrence match/weighted support is used to check for the associations of items within this tree. Transaction support could have also been used only if we were to pre-process the data so that information of each person (actor, producer) involved was separated as a separate transaction in the database. Hence, when we talk about a record in the context of this dataset, we are

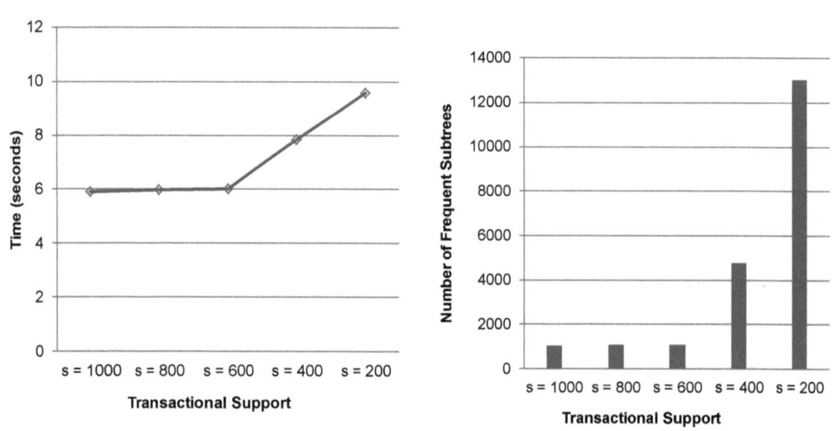

(a) IMB3 time performance (b) Number of frequent subtrees

Fig. 9.14 LOGML data

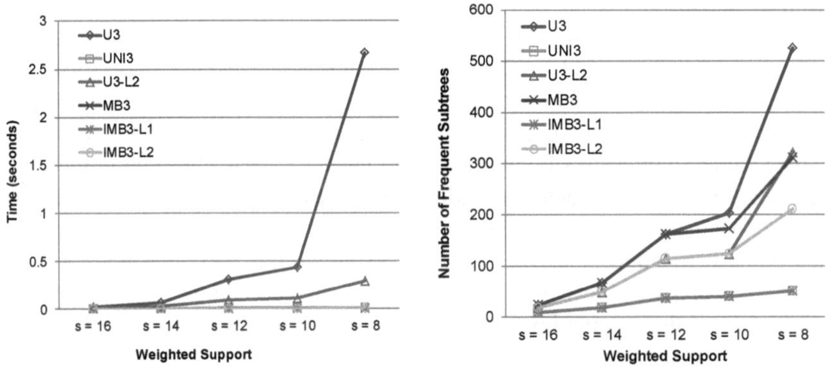

(a) Time performance (b) Number of frequent subtrees

Fig. 9.15 Godfather.xml

referring to the 'filmography' information about a person. In Fig. 9.15, we show the time performance (a) and the number of frequent subtrees detected (b) by various algorithms. We have also performed an experiment using the level of embedding constraint (δ) set to 2 for both ordered IMB3-L2 and unordered (U3-L2) subtrees. This serves as an analysis of the type of patterns extracted that are between embedded (no limit on δ) and induced (δ = 1) subtree extreme. From the figure, we can see that the U3 algorithm extracts the most subtrees as frequent (Fig. 9.15(b)) which adds to its processing time. Let us now look at some interesting patterns and the difference between the results obtained by using different tree mining parameters.

For higher support thresholds (s = 16 or 14), the frequent subtrees found by algorithms for mining induced subtrees (UNI3 and IMB3-L1), do not indicate associations between attribute values but rather, are comprised of general nodes occurring in all records. In other words, the patterns reflect only the common structure of the document without associating too many values together. At lower support threshold, the frequent subtrees set from UNI3 and IMB3-L1 contained the following pattern "*person filmography Actor title(Godfather Trilogy: 1901-1980, The (1992) (V)) / title(Godfather, The (1972))* s = 12). It indicates that 12 people who have acted in the first part of The Godfather movie (1972) have also acted in the two subsequent Godfather movies. As seen from Fig. 9.15(b), both UNI3 and IMB3-L1 detect the same subtrees as frequent, but as the maximum level of embedding (δ) is increased, we start to see some differences. For example, increasing the δ to 2 (i.e. UNI3-L2, IMB3-L2), results in the UNI3-L2 algorithm enumerating 111 more patterns in comparison to IMB3-L2, at s = 8. The reason for this is that some of the 'filmography' information is not ordered in the same manner for all the people about whom the information has been stored. Generally speaking, this is one of the main aspects that causes the unordered subtree mining to be more applicable than ordered subtree mining, since in some datasets, the databases may contain data from disparate sources and the order is unknown or not considered to be important to the user. The reason for this difference not being observed in an induced case is that all the nodes connecting other nodes had to be frequent in order for such a pattern to be frequent itself. Hence, by mining induced subtrees, we could not detect the associations that were embedded at more than one level within the dataset. From another perspective, one can use the δ to skip/ignore a number of intermediate nodes separating attribute value relationships lying at different levels. As was observed in these experiments, setting δ to 2 detected additional patterns because the node separating the relationships at different levels was not frequent enough for it to be detected when $\delta = 1$ (i.e. induced subtree case). For example, this particular node was the 'Actor' node that was present in the pattern mentioned previously, which gives us the information that people involved in this movie were actors. At s = 8 UNI3-L2 and IMB3-L2 detected many more associations between movies, but these associations are general for anyone involved in the movie. For example, the pattern "*person filmography title(Godfather: Part III, The (1990)) / title(Conversation, The (1974)) / title(Godfather, The (1972))* s = 19" was found by both UNI3-L2 and IMB3-L2 algorithms, indicating that 19 people who were involved in 'Godfather III' movie were also involved in 'Conversation' and 'Godfather I' movie. However, the roles that these people played in the movie are not known. At the same time, this illustrates the drawback of mining subtrees where $\delta > 1$. While one can find additional associations, there is also the possibility of losing some context information, and it may be possible that out of those 19 people, one had two different roles in one of the movies (e.g. director and producer).

In the last scenario, when no constraint on δ was imposed, the number of frequent subtrees extracted was much larger, as expected. In this scenario of mining of embedded subtrees, any attribute values that are frequently associated together will be detected. Generally speaking, the MB3 algorithm detects patterns similar to those discussed above detected by IMB3-L2. The information found is mainly concerned with the movies that are frequently associated together. At the given support, the movie 'Conversation' is the only movie that is frequently associated with all the parts of 'Godfather' movies. The largest pattern extracted by the MB3 algorithm (s = 8) was "*person date_of_birth / filmography title(Godfather Trilogy: 1901-1980, The (1992) (V)) / title(Godfather: Part III, The (1990)) / title(Godfather: Part II, The (1974)) / title(Conversation, The (1974)) / title(Godfather, The 1972))* s = 41". This time it indicates that the found association occurs 41 times when we have no limit on δ, because all the instances of occurrence are counted including those that are not associated with the same role, as is the case in induced subtree mining. Again, the large jump in occurrence of associations has occurred in comparison to δ being set to 2, which confirms the extreme difference between induced and embedded subtree mining. Similarly, at s = 8, the largest pattern detected by the U3 algorithm was "*person date_of_birth / filmography title(Godfather Trilogy: 1901-1980, The (1992) (V)) / title(Godfather Trilogy: 1901-1980, The (1992) (V)) / title(Godfather: Part III, The (1990)) / title(Godfather: Part III, The (1990)) / title(Godfather: Part II, The (1974)) / title(Godfather: Part II, The (1974)) / title(Conversation, The (1974))* s = 82". Here we can see both 'Godfather II' and 'III' repeating in the pattern, which indicates that in 82 cases a person had multiple roles in those movies. This association was not detected by the MB3 algorithm because of the different order in which they occurred, while the U3 algorithm has grouped all those occurrences into one entity. While in this domain, these kinds of patterns may not be considered as useful because of the loss of context information (i.e. role); in other domains, associations lying at different levels in the tree are valid and should be considered. Hence, depending on the application, one would know the difference in the level that is acceptable and whether the intermediate node(s) are necessary to keep the information in context. One can then set the value for δ to be suitable for the application and not to be at such an extreme of induced or embedded subtrees. Another option would be to use the distance constraint to further distinguish the embedded subtrees depending on the different distances between the nodes. This distance information would then indicate the number of nodes that have not been taken into account by the pattern, and the user would then decide whether such patterns are to be considered as valid with respect to the specific application. However, in this particular dataset, all the frequent embedded subtrees extracted have the same distance between the nodes in the original tree. This is confirmed by the fact that when we ran the U3Razor algorithm on this dataset, the frequent subtree set was exactly the same as for U3, which is the reason for not including U3Razor in the results presented in Fig. 9.15.

This section has presented an approach for Web content and Web usage mining, by recasting it as the problem of mining frequent subtree patterns from tree-structured data. A number of experiments on real-world and synthetic data were performed to illustrate and discuss the important differences and implications of mining different subtree types under different supports/constraints. This has demonstrated how frequent subtree mining algorithms can be a very useful tool for the mining of Web logs and content that is represented in a semi-structured form.

9.6 Application for the Knowledge Matching Task

Matching of heterogeneous knowledge sources is of increasing importance in areas such as scientific knowledge management, e-commerce, enterprise application integration, and many emerging Semantic Web applications. For example, many web services use XML as a unified exchange format, since it provides the extensibility and language neutrality that is the key for standards–based interoperability between different software applications (Fensel et al. 2007; Alesso & Smith 2006). The process of discovering particular web services and composing them together in order to accomplish a specific goal is an important step toward the development of 'semantic web services' (Fensel et al. 2007; Alesso & Smith 2006; Paolucci, Kawamura, Payne & Sycara 2002). In this process, being able to detect common knowledge structures between the information presented by the services to be integrated will be a useful step toward automation.

The knowledge matching method presented in this section is based on the use of our previously developed tree mining algorithms in order to automatically extract shared document structures (Hadzic, Dillon & Chang 2007). By using the tree mining approach, many of the structural differences among the knowledge representations can be detected and the largest common structure is automatically extracted. The implications of mining different subtree types are discussed and the most suitable subtree type within the current tree mining framework is indicated. The experiments are performed on a set of XML schemas used for describing organization specific information. In general, the method is applicable for matching any tree-structured knowledge representations. The focus is solely on knowledge structure matching, and the problem of matching at the conceptual level has not yet been considered. Hence, the main purpose of the work is to demonstrate the potential of applying tree mining methods to the problem of knowledge matching which brings it a step closer towards automation. In earlier chapters, the developed tree mining algorithms were applied on large and complex tree structures and their scalability was experimentally demonstrated. In the experiments presented later in this section, smaller trees are considered in order to illustrate the underlying concept and issues in a more comprehensible manner.

9.6.1 Method Description

In this section, an overview of the proposed method for knowledge structure matching is described. The approach described is motivated by a real-world scenario of different organizations using different knowledge structures for representing their domain-related information. The organizations may want to discover the general knowledge of the domain that is shared in all knowledge representations in order to obtain a shared understanding of the domain. Furthermore, the shared knowledge structure can be used as the knowledge basis to be used by new organizations. In the real world, it is common for different organizations to use different names for the same concepts. This is a problem of concept matching and as mentioned earlier, the focus of this work is on matching at the structural level. Hence, the current assumption is that all concept names are the same, or some concept matching algorithm has already been applied to find the corresponding mappings.

The general approach is shown here in Fig. 9.16 which at the same time describes the experimental setup. While only two knowledge structures (KSs) are displayed in the figure, the approach is valid when the number of available KSs is larger. Suppose that we have two document structures used by different organizations for representing the knowledge from the same domain. The aim is to merge them into one representation which captures the general knowledge for that specific domain. Each KS is most likely to differ in the way the knowledge is represented and the amount of concept granularity. However, since they are describing the same domain, there will be some common parts of knowledge. This is where a tree mining approach will prove useful since sub-structures from large tree databases can be automatically extracted.

The way that the experiment is set up can be explained as follows. Given n number of tree structured documents D_i (where $i = \{1, \ldots, n\}$) to be matched which can be modeled as rooted ordered labeled trees KS_i (where $i = \{1, \ldots, n\}$) set up a tree database (T_{DB}) so that each KS_i is represented as an independent subtree (i.e. a

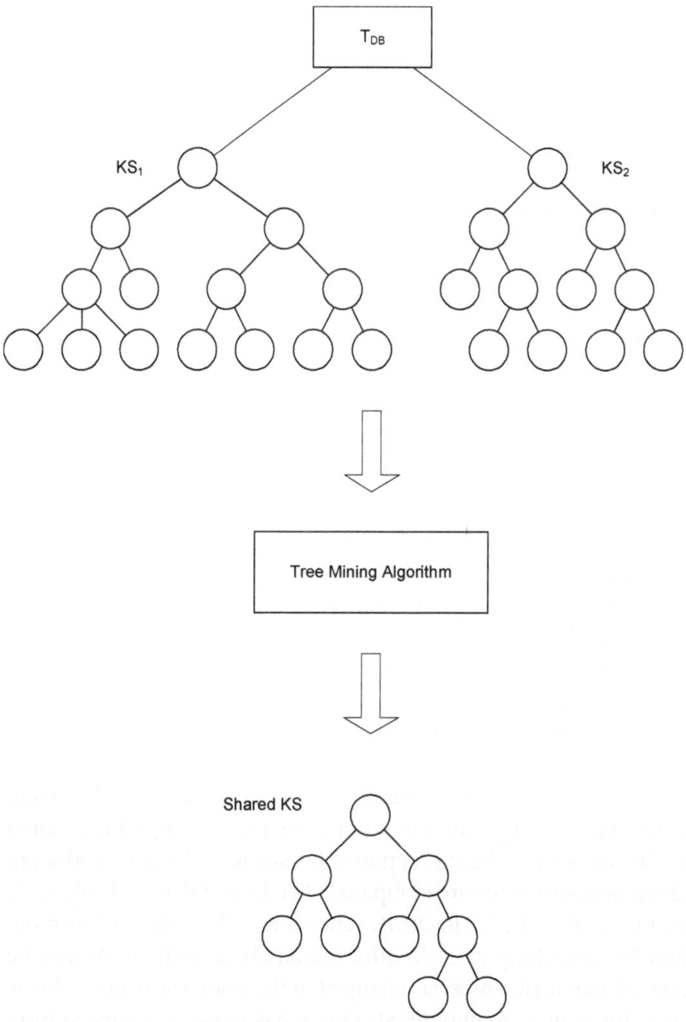

Fig. 9.16 Proposed knowledge matching approach

separate transaction) in T_{DB}. The root of each KS_i will be a child of the root node of T_{DB} as is illustrated in Fig. 9.16.

Since the order of sibling nodes is irrelevant, and the knowledge structures can differ in the level of detail stored about a concept, it was decided that the application of an algorithm for the mining of unordered embedded subtrees is most suitable for this problem. Hence, the U3 algorithm (Chapter 6) was applied to the database using the transaction-based support definition of n (i.e. the number of documents to be matched). The detected set of frequent patterns corresponds to the common substructures of the documents being compared. The largest frequent k-subtree (i.e. largest common unordered embedded subtree among all KS_i in T_{DB}) is selected to represent the shared knowledge structure among the documents compared. It should be noted that this subtree corresponds to the frequent maximal subtrees and hence, the mining of maximal unordered embedded subtrees is more suitable for this task.

The shared KS detected could be less specific than the KS from a particular organization, but it is nevertheless valid for all the organizations. Furthermore, each of the different organizations could have its own specific part of knowledge which is valid only from their perspective, and which can be added to the shared KS so that every aspect for that organization is covered. Hence, the shared KS can be used as the basis for structuring the knowledge for that particular domain and different communities of users can extend this model when required for the specific purposes of their own organization.

9.6.2 Experiments

This section describes the knowledge models used in the experiments and shows the resulting common document structure as detected by the U3 algorithm. The example XML (schema) documents were obtained from the ontology matching website (www.ontologymatching.org). The documents correspond to a representation of specific domain information used by different organizations. Please note that in the examples, the node labels concepts describing the same concepts have been replaced where necessary by a common name, since the problem of concept matching is beyond the scope of this chapter.

In Fig. 9.17, we can see the underlying tree structures form the XML schema documents used by Amazon (S1) and Yahoo (S2) for describing a sales item. Each of these tree structures was represented as a separate transaction in the XML document that we used as input to the U3 algorithm in order to extract the largest unordered embedded subtree. The transaction-based support was used with the threshold set to 2 since only two document structures are compared. The largest detected subtree is presented at the bottom of Fig. 9.17. The XML schema used by Amazon has more specific information for describing the sale information about an item. As can be seen in Fig. 9.17, the sibling node order has changed in the shared structure. This is because an algorithm for mining an unordered subtree has to use a canonical form of a subtree, according to which candidate subtrees will be converted and grouped. As discussed in Chapter 6, in the canonical form used by the U3 algorithm, the sibling nodes are ordered according to the alphabetical order and the node with the smallest label is placed to the left of the subtree (as shown in shared structure of Fig. 9.17). This process is required so that all the subtrees with a different order of sibling nodes describing the same concept, are still grouped to one candidate.

In the scenario depicted by Fig. 9.17, it would have even been sufficient to mine ordered induced subtrees since no specific concept information was stored at different levels of the tree and all the common concepts were presented in the same order. However, one cannot assume that the compared knowledge representations will have their common concepts ordered in the same way, and that a particular set of concepts will not be grouped under a certain criterion in different representations. In this experiment, smaller examples were used for a clearer illustration of

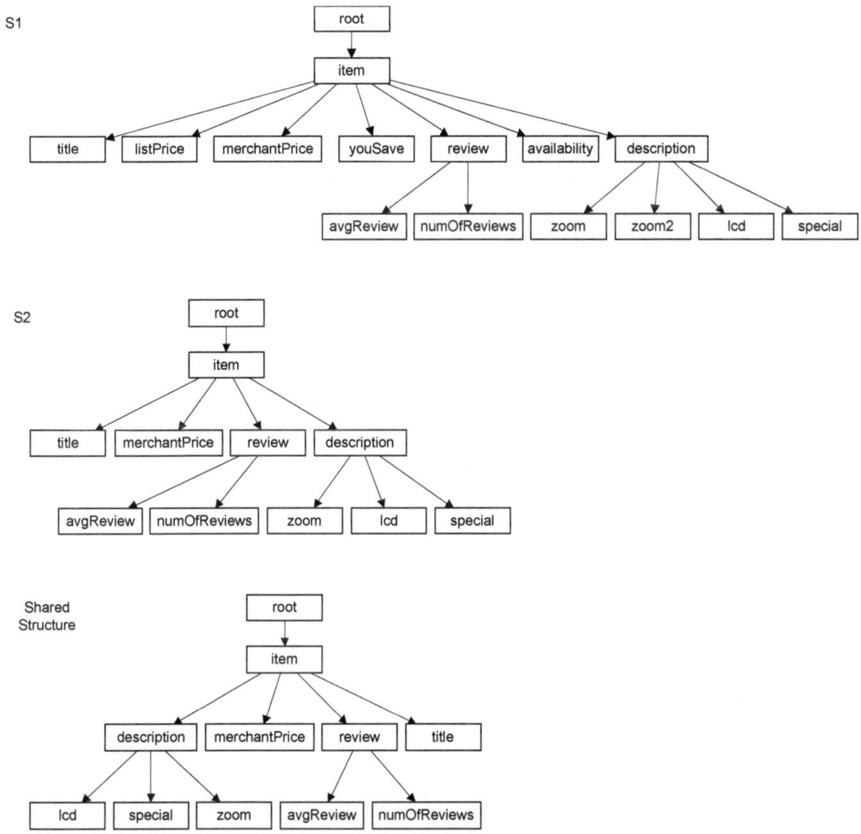

Fig. 9.17 XML schema structures for sale item description by Amazon (S1) and Yahoo (S2), and their shared document structure

the underlying concept. However, when document structures being compared are much larger, it is more likely that the same concept information will be stored in a different order and at different levels in the tree. The next scenario illustrates that it is important to relax the order of sibling nodes and to extend the relationships between the nodes from parent-child to ancestor-descendant (i.e. extracting embedded as opposed to induced subtrees).

Fig. 9.18 displays tree structures that correspond to the XML schemas used by different organizations for describing their purchase order (PO) commercial document. The largest common document structure is displayed last and by comparing it to S1 and S2, one can see that it was necessary to extract the largest unordered embedded subtree. First of all, the order of nodes describing the address information was different, and the nodes were stored at different levels in the tree. Hence, the common relationship between the 'DeliverTo' nodes and the address details would not be detected if the largest induced subtree were extracted, since the level of embedding is equal to 2, while the allowed level of embedding in an induced subtree is limited to 1. The largest common induced subtree would consist of only the root node 'PO' and two nodes emanating from the node 'PO' i.e. 'DeliverTo' and 'BillTo'. The remaining common structures would be missed altogether and hence, mining embedded subtrees in these scenarios is a necessity.

It is worth noting that it could be possible for some knowledge structures to have a few nodes with the same label located deeper in the tree. In this case, if embedded subtrees are mined, misleading results could be returned since the level of embedding allowed in the subtrees is not limited. Making the major change from mining embedded to induced subtrees runs the risk of missing many other common structures where the level of embedding among the nodes is different. In these cases, the maximum level of embedding (δ) constraint could be used to impose a limit on the allowed level of embedding in the extracted embedded subtrees. The δ could also be progressively decreased until some differences are resolved. In this scenario where multiple nodes exist with the same label, it probably would not provide only one matching document structure as for the induced case, but many common structures

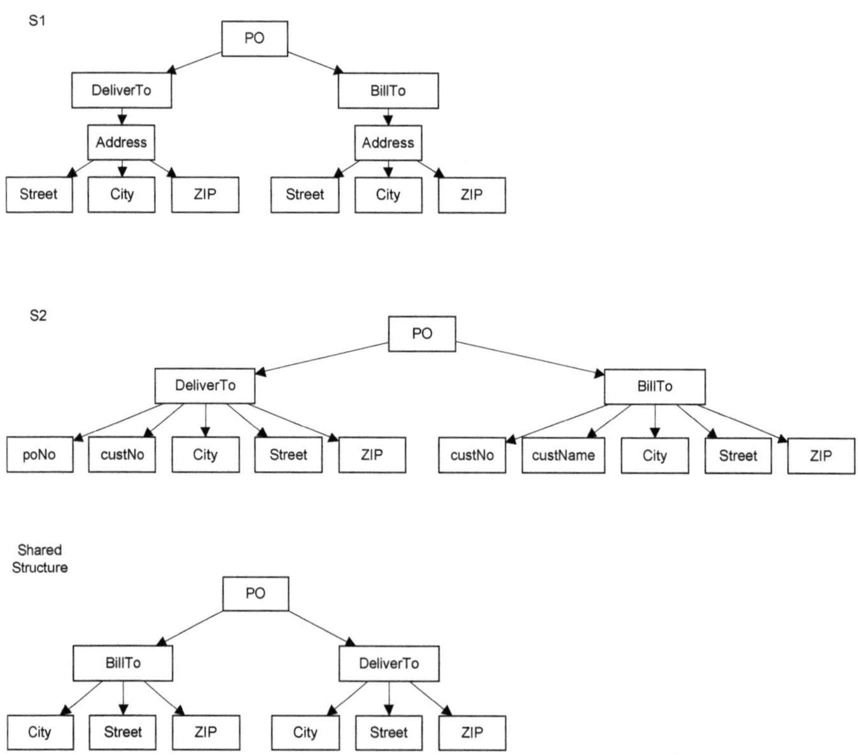

Fig. 9.18 XML schema structures describing different post order document (S1 and S2) and their shared document structure

of same size could be detected. The choice of method to adopt is again dependent on the type of knowledge that is being matched as, for some applications, induced subtrees may be sufficient and the level of embedding can be ignored, while for others it is important as it indicates that extra specific information is stored for a particular concept in a document structure. Even if the user is not a domain expert, different options can be tried with respect to δ and this should in itself reveal some more detail about the similarities and differences among the compared document structures.

9.7 Mining Substructures in Protein Data

As the final application area considered in this chapter, this section indicates the potential of tree mining algorithms as a means of providing interesting biological information when applied to tree-structured biological data. Prion (short for pertinacious infectious particle) is a type of infectious agent. Prions are abnormally structured forms of a host protein, which are able to convert normal molecules of protein into an abnormally structured form. The Prions dataset describes a Protein Ontology (Sidhu et al. 2005) database for Human Prion proteins in XML format.

As mentioned in Chapter 5, the first step performed for optimization purposes is to map the XML tags to integer indexes. Since the maximum height of the underlying tree structure of the Prions dataset is 1, all candidate subtrees generated are induced subtrees. Furthermore, the order in which the information is represented is consistent and known beforehand. This makes the application of an algorithm for mining of ordered subtrees sufficient for the aim of this application. The experiments were run on 3Ghz (Intel-CPU), 2Gb RAM, Mandrake 10.2 Linux machine and compilation was performed using GNU g++ (3.4.3) with the –O3 parameter. Occurrence-match support definition was used. The total run-time and memory usage of the IMB3-Miner algorithm is displayed in Fig. 9.19, for varying support thresholds.

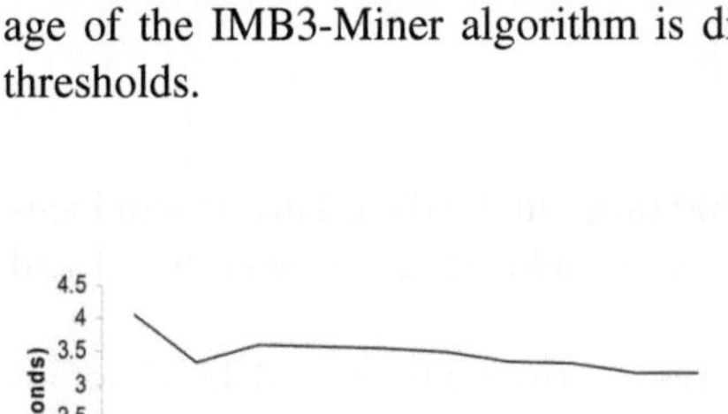

(a) Run-time

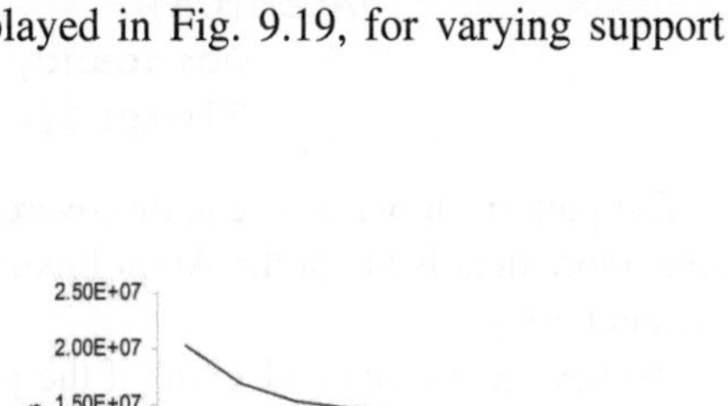

(b) Memory usage profile

Fig. 9.19 IMB3-Miner

In bioinformatics, the discovery of structural patterns by matching data representation structures is essential for the analysis and understanding of biological data. If a structural pattern occurs frequently, it ought to be important in some way. On the other hand, infrequent patterns may also provide meaningful information. Next, we discuss some interesting patterns detected by the IMB3-Miner algorithm.

```
ATOMSequence
    _ATOM_Chain[L]
        _ATOM_Residue[ALA]
```

This pattern was discovered 40 times in the dataset. Here, ATOMSequence refers to the chain of residue sequence for the atom list in question (_ATOM_Chain = L). The description of the structure of the Chain refers to numerous instances of Residues defined as (_ATOM_Residue = ALA). Each residue has a number of Atoms linked to it, which make up the atom list for the residue. Similarly, the collection of atom lists for all residues in the chain describes the entire ATOMSequence.

```
ATOMSequence
    _ATOM_Chain[A]
        _ATOM_Residue[ALA]
            Atom[H]
```

This pattern was discovered 101 times in the data. The pattern is similar to the pattern discussed above, with the inclusion of identifying the Atom (Atom = H) linked to the residue.

```
ATOMSequence
    _ATOM_Chain[H]
        _ATOM_Residue[THR]
            Atom[CA]
                Occupancy[1]
                Element[C]
```

The pattern shown above is discovered 100 times in the data. This pattern identifies more details about the Atom linked to the Residue (like: Occupancy = 1 and Element = C).

We have just discussed some of the patterns discovered by the IMB3-Miner algorithm. Generally speaking, a frequent subtree mining algorithm can aid in the process of discovering useful pattern structures in Protein datasets. This makes it useful for comparison of protein datasets taken across protein families and species, and it can help to discover interesting similarities and differences.

9.8 Conclusion

This chapter has provided a general overview of some important applications of tree mining. The motivation behind each development was illustrated with example scenarios. An overview of possible tree mining parameters that can be used within the current tree mining framework was provided. These parameters include the support definitions, types of subtrees that can be mined, and types of constraints that can be imposed on the subtrees. Possible application areas and the implications of using different tree mining parameters were discussed in the context of knowledge management related tasks.

Furthermore, the chapter discussed some independent applications of tree mining to the problems of analysis of mental health information, Web log/content mining, knowledge matching and protein structure mining, As a whole, the chapter is useful for those not so familiar with the area of tree mining as it can reveal useful applications within their domain of interest. It gives guidance as to which type of tree mining parameters are most suitable for the application at hand, and in general how tree mining can be utilized as a tool for domain-specific tasks.

References

1. Alesso, H.P., Smith, C.F.: Thinking on the Web: Berners-Lee, Godel, and Turing. John Wiley & Sons, Inc., Hoboken (2006)
2. Asai, T., Arimura, H., Uno, T., Nakano, S.-i.: Discovering Frequent Substructures in Large Unordered Trees. In: Grieser, G., Tanaka, Y., Yamamoto, A. (eds.) DS 2003. LNCS (LNAI), vol. 2843, pp. 47–61. Springer, Heidelberg (2003)
3. Borges, J., Levene, M.: Mining association rules in hypertext databases. Paper presented at the Proceedings of the 4th International Conference on Knowledge Discoevry and Data Mining, New York City, NY, USA, August 27-31 (1998)
4. Buchner, A.G., Baumgarten, M., Anand, S., Mulvenna, M.D., Hughes, J.: Navigation pattern discovery from internet data. Paper presented at the Proceedings of the WEBKDD 1999 Workshop on Web Usage Analysis and User Profiling, Boston, MA, USA, August 15 (1999)
5. Cooley, R., Mobasher, B., Srivastava, J.: Web mining: Information and pattern discovery on the World Wide Web. Paper presented at the Proceedings of the 9th International Conference on Tools with Artificial Intelligence (ICTAI 1997), Newport Beach, CA, USA (1997)
6. Cooley, R., Mobasher, B., Srivastava, J.: Data preparation for mining World Wide Web browsing patterns. Knowledge and Information Systems 1(1), 5–32 (1999)
7. Dillon, T., Tan, P.L.: Object Oriented Conceptual modelling. Prentice Hall of Australia Pty Ltd (1993)
8. Erinaki, M., Vazirgiannis, M.: Web Mining for Web Personalization. ACM Transactions on Internet Technology (TOIT) 3(1), 1–27 (2003)
9. Facca, F.M., Lanzi, P.L.: Mining interesting knowledge from Weblogs: a survey. Data and Knowledge Engineering 53(3), 225–241 (2005)
10. Feng, L., Dillon, T.S., Weigand, H., Chang, E.: An XML-Enabled Association Rule Framework. In: Mařík, V., Štěpánková, O., Retschitzegger, W. (eds.) DEXA 2003. LNCS, vol. 2736, pp. 88–97. Springer, Heidelberg (2003)
11. Fensel, D., Lausen, H., Polleres, A., Brujin, J.D., Stollberg, M., Roman, D., Domingue, J.: Enabling Semantic Web Services: The Web Service modelling Ontology. Springer, Berlin (2007)

12. Hadzic, F., Dillon, T.S.: Using the Symmetrical Tau (t) Criterion for Feature Selection in Decision Tree and Neural Network Learning. Paper presented at the Proceedings of the SIAM SDM 2nd Workshop on Feature Selection for Data Mining: Interfacing Machine Learning and Statistics Bethesda, MA, USA, April 20-22 (2006)
13. Hadzic, F., Dillon, T.S., Sidhu, A.S., Chang, E., Tan, H.: Mining Substructures in Protein Data. Paper presented at the IEEE Workshop on Data Mining in Bioinformatics (DMB 2006) in conjunction with IEEE ICDM 2006, Hong Kong, December 18-22 (2006)
14. Hadzic, F., Dillon, T.S.: Application of tree mining to matching of knowledge structures of decision tree type. Paper presented at the OTM Confederated international conference on 'On the move to meaningful internet systems' (SWWS workshop), Vilamoura, Portugal (2007)
15. Hadzic, F., Dillon, T.S., Chang, E.: Tree Mining Application to Matching of Heterogeneous Knowledge Representations. In: Proc. IEEE Intl. Conf. on Granular Computing (GRC), California, USA, pp. 351–351 (2007)
16. Hadzic, F., Tan, H., Dillon, T.S.: UNI3 - Efficient Algorithm for Mining Unordered Induced Subtrees using TMG Candidate Generation. In: Proceedings of IEEE Symposium on Computational Intelligence and Data Mining (CIDM), Honolulu, Hawaii, USA, April 1-5, pp. 568–575. IEEE, Los Alamitos (2007)
17. Hadzic, F., Tan, H., Dillon, T.S.: U3 mining unordered embedded subtrees using TMG candidate generation. In: Proceedings of the IEEE / WIC / ACM International Conference on Web Intelligence, Sydney, Australia, December 9-12, pp. 285–292 (2008a)
18. Hadzic, F., Tan, H., Dillon, T.S.: Mining Unordered Distance-constrained Embedded Subtrees. In: Boulicaut, J.-F., Berthold, M.R., Horváth, T. (eds.) DS 2008. LNCS (LNAI), vol. 5255, pp. 272–283. Springer, Heidelberg (2008b)
19. Hadzic, M., Hadzic, F., Dillon, T.S.: Tree mining in mental health domain. Paper presented at the Proceedings of the 41st Hawaii International Conference on System Sciences (HICSS-41), Waikola, Hawaii, USA, January 7-10 (2008)
20. Menasalvas, E., Millan, S., Pena, J.M., Hadjimichael, M., Marban, O.: Subsessions: a granular approach to click path analysis. Paper presented at the Proceedings of the IEEE International Conference on Fuzzy Systems, Honolulu, Hawaii, USA, May 12-17 (2002)
21. museum, A.E.t.n.h, Understanding gene testing: What does a predictive gene tell you (2007), `http://www.accessexcellence.org/AE/AEPC/NIH/gene14.html`
22. Paolucci, M., Kawamura, T., Payne, T., Sycara, K.: Semantic matching of web services capabilities. In: Horrocks, I., Hendler, J. (eds.) ISWC 2002. LNCS, vol. 2342, p. 333. Springer, Heidelberg (2002)
23. Prodromidis, A., Chan, P., Stolfo, S.: Meta-learning in distributed data mining systems: Issues and approaches. In: Kargupta, H., Chan, P. (eds.) Advances of Distributed Data Mining, AAAI/MIT Press, Cambridge, MA (2000)
24. Punin, J., Krishnamoorthy, M., Zaki, M.J.: LOGML: Log markup language for Web usage mining. Paper presented at the Proceedings of the 3rd ACM SIGKDD Workshop on Mining Log Data Across All Customer Touch Points, San Francisco, CA, USA, August 26-29 (2001a)
25. Punin, J., Krishnamoorthy, M., Zaki, M.J.: Web usage mining: Languages and algorithms. In: Studies in Classification, Data Analysis, and Knowledge Organization. Springer, Heidelberg (2001b)
26. Shasha, D., Wang, J.T.L., Shan, H., Zhang, K.: ATreeGrep: Approximate Searching in Unordered Trees. Paper presented at the Proceedings of the 14th International Conference on Scientific and Statistical Database Management, Edinburgh, Scotland, UK, July 24-26 (2002)

27. Shasha, D., Wang, J.T.L., Zhang, S.: Unordered Tree Mining with Applications to Phylogeny. Paper presented at the Proceedings of the 20th International Conference on Data Engineering, Boston, MA, USA, March 30-April 2 (2004)
28. Sidhu, A.S., Dillon, T.S., Chang, E., Sidhu, B.S.: Protein ontology: vocabulary for protein data. Paper presented at the Proceedings of the 3rd International Conference on Information Technology and Applications (ICITA 2005), Sydney, Australia, July 4-7 (2005)
29. Stolfo, S.J., Prodromidis, A.L., Tselepis, S., Lee, W., Fan, D.W., Chan, P.K.: Jam: Java agents for meta-learning over distributed databases. Paper presented at the Proceedings of the 3rd International Conference on Knowledge Discovery and Data Mining, Newport Beach, CA, USA, August 14-17 (1997)
30. Tan, H., Dillon, T.S., Hadzic, F., Chang, E., Feng, L.: IMB3-Miner: Mining Induced/Embedded Subtrees by Constraining the Level of Embedding. In: Ng, W.-K., Kitsuregawa, M., Li, J., Chang, K. (eds.) PAKDD 2006. LNCS (LNAI), vol. 3918, pp. 450–461. Springer, Heidelberg (2006)
31. Tan, H., Hadzic, F., Dillon, T.S., Feng, L., Chang, E.: Tree Model Guided Candidate Generation for Mining Frequent Subtrees from XML. ACM Transactions on Knowledge Discovery from Data 2(2) (2008)
32. UCI Repository of Machine Learning Databases. University of California, Department of Information and Computer Science, Irvine, CA (1998), http://www.ics.uci.edu/~mlearn/MLRepository.html
33. Wang, J.T., Zhang, K., Jeong, K., Shasha, D.: A System for Approximate Tree Matching. IEEE Transactions on Knowledge and Data Engineering 6(4), 559–571 (1994)
34. Wang, K., Liu, H.: Discovering Structural Association of Semistructured Data. IEEE Transactions on Knowledge and Data Engineering 12(3), 353–371 (2000)
35. Zaki, M.J.: Efficiently Mining Frequent Trees in a Forest: Algorithms and Applications. IEEE Transactions on Knowledge and Data Engineering 17(8), 1021–1035 (2005)
36. Zhang, Y., Street, W.N., Burer, S.: Sharing Classifiers among Ensembles from Related Problem Domains. Paper presented at the Proceedings of the 5th IEEE International Conference on Data Mining (ICDM), Houston, Texas, USA, November 27-30 (2005)

Chapter 10
Extension of TMG Framework for Mining Frequent Subsequences

10.1 Introduction

Data mining strives to find frequent patterns in data. The type of data, structures and patterns to be found characterizes a task in data mining. Advancements in data collection technologies have contributed to the high volumes of sales data. Sales data typically consists of transaction timestamps and the list of items bought by customers. If one is interested in inter-transactional patterns, sales data is a good example of a sequential pattern. In addition, sequential patterns exist in data such as DNA string databases, time series, occurrences of recurrent illness, etc.

In this chapter, we are describing an algorithm, referred to as SEQUEST (Tan et al. 2006b; Tan 2008), to mine frequent subsequences from a database of sequences as another extension of the TMG framework for tree mining. The main aim of this chapter is to show how the framework utilized to tackle the problem of mining frequent ordered subtrees described in Chapter 5 can be reused and extended to cover the problem of mining frequent subsequences.

There are two different sub-problems in mining frequent subsequences, i.e. problems of addressing (1) sequences of items and (2) sequences of itemsets. Addressing sequences of itemsets, i.e. sequences whose elements consist of multi items, is a more difficult problem and poses more challenges (Agrawal & Srikant 1995; Zaki 2000). SEQUEST is designed to tackle both sub-problems of mining frequent subsequences.

There are very efficient techniques for mining frequent subsequences such as GSP (Agrawal & Srikant 1995), SPADE (Zaki 2000) and many others. In this study, we focus mainly on showing how the proposed framework previously used for mining frequent ordered subtrees can also be used to address other problems that are in the similar domains. As a basis for comparison, we choose a recently developed algorithm, PLWAP (Ezeife & Lu 2005). The PLWAP algorithm uses preorder intermediate WAP trees during sequential mining as done by WAP-tree based techniques. (Ezeife & Lu 2005) have conducted a study comparing PLWAP against the GSP and reported some performance improvements over the GSP algorithm.

F. Hadzic, H. Tan, and T.S. Dillon: Mining of Data with Complex Structures, SCI 333, pp. 249–286.
springerlink.com

We will compare the SEQUEST against the PLWAP by running some experiments utilizing relatively large databases of sequences to demonstrate the effectiveness of the proposed framework. We leave the comparison of our approach against other techniques as possible future work.

The chapter outline is as follows. We start with some formal definitions of general sequence mining related concepts in Section 10.2. In Section 10.3, an overview of some existing work in the field of sequential pattern mining is provided, while in Section 10.4 we present a detailed description of a popular tree-based approach to sequence mining. In Section 10.5, we give an overview of the proposed solution using an extension of TMG framework. We present the algorithm in detail in Section 10.6. In Section 10.7, we empirically evaluate the performance and scale-up properties of the described technique. We summarize the chapter in Section 10.8.

10.2 General Sequence Concepts and Definitions

A sequence S consists of ordered elements $\{e_1,\ldots,e_n\}$. Sometimes elements of a sequence are referred to as *events*. Each element in a sequence can be either an atomic or a complex type. For a sequence with atomic elements, there is one entity or item per element. If the element is of a complex type, each element can have multiple items (itemsets). The itemsets can be either ordered or unordered. In this chapter, we will show how our approach is designed to mine both *sequences of ordered itemsets* and *sequences of ordered items*. We refer to sequences of ordered itemsets or items as simply sequences of itemsets or items. A sequence α is denoted as $e_1 \to e_2 \to \ldots \to e_n$ and $|e_i|$ refers to the size of itemsets. We refer to the first item of the sequence as its *root*. In case of sequences of itemsets, the root is the first item of the first itemset in the sequence. Zaki (2000) described $e_1 \to e_2 \to \ldots \to e_n$ to refer to the same notation. A sequence can be identified uniquely through its encoding which is a string of element labels separated by an arrow $\to$. In a case of a sequence of items, we assume that each event e_i is an atomic item. Encoding of a sequence α is denoted by $\varphi(\alpha)$. An element with j items is denoted by e:$\{\lambda_1,\ldots,\lambda_j\}$ where λ_i is a label of item i. A label of an element with j items, denoted by $\lambda(e)$, is a concatenation of $\lambda_1 + \ldots + \lambda_j$. Say a sequence α:$\{e_1,e_2\}$ where e_1:$\{A,B\}$ and e_2:$\{C,D\}$, the label of e_1 , $\lambda(e_1)$, is '*AB*' and the label of element e_2, $\lambda(e_2)$, is '*CD*' and the encoding of sequence α, $\varphi(\alpha)$, is '*AB*→*CD*'. This labelling function translates equivalently for an element that consists of an atomic item. We normally refer to a sequence by its encoding without quotes. Consider a sequence with n elements. We refer to a sequence with k items as k-sequence, i.e. k equals to the sums of $|e_i|(i = 1,\ldots,n)$. For example, *AB*→*CD* is a 4-sequence and *AB*→*DF*→*G* is a 5-sequence. A subsequence is a subset of a sequence such that the order of elements in the subsequence is the same as the order of the elements in the sequence. Accordingly, a subsequence with k elements is called a k-subsequence. We say that a subsequence S is a subset of sequence T if the elements of subsequence S are a subset of elements of sequence T and the ordering of items is preserved ($S \leq T$). *A*→*C* is an example of subsequence of $AB{\rightarrow}CD$ whereas AC is not. A is a subset of AB and C is a subset of CD, whereas AC is not a subset of either AB or CD. We denote that element α is a subset of element β as $\alpha \subseteq \beta$. Further, the ordering of elements in a sequence follows a certain ordering function. The most common ordering function is a time function. If an element α occurs at t_1 and element β occurs at t_2 where $t_1 < t_2$, the position of α and β in a sequence is i and j respectively and $i < j$. A *gap constraint* (Srikant & Agrawal 1996) is very similar to the notion of level of embedding Δ (Tan et al. 2006a) between two nodes in a tree with an ancestor-descendant relationship. The gap between two elements in a sequence is determined by the *gap distance* (∇) between two elements at position i and j, i.e. ∇ is a delta between i and j. A subsequence is said to be contiguous if there is no gap between every two consecutive elements. In other words, for a *contiguous subsequence,* every two consecutive elements have a gap distance of 1. Subsequences that contain any two consecutive elements with $\nabla > 1$ are called *non-contiguous subsequence*. Suppose we have a sequence φ:$A{\rightarrow}BCD{\rightarrow}F{\rightarrow}GH{\rightarrow}IJ{\rightarrow}JK$. $A{\rightarrow}B{\rightarrow}F$ and $A{\rightarrow}F{\rightarrow}K$ are both subsequences of φ. $A{\rightarrow}B{\rightarrow}F$ is a contiguous subsequence of φ since the gap distance of every two consecutive elements is 1, whereas $A{\rightarrow}F{\rightarrow}K$ is a non-contiguous subsequence of φ because the gap distance between A and F in $A{\rightarrow}F{\rightarrow}K$ is 2 and between F and K is 3. The database D for sequence mining consists of input-sequences. Each input-sequence can contain one or many itemsets. Each input-sequence is identified through a unique sequence-id (*sid*) whereas each element (itemsets) is identified through a unique event-id (*eid*). If D contains sales data, the *sid* can be a customer-id. A common event identifier in transactional databases is a timestamp.

10.2.1 Problem of Mining Frequent Sequences from a Database of Sequences

Mining frequent subsequences from a database of sequences can be formulated as discovering all frequent subsequences whose support count γ is greater or equal to the user specified minimum support threshold σ. We say that a sequence α has a support count γ if there is a γ number of input sequences that support it. An input sequence supports sequence α if it contains at least one occurrence of α.

10.3 Overview of Some Existing Techniques for Mining Sequential Data

This section provides an overview of a number of existing sequence mining techniques. One of the first works that investigated the different learning problems and algorithms for knowledge discovery from sequential data has been presented in (Laird 1993). In this section, we focus on the more mature problems of sequential data mining which essentially exist within the general frequent pattern mining problems. We do not describe in great detail the low level characteristics and the data structures used in the approaches overviewed, but refer the interested reader to

(Zhao & Bhowmick 2003; Masseglia, Teisseire & Poncelet 2005, Kum et al. 2005; Han & Kamber 2006), where such information is provided together with a general discussion of sequence mining parameters, comparison of existing approaches and their advantages/disadvantages. To date, the sequence mining problem has been extensively studied and a large number of approaches have been proposed in the literature. Similar to the other frequent pattern mining problems, the two main groups are the Apriori-like and the pattern-growth/extension approaches. There is another category of approaches, often referred to as WAP-tree based techniques, and these are overviewed in Section 10.4, where we describe the WAP-mine (Pei et al. 2000) algorithm in detail. At the end of this section, we look at some other methods for sequence mining developed for particular application aims or scenarios, database characteristics and others motivated by the complexity problems that often arise when mining sequential databases.

10.3.1 Apriori-Like Approaches

The problem of sequential pattern mining was first introduced in (Agrawal and Srikant 1995) where a number of general Apriori-like algorithms were presented. In (Agrawal & Srikant 1996), they extended their notions and proposed the GSP algorithm which adopts a level-wise enumeration strategy where the sub-sequences of length k are used to generate the sub-sequences of length $k+1$, and the pruning strategy employed exploits the apriori characteristics, as no frequent sequence of size k can have an infrequent subsequence with size $< k$. The algorithm allows for incorporation of time constraints, a sliding time window and user-defined taxonomies. The time constraint allows one to specify a minimum/maximum time period between adjacent elements in a sequential pattern. Using the sliding time window, one can specify the transaction times within which the elements of a sequential pattern can be present, thereby relaxing the constraint that all elements must come from the same transaction. With a user-defined taxonomy, the discovered sequential patterns will include elements across all levels of taxonomy. An improvement to the efficiency of the GSP based approach, has been presented in (Masseglia, Cathala & Poncelet 1998) in the so-called PSP approach for sequential pattern mining. The fundamental principles of the GSP approach are kept, with the main difference that a more efficient hierarchical (prefix-tree) data structure is used for manipulating candidate sequences. In this hierarchical structure, any branch of the tree stands for a candidate sequence and labeled edges are used to store transactional (individual sequence records) information. It thereby allows for faster retrieval of candidate sequences by traversing the tree and incrementing the support values once a leaf node has been reached. The MFS algorithm (Zhang et al. 2001) is another extension of the GSP algorithm which reduces the I/O requirement during candidate generation. The main idea is that a sample of database is first mined to obtain a rough estimate of the frequent sequences, and the solution is then refined. It takes a set of frequent sequences of various lengths and generates candidate sequences of various lengths. These candidates are checked against the database for their frequency and the process continues until no further frequent sequences can be generated. Another

GSP-based algorithm has been proposed in (Joshi, Karpys & Kumar 2001), which can discover *universal* sequential patterns. This universal formulation of sequential patterns extends the previous approaches to make them more general in terms of representation, constraints and different support counting methods that are suitable for various applications. The modifications allow for the incorporation of newly defined timing constraints and pattern restrictions, mining for sequential patterns to be used for prediction purposes, as well as some performance improvements over the general GSP-based approach. Zaki (1997, 2001) proposed the SPADE algorithm which adopts a lattice-theoretic approach whereby it decomposes the original search space into smaller pieces which are processed independently in main memory using efficient lattice search techniques. Each sequence has a list (*id-list*) associated with it, which contains data objects in which it occurs and respective timestamps. The set of frequent sequences are enumerated using temporal joins on the frequent sequence lists, and two search strategies (breadth-first and depth-first) are proposed to enumerate frequent sub-sequences within each sub-lattice. The SPADE algorithm has been used in (Zaki, Lesh & Ogihara 1999) for mining sequences of plan executions for patterns that predict plan failures. The main modification is the combination of several techniques to remove non-predictive and redundant patterns in the application in real-world planning systems. These pruning strategies are performed in three phases: (1) rules consistent with background knowledge are pruned as they do not contribute any new information; (2) removal of sequential patterns that have the same frequency as at least one of their super-sequences (redundancy) and (3) removing patterns that are less predictive than any of its proper super-sequences. An extension of the SPADE algorithm, GO-SPADE, has been presented in (Leleu et al. 2003), with the motivation that many sequential databases can contain repetitions of items, that cause performance degradation in the traditional SPADE approach. The authors introduce the concept of *generalized occurrences* which are compact representations of several occurrences of a pattern, and describe corresponding primitive operators to manipulate them. More recently, the PRISM (**PRI**me-Encoding Based **S**equence **M**ining) algorithm has been proposed in (Gouda, Hassan & Zaki 2007). Similar to the SPADE algorithm (Zaki 1997, 2001), PRISM utilizes a vertical approach for enumeration and support counting, but uses novel *prime block encoding*, which is based on prime factorization theory, to represent candidate sequences. The join operations over the primal blocks are used to determine the frequency of each candidate.

10.3.2 Pattern Growth Based Approaches

The idea of using a pattern-growth approach to sequence mining was first presented in (Han et al. 2000), and it integrates the mining of frequent sequences with that of frequent itemsets. The general idea behind the proposed FreeSpan algorithm is that the sequence database is projected into a set of smaller databases and the subsequence fragments are grown in each of the projected databases. In this manner, the search space is confined and the number of candidate sequential patterns to be enumerated is reduced. This work has been extended into the PrefixSpan algorithm (**Prefix**-projected **S**equential **pa**ttern mining) (Pei et al. 2001) where the general idea is to project only frequent prefixes and enumerate all frequent subsequences by a pattern-growth approach to the projected frequent prefixes. The items within an element of a sequence are listed in alphabetical order and the algorithm starts by finding all frequent items in sequences of length 1.The search space is then divided into subsets corresponding to the different prefixes from the items in sequences of length 1. Each of these prefixes is grown with the items in the sequence starting with that particular prefix. The algorithm then recursively grows prefixes for each of the subsets and repeats the process progressively for longer prefixes. Using this approach, the PrefixSpan algorithm does not generate candidate sequences but rather grows longer sequential patterns from the shorter frequent sequences. Hence, the search space is much smaller in comparison to the search space that needs to be exploited through candidate generation and frequency testing. As the major cost of PrefixSpan is in the construction of the projected databases, the authors have also proposed two strategies to alleviate the cost, namely bi-level and pseudo-projection. The SPAM algorithm (Ayres, Flannick, Gehrke & Yiu 2002) is based on a lexicographic tree of sequences and assumes that the entire database fits into main memory. Each node in the lexicographical tree stores sequences from the database starting where the level of the tree corresponds to the length of the sequential pattern. Hence, the root of the tree contains no elements and then the nodes at level 1 correspond to sequences of length 1, nodes at level 2 to sequences of length 2, etc. The nodes at a particular level are ordered lexicographically from left to right. The children of a particular node in the tree correspond to the super-sequences of the sequence at that node. However, a distinction is made between the sequence-extended and itemset-extended sequences generated from a particular parent node. The tree is traversed in a depth-first manner and the frequency of each sequence-extended and itemset-extended sequence is tested so that only frequent sequences are further extended in a recursive manner. Hence, the pruning strategy employed is Apriori-based as if the nodes in the tree (sequence-extended and itemset-extended) are not frequent they are pruned as none of their super-sequences can be frequent. A vertical bitmap representation of the data is used to allow for simple and efficient support counting. The SPAMSEPIEP algorithm (Yusheng et al. 2008) is an optimization of the SPAM algorithm which employs sequence extension pruning (SEP) and item extension pruning (IEP). While the SPAM algorithm needs to check for frequency of every sequence-extended or itemset-extended sequence, the SEP and IEP pruning strategies will avoid the generation of infrequent sequence-extended and itemset-extended sequence, by pruning the inappropriate candidate sequences and items for extension, respectively. These are inappropriate candidates in the sense that they do not comply with the allowed sequences in that domain, and would result in an invalid and/or infrequent sequence which would need to be checked for frequency and pruned thereafter. The SEP and IEP share lists of frequent 2-sequences to determine the inappropriate candidate sequences and items for extension. Hence, the main optimization is in the avoidance of unnecessary frequency counting, with little more memory overhead. This work has been extended in (Yusheng, Lanhui, Zhixin, Lian & Dillon 2008) to dynamic SEP and IEP pruning strategies, DSEP and DIEP, which as opposed to sharing a frequent 2-sequence list, dynamically build private sequence and items lists to prune out the inappropriate frequent 2-sequences.

A strategy referred to as DISC (**DI**rect **S**equence **C**omparison) that can find frequent sequences without computing the support of infrequent sequences, has been presented in (Chiu, Wu & Chen 2004). The sequences are ordered based upon the alphabetical order of their items and their transactional identifiers. The smallest k-sequence according to this order is referred as k-minimum subsequence. Using this ordering scheme, all sequence transactions containing the same k-minimum subsequence are listed together, which allows for easy support calculation. Given the set of frequent minimum k-subsequences, the approach then iteratively finds the minimum k-sequence greater than a given frequent minimum k-subsequence and updates the sorted database, until all the frequent k-sequences are found.

Generally speaking, the pattern-growth or extension-based techniques for sequence mining usually outperform the Apriori-like approaches as their search space is much smaller due to the directed candidate pattern generation. The memory and run-time performance difference between the Apriori-based (GSP) and pattern-growth (PrerixSpan) based approaches has been studied in (Antunes & Oliveira 2004), aiming at defining a way by which Apriori-like approaches can be optimized to match the execution times of pattern-growth approaches which usually require more memory. They presented a new algorithm, SPaRSE, which combines the generate and test procedures of the GSP algorithm (Agrawal & Srikant 1996) with the restriction of the search space using the idea of projected databases in a manner similar to that of PrefixSpan (Pei et al. 2001). The experimental results have shown that PrefixSpan will outperform SPaRSE in sparse datasets and have similar performance on dense databases, and PrefixSpan will consume more memory then SPaRSE and GSP. Another approach that overcomes the memory issue caused by the projection of many intermediate databases is MEMISP (**MEM**ory **I**ndexing for **S**equential **P**attern mining) (Lin & Lee 2002). In addition to no candidate generation, the MEMISP requires only one scan of the database to read the data sequences into memory. A recursive find-then-index strategy is used to discover all sequential patterns from stored data sequences. The frequent 1-sequences (consisting of 1 item) are counted during the database scan, and the frequent ones form the basis for further extension. MEMISP then progressively discovers all frequent patterns of increasing size by searching the set of data sequences that contain the prefix of the frequent pattern being extended (i.e. have the pattern as the common subsequence). The compact indexing and find-then-index technique has a good linear scalability and outperforms PrefixSpan (Pei et al. 2001), as it avoids the generation of intermediate projected databases. A scalable two-phase approach to handle the complex biological sequence mining task has been presented in (Wang, Xu & Yu 2004). In the first phase, all frequent small patterns which contain no gaps are enumerated, and an in-memory index is used for the positions of these base segments. The second phase enumerates longer patterns by growing each of the base segments one at a time.

The search space is efficiently organized in a pattern tree structure and a depth-first tree traversal is adopted to progressively enumerate longer patterns from frequent base segments. By growing frequent base segments rather than singular items, the search space is reduced and a more effective pruning and pattern matching strategy is enabled.

10.3.3 Other Types of Sequential Pattern Mining

A number of specialized sequential mining methods have been developed, motivated by specific application aims or sequential database characteristics. For example, the problem of finding correlations and dependencies in multivariate, discrete-valued time series data have been presented in (Oates, Schmill, Jensen & Cohen 1997). A framework for discovering frequent episodes (partially ordered sets of events) in event sequences has been proposed in (Mannila, Toivonen & Verkamo 1999). Another approach for mining sequential patterns that include time-intervals has been described in (Chen, Chiang & Ko 2003), presenting Apriori-based and PrefixSpan-based algorithms for the task. The problem of discovering long sequential patterns in noisy environments has been studied in (Yang, Wang, Yu & Han 2002). This work has been motivated by the domains where it is likely that the recorded symbol in the sequence from the database may differ from the true symbol. The authors have introduced the concept of a compatibility matrix that provides a probabilistic connection between the recorded values and the underlying true value, in order to capture the expected support of a pattern if a noise-free environment was assumed. In the following subsections we discuss other types of sequential pattern mining which have received more research attention.

10.3.3.1 Incremental Mining of Sequential Patterns

As the sequential databases grow incrementally, it is undesirable to mine the sequential patterns from scratch on every database update. Initial work in this area started with the algorithm presented in (Wang & Tan 1996; Wang 1997) which uses a dynamic suffix tree data structure for indexing strings where general updates are allowed at any specified position in the string. The dynamic suffix tree structure supports mappings between string positions and disk addresses. By referencing substring addresses rather than positions, updates are made only to a small portion of the dynamic suffix tree, and this achieves efficient incremental mining of string databases where updates are allowed. However, the algorithm presented is developed for databases of a singular long sequence rather than databases consisting of many sequences (transactions). ISM (Incremental Sequence Mining) algorithm (Parthasarathy, Zaki, Ogihara, & Dwarkadas 1999) is based on the previously mentioned SPADE algorithm (Zaki, Lesh & Ogihara 1999). The main modification is the use of an efficient memory management scheme that indexes the database and generates an Increment Sequence Lattice (ISL), whose properties are exploited to prune the search space for potential new sequences. The ISL consists of all the frequent sequences and the sequences from the *negative border* (non-frequent sequences whose generating subsequences are frequent) of the original database. The support of each sequence is kept in the ISL and as new sequences arrive, the ISL is traversed to update the corresponding node and recalculate the support count. The authors also propose techniques for maintaining data structures to mine sequences in the presence of user interaction.

FASTUP (Lin & Lee 1998) is an algorithm for incremental sequence mining based on the GSP (Agrawal & Srikant 1996) sequence mining algorithm mentioned earlier. Rather than mining the updated database as a whole, it uses frequent sequences and support information from the original database and updates them with the support count and promising candidates obtained by scanning only the newly appended data. The ISE (Incremental Sequence Extraction) (Masseglia, Poncelet & Teisseire 2003) takes as input the frequent sequences from the original database, and for candidate generation uses a GSP-based approach (Agrawal & Srikant 1996). It counts the support of singular items in the new data and validates them with the support count of singular items in the original data so that only the frequent ones are kept. These are then self-joined to obtain the set of candidate 2-sequences and the new data is scanned so that only those candidate 2-sequences are kept that occur in the new data. These are then checked for the frequency in the databases as a whole (union of the old and new database). To find all the frequent k-sequences, an iterative process starts as follows. Each frequent 1-sequence from the new database is extended with a frequent 1-sequence (from the database as a whole), and this set is referred to as *freqSeed*. New candidates are generated by appending the already found frequent sequences to those from *freqSeed* where the last item of a sequence from *freqSeed* is the first item from the already found frequent sequence. The support count of these candidates is obtained by scanning the whole database to determine the set of frequent k-sequences.

IncSpan (Cheng, Yan & Han 2004) is another incremental sequence mining algorithm based on the PrefixSpan algorithm (Pei et al. 2001). IncSpan initially enumerates all frequent sequences (FS) and uses a buffer ratio to lower the minimum support and keep a set of semi-frequent patterns (SFS) from the original database. The intuition behind this approach is that the patterns in SFS are almost frequent and as new sequential data is incrementally appended, the new set of frequent patterns will either come from SFS or already be in FS. It is based on the assumption that the newly appended data has a uniform probability distribution on items. More recently, another approach for incremental sequence mining that is also based on the PrefixSpan algorithm (Pei et al. 2001) has been presented in (Chen, Wu & Zhu 2005). In addition to the extension technique used by PrefixSpan, the proposed MILE (**MI**ning in mu**L**tiple str**E**ams) algorithm also uses a *suffix-append* technique to speed up the candidate generation. It avoids the expensive depth-first search, by caching frequent suffixes for future appending. The algorithm can also incorporate some prior knowledge about the data distribution in the data stream to speed up the processing, and the authors also propose a memory and execution time balancing technique.

10.3.3.2 Parallel Sequence Mining

As the sequence mining problem can be very complex, and the memory of a single processor can be easily exhausted when large and complex sequential databases are in question, some research has investigated the use of a parallel multiprocessor system to accommodate this task. This requires the development of parallel sequence mining algorithms that can decompose the original search space into smaller ones that can be solved independently on each processor. A hash-based approach for parallel sequence mining has been explored in (Shintani & Kitsugerawa 1998). Three algorithms were proposed, NPSPM (Non Partitioned Sequential Pattern Mining), SPSPM (Simply Partitioned Sequential Pattern Mining) and HPSPM (Hash Partitioned Sequential Pattern Mining). In NPSPM, the candidate sequences are copied among the processors and each processor determines the frequent sequences by exchanging the support information among all the processors. In this approach, each processor still needs to examine all the candidate sequences. In SPSPM, the candidate sequences are partitioned equally over the memory space of all the processors, which thereby exploits the aggregate memory of the system. However, each processor needs to broadcast the support information of the candidate sequences it processes. In HPSPM, the candidate sequences are partitioned among the processors using a hash function. This approach eliminates the need for broadcasting the support information and can reduce the comparison workload significantly.

An extension of the original SPADE algorithm (Zaki 1997, 2001) has been presented in (Zaki 1999), where the parallel sequence mining algorithm, pSpade, was presented. It shares many of the features of the SPADE algorithm, such as the vertical list storing list of objects in which sequences occur, along with timestamps, candidate enumeration via vertical lists intersections, and a lattice-theoretic approach to decompose the original search space into smaller pieces, referred to as suffix-based-classes. Each processor works on separate classes, and a number of load balancing strategies are proposed where any free processor will join a busy processor to speed up task execution. Similarly, a study of a variety of distributed-memory parallel algorithms for tree-projection based sequence mining algorithms has been presented in (Guralnik & Karpys 2004). The authors presented three algorithms for distributing the work load among the processors to minimize various overheads caused by idle processors, overlapping databases and interprocessor communication. Two static distribution algorithms where one explores data-parallelism (distribute the database across the processors), and the other task-parallelism (decompose the task across processors to partition the lattice of frequent patterns). The third proposed algorithm improves upon the task-decomposition by dynamically balancing the load of the overall computation when load imbalance occurs. The task decomposition schemes use bipartite graph partitioning to balance the computations and minimize the positions of the database that need to be shared across different processors.

10.3.3.3 Multi-dimensional Sequential Pattern Mining

In certain applications, the sequential patterns discovered can and often are, associated with other information regarding different circumstances that come from a multi-dimensional space. For example, customer transactions are often associated with time, region, product category, customer segments, etc. This problem has given rise to research into multi-dimensional sequential pattern mining, initially proposed in (Pinto et al. 2001) with a framework that combines multi-dimensional analysis and sequential pattern mining. In this framework, the PrefixSpan algorithm (Pei et al. 2001) was taken as the basis and three algorithms were presented for the task, namely, UniSeq, Dim-Seq and Seq-Dim. UniSeq simply considers all dimensional values as additional items in the sequence and applies the PrefixSpan algorithm to find all sequential patterns. The Dim-Seq algorithm, first looks for frequent dimensional value combinations, and then extracts all sequences that satisfy each of these combinations, whereas Seq-Dim uses PrefixSpan to extract all the frequent sequences from the complete database, and then looks for corresponding frequent dimensional patterns to be associated with each frequent sequence. In (de Amo, Furtado, Giacometti & Laurent 2004) the multi-dimensional information is captured using first-order temporal logic and two Apriori-based algorithms were presented for the task. Multi-dimensional sequence mining in context of Web Usage Mining has been studied in (Yu and Chen 2005). The authors propose two algorithms AprioriMD (based on GSP (Agrawal & Srikant 1996)) and PrefixMDSpan (based on PrefixSpan (Pei et al. 2001)) for discovering multi-dimensional sequential patterns. In the AprioriMD algorithm, the dimensional information is associated with each item name in the data structure used, so that when candidates are generated through join, the pairs for all possible dimensional value relations are considered. On the other hand, in PrefixMDSpan, a lookup table is used, where for each item the number of transactions containing the item is stored together with the corresponding dimensional value. The extension based technique of PrefixSpan is used taking the dimensional values into account. Another method that incorporates multi-dimensional information into sequential pattern mining when applied to Web usage mining has been presented in (Stefanowski and Ziembinski 2005). In their method, context attributes may be introduced for both describing the complete sequence and for each element in the sequence, and they can be on a nominal, numerical or ordinal scale. The M2SP algorithm (Plantevit et al. 2005) also mines for multi-dimensional patterns which in addition can contain associations inferred across transactions (inter-pattern sequences). The so-called *jokerized* patterns are introduced, which are patterns in which some of the dimensional values have not been instantiated. The algorithm can take these kinds of patterns into account but will consider the patterns having too many dimensional values not instantiated, as infrequent. This work has been extended into the more efficient M3SP algorithm (Plantevit 2010), which takes hierarchies into account, thereby adding more granularities to the available dimensional information. More recently Esposito et al. (2009), have provided a framework and an Inductive Logic Programming algorithm for mining complex sequential patterns. The multi-dimensional patterns are defined as a set of atomic first order formulae where events are represented by a variable and a set of dimensional predicates captures the relations between the events. The ApproxMGMSP (Approximate Mining of Global Multidimensional Sequential Patterns) method (Zhang, Hu, Chen, Chen & Dong 2007) was one of the first approaches to use an approximate method in multi-dimensional sequence mining problem. In their approach, the multi-dimensional information is embedded into the corresponding sequences. The sequences are then clustered, summarized and analyzed on the local sites, to discover local patterns using an approximate mining method. The global multi-dimensional sequential patterns are then obtained using a high vote sequential pattern model on all the discovered local patterns from distributed sites. Another clustering based approach for multi-dimensional sequence mining has been presented in (Zeitouni 2009). An algorithm using an indexed structure associated with a bit map representation is presented for standard sequence mining. To allow for multi-dimensional information to be incorporated, the method first groups similar sequences together in a cluster and then characterizes each cluster using associated multi-dimensional information.

10.3.4 Constraint-Based Sequential Mining

In this section, we overview some of the existing work done toward the incorporation of constraints in the sequence mining problem. In situations where the data set is of such characteristics that the generation of the complete set of frequent patterns would be infeasible, constraints can serve a good purpose since, when incorporated during the pattern generation phase, the result can be a significant memory and execution time saving. In other scenarios, it can allow the user to focus the search so that the resulting patterns contain characteristics desirable for a particular application aim.

For example, the use of regular expression constraints in sequence mining has been proposed in (Garofalakis, Rastogi & Shim 1999, 2002), with a family of algorithms (referred to as SPIRIT – **S**equential **P**attern m**I**ning with **R**egular express**I**on cons**T**raints). The presented algorithms exploit the equivalence of regular expressions to deterministic finite automata to push the constraints in deeply during the pattern generation phase and thereby save on execution time and memory through effective pruning strategies. It enables user-controlled focus on the mining process which yields results that are more focused on user/application needs and thereby prevents the user from being overwhelmed by a large number of frequent patterns that are useless for application purposes. The cSPADE algorithm presented in (Zaki 2000) is an extension of the general SPADE algorithm (Zaki 1997) for discovering frequent sequences where the following constraints can be incorporated: (1) maximum allowable width or length of the sequential pattern; (2) minimum and maximum gaps (time intervals) between sequence elements; (3) time window (time-frame restriction on the whole sequence); item constraints (controlling the presence of items in the sequence) and (4) distinctive sequences with respect to a particular class (attribute-value pair that the user is interested in predicting). A variant of the cSPADE algorithm to incorporate similarity constraints during sequential pattern mining has been presented in (Capelle, Masson & Boulicaut 2002). A similarity measure was proposed so that only those patterns are extracted whose similarity distance to a specified pattern is within a chosen threshold. The work in (Wojciechowski 2001) is focused on exploiting the results from previous sequential pattern queries to enable one to answer a set of new queries without having to redo the whole sequence mining process. It is motivated by the claim that a user is likely to execute a set of similar sequential pattern queries in one session. A set of criteria is provided for determining whether a new query can be answered from the results of previous related queries and an algorithm is presented for sequential pattern query processing by exploiting cached results of previous mining queries. A framework for constraint-based pattern-growth sequential mining, *Prefix-growth*, has been proposed in (Pei, Han & Wang 2002). The approach uses the PrefixSpan (Pei et al. 2001) algorithm as a base and pushes monotonic, anti-monotonic and regular expression constraints deep into the mining process. In addition, more complex aggregate constraints such as sum and average can be handled. The constraints used were demonstrated to be prefix-monotone, which enabled an effective approach for incorporating the constraints deep into the mining process, and reducing the search space significantly through effective pruning. The SLPMiner algorithm (Seno & Karypis 2002) uses a length-decreasing support constraint during sequence mining. This work is motivated by the observation that small patterns tend to be interesting if they have a large support, whereas long patterns remain to be interesting even when their support is relatively small. Hence, the algorithm finds all frequent patterns whose support decreases as a function of their length. It follows a database-projection-based approach, and significant run-time reduction is achieved through effective pruning strategies, that can discard items within sequences, entire sequences, or entire projected databases. The concept of recency and compactness in sequential patterns has been studied in (Chen & Hu 2006), and an extension of the PrefixSpan (Pei et al. 2001) algorithm was presented to incorporate these constraints. Timestamps are associated with the data and the concept of recency ensures that the discovered patterns have also occurred in the more recent subset of the data, while compactness ensures that the time span from the first to the last item in the discovered patterns must be no longer that a pre-specified threshold. A comparison and evaluation of the different constraints that can be incorporated into a pattern-growth sequence mining based approach, has been provided in (Pei, Han & Wang 2007).

Another stream of research that can be to some extent viewed as constraint-based sequence mining is the development of sequence mining methods that provide approximate solution, and often employ fuzzy-based or clustering-based approach to sequence mining. A clustering based method, CLUSEQ, has been proposed in (Yang & Wang 2003), that explores various statistical properties of the sequences. The similarity measure is based on the conditional probability distribution of the next symbol given a preceding sequence segment. A probabilistic suffix tree is used for efficient organization of statistical properties of a sequence cluster, and the designed algorithm discovers clusters of sequences allowing for an automatic adjustment of the number of clusters and the amount of outliers. A fuzzy-based approach to sequence mining has been presented in (Hu, Chen, Tzeng & Sheh 2003), where fuzzy sequential patterns described by the natural language are considered. The proposed FGBSPMA (Fuzzy Grid Based Sequential Patterns Mining Algorithm) scans the database once and applies a sequence of Boolean operations to generate fuzzy grids and sequences among attributes. Quantitative attributes are divided into many linguistic values, and each linguistic value is viewed as a candidate 1-dimension fuzzy grid, and progressively fuzzy grids of larger dimensions are generated from smaller ones, and their fuzzy support with respect to their assigned grid is determined. ApproxMAP (Kum, Pei, Wang & Duncan 2003; Kum, Chang & Wang 2006) is another clustering-based approach that deals with the problem of finding patterns that are approximately shared by many sequences. These kinds of patterns can represent the local data in a compact manner that allows for efficient global sequential pattern mining. The first step consists of clustering the sequential patterns using a distance measure based on sequence (string) edit operations. A strategy based on the multiple alignment of sequences is used to identify the longest approximate pattern for each cluster. These are referred to as *consensus patterns*, and are essentially a more expressive and general representation of the information conveyed by the cluster. In addition, a post-processing model was presented in (Kum, Chang & Wang 2006) to find both high vote and exceptional sequential patterns from the locally discovered consensus patterns.

Similar to other frequent itemsets mining problems, another direction for constraint-based sequence mining is the task of discovering frequent closed sequences (a frequent sequence of which no other proper super-sequences have the same support), and frequent maximal sequences (sequences of which no other proper super-sequences are frequent). Some algorithms for mining frequent closed sequences are BIDE (Wang & Han 2004) and CloSpan (Yan, Han & Afshar 2003). TFP algorithm (Tzvetkov, Yan & Han 2004) mines top-k frequent closed patterns with specified minimum length. Rather than using the minimum support threshold, the algorithm uses the minimum length constraint and the properties of the top-k closed patterns to perform dynamic support increase and projected database pruning. In regards to mining of maximal frequent sequences, a number of algorithms have been developed. The MSPS (Maximal Sequential Patterns using Sampling) algorithm (Luo & Chung 2004, 2008) uses a sampling technique to identify long frequent sequences and then uses a bottom-up level-wise candidate enumeration strategy. A signature-based and a hash-based method for partial subsequence infrequency-based pruning and a new prefix tree structure are used for efficient candidate sequence counting. Another approach, TSETMAX, has been proposed in (Guil, Bosch & Marin 2004), and it starts by creating a *temporal set-enumeration tree* data structure that stores all the frequent singular items (i.e. sequences of length 1). It then recursively uses a look-ahead technique and a depth-first search on this data structure, and with the help of a queue structure, it enumerates all candidate k-sequences and prunes out the infrequent ones. A maximal sequence mining algorithm based on the WAP-mine (Pei, Han, Mortazavi-asl & Zhu 2000) has been proposed in (Xiaoqiu, Min and Jianke 2006). The authors present an extension of the WAP-tree data structure initially presented in (Pei, Han, Mortazavi-asl

& Zhu 2000), referred to as an IWAP-tree, which introduces a restrained subtree structure to capture all maximal frequent sequences. The MSMA (**M**aximal **S**equential Pattern Mining Based on Simultaneous **M**onotone and **A**nti-monotone constraints) algorithm (Ren, Sun & Guo 2007) allows for the incorporation of monotone and anti-monotone constraints during the discovery of frequent maximal sequential patterns. It transforms the database into a tree-based data structure according to the order of items in each of the sequence transactions. It then traverses the branches of the tree to construct the maximal patterns, during which pruning is applied to reduce the search space. In (Armua & Uno 2007), a polynomial algorithm for the problem of mining flexible maximal sequential patterns is presented. A flexible sequence is defined as the sequence that can have a gap between its constituting elements, which preserves the order of the frequent items of the sequence but indicates that a number of infrequent items occurred in the gap. A Sequence Mining Automata (SMA) for mining sequences under regular expression constraints has been proposed in (Trasarti, Bonchi & Goethals 2008). In contrast to other approaches, which focus on pushing the constraints into the search space to prune those candidates that do not satisfy the regular expression, the authors present an approach that starts checking the constraint during initial database scanning. The SMA used is a specialized kind of Petri Net that for each input sequence produces all constraint-satisfying sequential patterns. An interestingness measure for sequential association rules that combines cohesion and frequency has been introduced in (Cule, Goethals & Robardet 2009). The combined interestingness measure evaluates candidate patterns based on a combination of how often the items comprising the candidate appear in the sequence and how close they appear to each other on the average. Thus, a pattern is considered interesting if its items occur close enough to each other in a sufficiently high number of sequences. This work has been further extended in (Cule & Goethals 2010), to be effectively applied in the association rule setting. The previously used pruning function is relaxed to make the computation easier (i.e. reduce run-time) and associations are discovered within the predetermined interesting itemsets. Itemsets and rules are mined in parallel, and the measure of association rule confidence indicates the likelihood that when a part of an itemset is encountered in the sequence, the rest of the itemset will be found nearby.

10.4 WAP-Mine Algorithm

Besides the general Apriori and pattern-growth based approaches overviewed in the previous section, another popular approach utilizes a compressed tree-based approach and performs an alternative counting approach without having to perform candidate generation. WAP-mine (Pei et al. 2000), PLWAP (Ezeife & Lu 2005) and many of its derivatives represent the other class of algorithms that utilize the compressed tree-based approach. In this section, we will describe the WAP-mine (Pei et al. 2004) algorithm.

WAP-mine is a non-Apriori algorithm, proposed in (Pei et al. 2000). It uses the highly compressed WAP-tree (Web Access Pattern tree) structure to store the web log data in a prefix tree format. A similar format is also used in the frequent pattern tree (FP-tree) (Han et al. 2005) for mining frequent subtrees from the database of trees. A WAP-tree is a highly compact and compressed data structure that in many ways resembles the Trie structure (Fredkin 1960) that was developed by Edward Fredkin in 1960. Both WAP-tree and Trie can only benefit from a smaller memory footprint when many of the patterns (sequences) share their prefixes or there are high numbers of patterns that are sub-patterns of many of the other patterns.

Unlike the Apriori-like algorithms that perform candidate generation and frequency counting, WAP-mine does not perform candidate generation and it derives the candidate frequency in a different way.

WAP-mine generally works as follows. It scans the database and computes all frequent 1-subsequences. Then it scans the database a second time to construct a WAP-tree from the frequent 1-subsequences of each transaction. WAP-mine then recursively mines the constructed WAP-tree using *conditional search* by building a conditional WAP-tree for each conditional suffix frequent pattern found. The process ends when it has only one branch or is empty. The WAP-mine pseudo-code is provided below.

WAP-Mine Pseudo-code

```
INPUT  : S (sequence database), σ (min support count)
OUTPUT : Complete frequent k-subsequences (k = 1,2,3,…,n)

   1. Scan S once, find all frequent 1-subsequences
   2. Scan S again, using the frequent 1-subsequences, construct WAP-tree
   3. Recursively mine the WAP-tree using conditional search
```

10.4.1 WAP-Tree Construction

Only the 1-subsequences from the sequence database S are used for constructing the WAP-tree. If two subsequences share a common prefix *P*, the prefix *P* can be shared in the WAP-tree. This sharing of common prefix saves space and facilitates the support counting of any subsequence of the prefix *P*.

Each node in a WAP-tree stores a label and frequency count. The tree has a special virtual root node with an empty label and 0 count. All other nodes are labeled accordingly and have an associated count, which denotes the number of occurrences of the corresponding prefix ended with the subsequence in the sequence database.

The WAP-tree is constructed by filtering out any non-frequent events, and inserting the frequent subsequences into the WAP-tree. The insertion of the frequent subsequence is done as follows. The insertion always starts from the root node. Consider the first event *e* of the frequent subsequence that will be inserted. If *e* already exists in the WAP-tree, only increment the count by 1. Otherwise, create a child node *e* with the frequency count set to 1. Then recursively process all the subsequent events from the same frequent subsequence started from the last inserted child node *e*.

The next phase is to construct auxiliary node linkage structures. The auxiliary node linkage structures are used to assist node traversal in a WAP-tree. The auxiliary node linkage structures are constructed by linking all the nodes in the tree with the same labels and inserting them into *event-node queue*. The head of each event-node queue is inserted into a global header table *H*. For any frequent event, all the frequent subsequences containing that event can be visited by iterating the event-node queue starting from the event entry in the header table *H*.

Once the WAP-tree has been constructed, database scanning is no longer required as everything that the subsequent operations need is already available on the constructed WAP-tree. The WAP-tree construction phase is summarized by the pseudo code below (Pei et al. 2000).

WAP-Tree Construction Pseudo-code

```
1. Create a root node for WAP-tree T
2. For each subsequence s in the sequence database S:
     a. Extract frequent subsequence S' from S by removing all
        infrequent events in S. Let S' = si where 1 ≤ i ≤ n are events in
        S'. Let γ be a current node points to the root of T.
     b. For i = 1 to n do, if γ has a child labeled si, increment the
        frequency count of si by 1 and set γ to point to si, else create
        a new child node with frequency count 1 (si:1) and set γ to point
        to the new node, and insert it into the event-node queue, in
        this case the si-queue.
3. Return T
```

The following example illustrates the process of constructing the WAP-tree from a database of web access subsequences shown in Table 10.1.

Table 10.1 A database of web access subsequences S

User ID	Web Access Sequence	Frequent Subsequence
#1	ABDAE	ABAE
#2	GAGBEAE	ABEAE
#3	BABFAGE	BABAE
#4	AFBAEFE	ABAEE

The table shows that there are 4 subsequences. With a minimum support count of 3, only event ‘A’, ‘B’ and ‘E’ are frequent 1-subsequences as they appear in at least 3 out of 4 total sequences. Event ‘F’ and ‘G’ are considered as infrequent 1-subsequences because 1 or more occurrences of an event from the same User Id# is only counted once. Hence, even though event ‘G’ appears twice in #2, only 1 is counted. Now that we have removed all the infrequent 1-subsequences from the original sequences, the next step is to generate the WAP-tree from the frequent 1-subsequences.

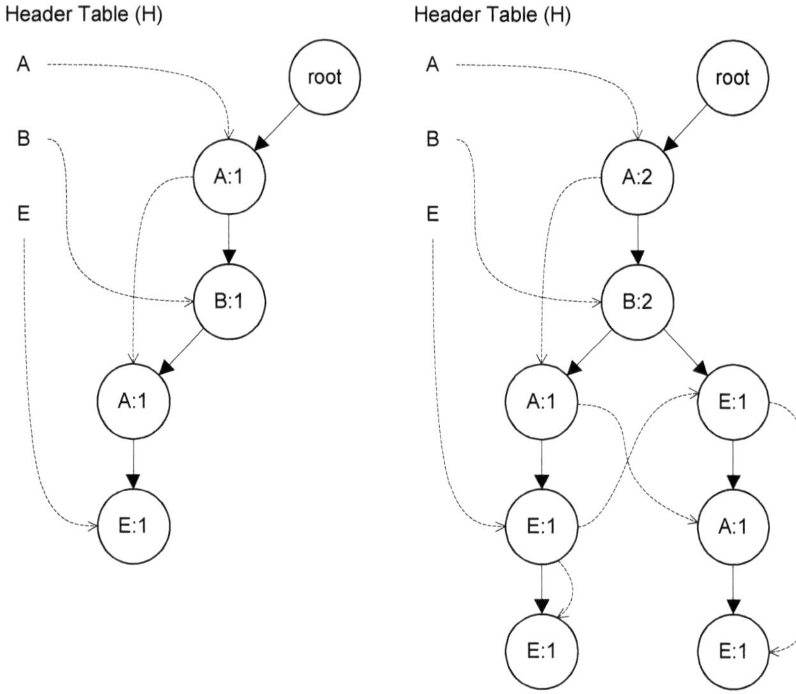

Fig. 10.1 (a) WAP-tree after inserting 'ABAE' (b) WAP-tree after inserting 'ABEAE'

The WAP-tree in Fig. 10.1 is constructed from the frequent 1-subsequences as follows. Start with the first subsequence 'ABAE', insert event 'A' into the initial empty tree with one virtual root node followed by the remaining events 'B', 'A', and 'E'. For each of the events, the frequency count is set to 1. After this point, we obtain the WAP-tree as shown in Fig. 10.1(a).

Next, we insert the sequence 'ABEAE' into the WAP-tree. Starting from the root, the first event 'A' is inserted into the WAP-tree. This time, since the WAP-tree already has the node A with frequency count 1, we simply need to increment the count by one without having to create a new node. Now we have the immediate child node after the root node updated from (A:1) to (A:2). A similar operation is then performed for the rest of the events. Note that we create 3 new nodes (E:1)→(A:1)→(E:1) as shown in the Fig. 10.1(b) and set the frequency count to 1. The auxiliary node linkage is updated to link nodes that share the same prefix.

Then, we insert the next two sequences 'BABAE' and 'ABAEE' to give us the final WAP-tree as shown in Fig. 10.2. For the 'ABAEE', a new node is created only for the last event 'E' whereas for the rest (first four events 'ABAE') no new nodes are created as they share the same prefix with the existing nodes previously created in Fig. 10.1(a). Only frequency count is updated. For the sequence 'BABAE', as no existing nodes share the same prefix with 'BABAE', new nodes are created for

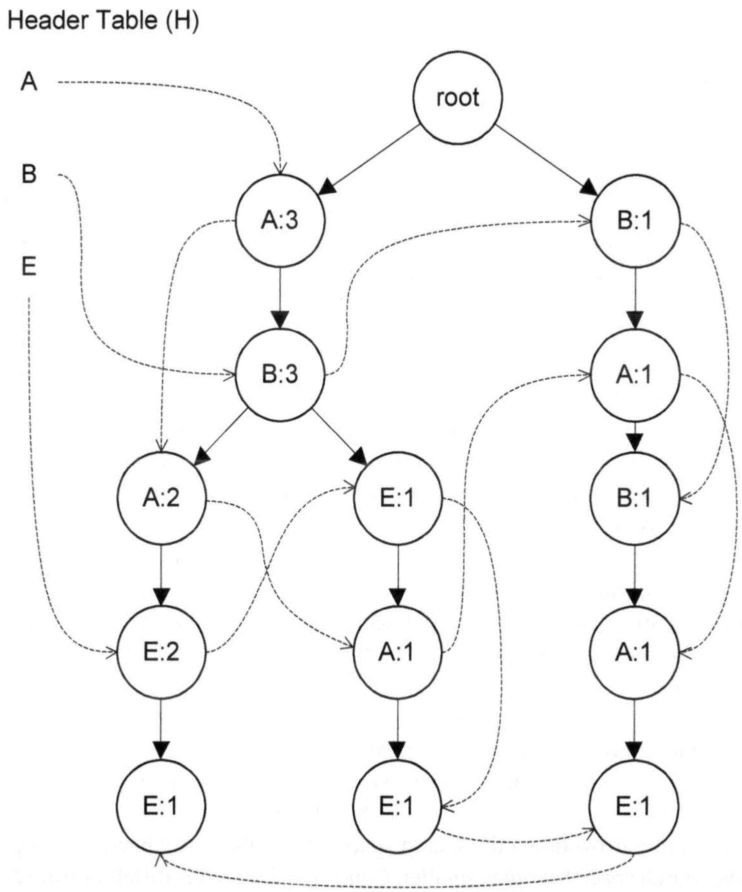

Fig. 10.2 The final WAP-tree *T*

each of the events and frequency count of 1 assigned. The auxiliary node linkage is updated to link nodes with the same prefix.

10.4.2 Mining Frequent Subsequences from WAP-Tree

After the WAP-tree construction has been completed, the next step is to enumerate and count subsequences from it. No candidate generation is required in the mining procedure, and only the patterns that have a support count greater than the specified minimum support will be considered.

The WAP-tree *T* has an important property as mentioned earlier in that for any frequent event, all the frequent subsequences that contain that event can be visited by following the event-queue, starting from the event entry in the header table of *T*. Therefore, all the patterns' information related to that frequent event can be obtained by following all the branches in T linked by the event-queue only once. Let P be a *prefix sequence* of an event e_i. P is defined as all nodes in the path from the root of the tree to this node labeled e_i (exclusive). The frequency count of this node labeled e_i is also the count of P.

Let P' be a *super-prefix sequence* and P be a *prefix sequence* of an event e_i. P must be the subsequence of the P'. Consider a sequence "*AEAE*". "*AEA*" is a *prefix sequence* of an event '*E*'. But, it is not the only *prefix sequence* of the event '*E*'. '*A*' is also a *prefix sequence* of the event '*E*'. In this case, "*AEA*" is called the *super-prefix sequence* of '*A*'. This property is important in computing frequency count of prefix-sequence of an event.

If a prefix sequence of an event e_i has no *super-prefix sequence*, the *unsubsumed count* of the prefix sequence is the same as the prefix sequence frequency count. If a prefix sequence of an event e_i has one or more super-prefix sequences, the *unsubsumed count* of that prefix sequence is defined as the frequency count of that prefix sequence minus the *unsubsumed counts* of all its *super-prefix sequences*. To illustrate this consider a sequence $s = (A{:}5) \rightarrow (E{:}4) \rightarrow (A{:}1) \rightarrow (E{:}1)$. The *unsubsumed count* of the first event '*A*', which is the prefix sequence of the event '*E*', should be 3 instead of 4, since one of them belongs to the super-prefix sequence of '*A*', that is "*AEA*". Thus, the frequency count of an event *e* is the sum of *unsubsumed counts* of all its prefix sequences which is a super-sequence of a sequence with event *e* as the last node. All the abovementioned properties allow a conditional search to be performed efficiently.

Instead of searching patterns level-wise as Apriori-like algorithms, a partition-based divide and conquer *conditional search* is used to narrow down the search space by looking for patterns with the same suffix, and count frequent events in the set of prefixes with respect to condition as suffix. In other words, the suffix is used as the condition to narrow down the search space. When the suffix becomes longer naturally, the search space becomes smaller. Conditional search is different from the bottom-up generation of combinations in that it can avoid generating large candidate sets more effectively than does the bottom-up approach.

In summary, the general flow of the mining frequent subsequences from WAP-tree is defined as follows (Pei et al. 2000):

```
INPUT  : WAP-tree T and a minimum support threshold σ
OUTPUT : Complete set of frequent subsequences (S)
   1. If T has only one branch, return all the unique combinations of nodes
      in that branch.
   2. Let S = ϕ. Insert every frequent event in T into S.
   3. For each event e in T
      a. Construct conditional sequence base of e, that is (T|e), by
         traversing the e event queue e-queue. At the same time, count
         conditional frequent events.
      b. If the set of conditional frequent events is not empty, build a
         conditional WAP-tree for event e (T|e). Mine the conditional WAP-
         tree recursively.
      c. For each subsequence returned from step 3(b), mine the conditional
         WAP-tree, append event e to it and insert it into S.

   4. Return S
```

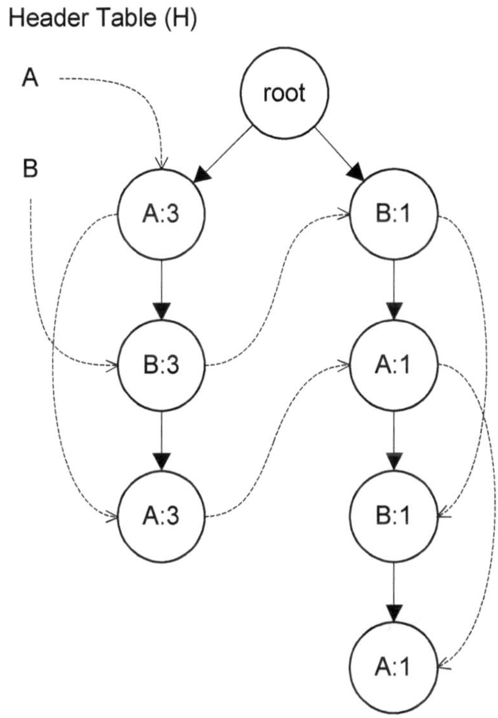

Fig. 10.3 Conditional WAP-tree ($T|E$)

The following example illustrates how the enumeration and counting is performed using conditional search. Using the same subsequences from the previous Table 10.1 and the minimum support count of 3, we start the conditional search from one of the frequent events *E*. The conditional sequence base of *E* is listed as follows:

{ABA:2, AB:1, ABEA:1, AB:-1, BABA:1, ABAE:1, ABA:-1}

The conditional sequence list of a suffix event is obtained by following the auxiliary link of the event and following the path from the root to each node (excluding the node). The conditional sequences based on *E* after removing infrequent event *E* are:

{ABA:2, AB:1,ABA: 1,AB:-1,BABA: 1,ABA:1,ABA : -1}

One event can only be a frequent conditional event if and only if the frequency count is greater or equal to the minimum support count. In this case, the minimum support count is 3. Thus, the conditional frequent events are 'A' and 'B' and both have frequency count 4. Using these conditional sequences, a conditional WAP-tree, $T|E$ is constructed as shown in Fig. 10.3.

The recursion continues with the suffix path E, AE. The algorithm keeps running, finding the conditional sequence bases of BAE, B, A. After traversing the whole tree, the discovered frequent subsequences are: {E, AAE, BAE, ABAE, AE, ABE, BE, B, AB, A, AA, BA, ABA}.

10.4.3 Other WAP-Tree Based Algorithms

Although in general, WAP-tree based algorithms may have an advantage over Apriori-like algorithms, in that they avoid the problem of generating explosive candidate subsequences, they have their own disadvantage(s). One of the main disadvantages of the WAP-mine algorithm is that it recursively re-constructs large numbers of intermediate WAP-trees and patterns during mining. This can result in serious performance degradation.

There are a number of proposed algorithms based on the WAP-tree that try to improve WAP-mine such as BC-WAP (Gomathi, Moorthi & Duraiswamy 2008), BC-WASPT (Vasumathi & Govardhan 2009), PLWAP (Ezeife & Lu 2005) and PL4UP (Ezeife & Chen 2004). PLWAP (Ezeife & Lu 2005) in particular is a derivative of the WAP-mine algorithm that uses a preorder linked, position-coded version of WAP-tree and eliminates the need to recursively re-construct intermediate WAP-trees during sequential mining. PLWAP produces a significant reduction in response time achieved by the WAP algorithm and provides a position code mechanism for remembering the stored database. Thus, it eliminates the need to re-scan the original database as would otherwise be necessary.

10.5 Overview of the Proposed Solution

In this section, we will provide an overview of the proposed solutions to address the problem of mining frequent subsequences from a database of sequences. The algorithm will be given in more detail in Section 10.6.

The challenges of mining a very large database of sequences are that it is computationally expensive and requires large memory space due to the fact that the order of items in sequences is important. When the order of entities in data is important, naturally the granularity inherent in the data is more refined and often contributes to the combinatorial explosion problem. One way to achieve efficient processing is to develop an intermediate memory construct that helps us to easily process and manipulate it. An example of an intermediate memory construct is the *embedding list* (*EL*) utilized for efficient enumeration of subtrees as explained in Chapter 4. However, the problem of using such an intermediate memory construct is that it demands some additional memory storage. For mining a very large database of sequences, this can be a serious problem.

The proposed algorithm uses a slightly modified version of the *recursive list* (*RL*) structure we discussed in Chapter 4. We refer to this structure as Direct Memory Access Strips (DMA-Strips). DMA-Strips allow generation of candidate subsequences to be done efficiently. The DMA-Strips structure provides direct memory access to sequences in the database. The DMA-Strips structure does not store hyperlinks of items. Instead, it segments the database of sequences systematically (per transaction) so that it allows direct access processing and manipulation. This contributes to better speed and space performance than using an intermediate memory construct like the *EL* that reserves extra storage in memory to store a corpus of hyperlinks of items. Nevertheless, the DMA-Strips structure resembles the *EL* in that it allows efficient enumeration using a model or structure guided approach. Furthermore, the DMA-Strips structure provides direct access to each item to be manipulated and thus is optimized for speed and space performance. In contrast to the *RL* structure, DMA-Strips has better transaction segmentation as each segmentation provides data locality to each transaction, in respect to the position of an item, for each transaction in the database. On the other hand, the *RL* requires more global data space so that the index of the item in the *RL* is local to the database of transactions but not to each transaction in the database.

In addition, the proposed technique uses a hybrid principle of frequency counting by the vertical join approach and candidate generation by the horizontal extension method. The horizontal extension method utilized here is adapted from the TMG enumeration approach used for enumerating ordered embedded subtrees as formulized in Chapter 5. We model the sequences as a vertical tree. By viewing a sequence as a vertical tree, we can reuse the *TMG* enumeration approach for enumerating sequences. A vertical tree is defined as a uniform tree with degree 1, i.e. each node has fan-out of 1.

As discussed in Chapter 5, the TMG enumeration approach is a candidate generation technique that utilizes the tree model to generate candidate subtrees. A unique property of the structure or model-guided approach is that it ensures that all enumerated patterns are valid patterns. These patterns are valid in the sense that they exist in the database, and hence by ensuring only valid patterns, no extra work is required to prune invalid patterns. This applies to tree-structured data as well as sequences.

Our strategy for efficient and more scalable frequency counting is as follows. We transpose the database from the horizontal format to the vertical format for counting frequent 1-subsequences. We combine efficient horizontal counting using a hashtable for counting 2-subsequences and the space efficient vertical join counting approach (Zaki 2000) for counting k-subsequences where $k \geq 3$. This hybrid approach overcomes the non-linear performance for counting 2-subsequences of the join enumeration approach due to the $O(n^2)$ complexity (Zaki, 2005; Han & Kamber 2006) and overall improves the memory space efficiency over the pure horizontal counting approach.

10.6 SEQUEST: Mining Frequent Subsequences from a Database of Sequences

In this section, we will present the SEQUEST algorithm (Tan et al. 2006b). The unique properties of the SEQUEST are as follows. It uses the DMA-Strips structure to allow efficient candidate subsequences generation. DMA-Strips is different from the *EL* in that it has better segmentation of each transaction. Therefore, the coordinates of items in the database will be local only to the transaction where they belong. In addition, SEQUEST utilizes a hybrid of horizontal candidate enumeration and vertical frequency counting approach. SEQUEST reuses the concept of structure-guided *TMG* by modelling sequences of itemsets as a vertical tree. This allows us to utilize the *TMG* approach seamlessly. Furthermore, by using a structure-guided concept, it does not generate any invalid sequences. At the same time, it takes advantage of the vertical counting approach that is more space efficient than the horizontal counting approach.

Given that SEQUEST reuses the TMG enumeration approach and it utilizes the DMA-Strips which is a variation of the *RL* structure that is utilized for describing tree-structured data, the general structure of the framework proposed in Chapter 5 can be easily adapted. Therefore, it is suggested that readers see the overview of the general structure of the framework again in Chapter 5 to get a clear picture. However, we will highlight some areas that are new and important to be discussed throughout this chapter, including but not limited to, the DMA-Strips generation, enumeration of subsequences using structure-guided approach, and vertical counting technique utilizing the vertical join counting approach. The pseudo-code of the SEQUEST algorithm is given later in Fig. 10.7, and in the explanation that follows, we make reference to a particular line of the pseudo-code that corresponds to the aspect being explained.

10.6.1 Database Scanning

SEQUEST scans the database S_{db} twice and generates a frequent database that contains only frequent 1-subsequences, S_{db}' after the first scan (Fig. 10.7 [1]). S_{db}' is obtained by intersecting S_{db} with L_1 (Fig. 10.7 [2]), the set of frequent 1-subsequences. The first scan and the second scan are executed in two separate runs. S_{db}' is used to construct DMA-Strips structure.

10.6.2 Constructing DMA-Strips

A DMA-Strip is constructed as follows (Fig. 10.7 [3]). For each item in S_{db}', an ordered list (*strip*) is generated which stores a sequence of items' *label*, *scope* and *eid*. Each strip is uniquely identified by a sequence-id *sid*. To allow mining sequences of itemsets, each item in a strip stores event-id *eid* and a notion of *scope*. The *eid* groups items in a strip based on their timestamps. The *scope* is used to determine the relationship between two consecutive items in a strip. The *scope* of an item A is determined by the position of the last item that belongs to the same event. If the preceding item's position is out of the prior item's scope, this tells us that for a sequence of itemsets, the two items (say *A* and *B*) occur at different time, i.e. *A*→*B*. Otherwise A and B occur at the same time that the event was generated, i.e. *AB*. Through the use of the *scope* DMA-Strips can be used for both mining sequences of itemsets or items.

Fig. 10.4 shows an example of a strip of a sequence *AC*→*AH*→*HJL* whose *sid* is 0. The above sequence consists of 3 different events. *AC*, *AH*, and *HJL* belong to event 0, 1 and 2 respectively. The scope of *A* is equal to the position of the last item whose *eid* is the same as *A*, which is *C*. Hence, *A*'s scope is determined to be 1. Similarly, using the same rule, *H*'s scope that belongs to *eid* 2 is determined to be 6. Each strip can be fetched in $O(1)$ by specifying its offset or its *sid*.

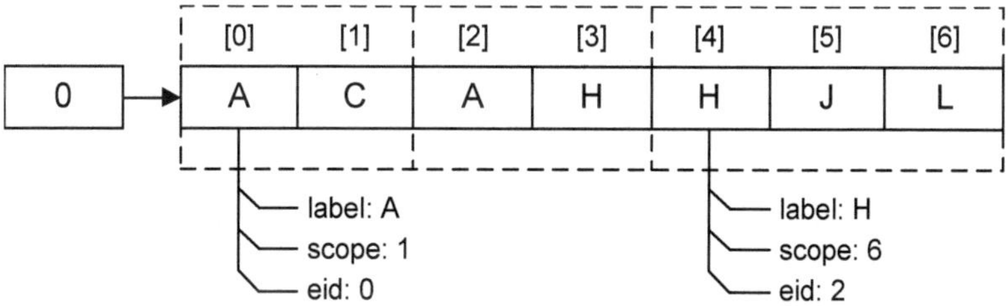

Fig. 10.4 Example of a strip of a sequence AC→AH→HJL

10.6.3 Enumeration of Subsequences

In previous chapters, we have explained the use of the structure-guided enumeration technique (TMG) to generate candidate subtrees. To reuse the TMG enumeration technique for enumerating subsequences, we need to first develop the analogy between a sequence and a tree such that we can define a sequence as a certain type of tree structure.

A sequence contains no hierarchical relationship but only horizontal (linear) relationships. Each item in a sequence has fan-out 1. By definition, the order of items in a sequence is important. A tree structure on the other hand has hierarchical relationships and horizontal relationships. But, a uniform tree with degree 1 has only hierarchical relationships. By definition, a hierarchical relationship implies that the order between nodes is vertically significant. By corollary, we can then view a sequence as a uniform tree with nodes degree 1. With this definition, then we can define the sequence of itemsets as a collection of uniform trees with nodes degree 1 rooted on the same root node, and we refer to this tree as a vertical tree. In this case, the root node is the *sid*. For instance, a sequence *AC→AH→HJL* with *sid* 0 can be represented as the vertical tree shown in Fig. 10.5.

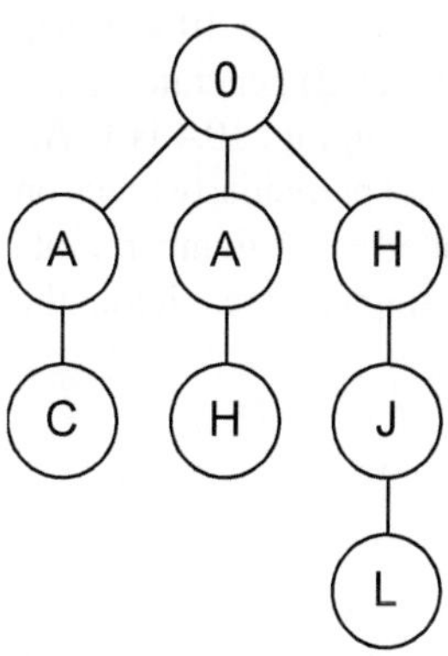

Fig. 10.5 Vertical tree representation of the strip in Fig. 10.1

The enumeration of subsequences is done by iterating a strip and extending one item at the time to the expanding k-subsequence.

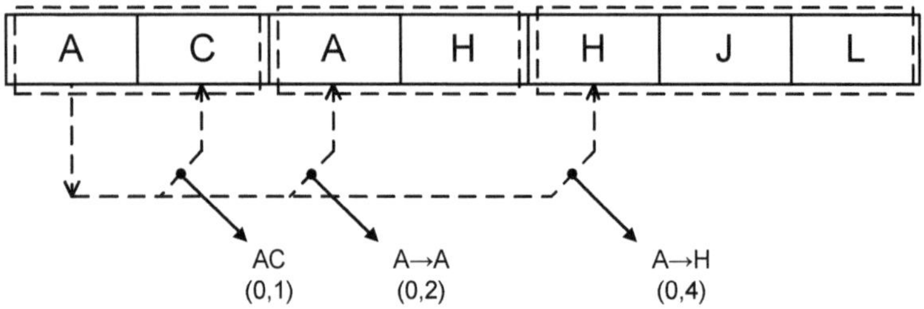

Fig. 10.6 Enumeration examples

Fig. 10.6 illustrates how the enumeration of AC, $A{\rightarrow}A$, and $A{\rightarrow}H$ is done through the strip. Given that we have 1-subsequence I at position r and the strip S with size $|S|$, 2-subsequences are generated by extending I with each item J at position t on its right to the end of the strip such that $r < t \leq |S|$ (Fig. 10.7 [4]). This can be generalized to enumerating $(k{+}1)$-subsequences from a k-subsequence (Fig. 10.7 [5]). For a k-subsequence it is sufficient to store only a pair of *sid* and the last item's position (*tail*) as its occurrence coordinate. Hence, given an occurrence coordinate of a $(k\text{-}1)$-subsequence, its occurrence coordinate will be in a form of $(sid,tail)$. Hence the enumeration of $(k{+}1)$-subsequences M using a strip S from a k-subsequence s whose encoding is $\varphi(s)$, and occurrence coordinate is (p,q), where p is its *sid* and q is the *tail* can be formalized as follows. Suppose that C is the set of occurrence coordinates of $(k{+}1)$-subsequences, $C{=}\{(p,t)|q{\leq}\, t = |S|\}$. The encoding of the $(k{+}1)$-subsequences is computed by performing a scope test. We define two functions, $\Psi(p,q)$ and $\lambda(p,q)$ that determine the *scope* and *label* of an item in a strip with *sid* p at position q. If $t > \Psi(p,q)$ then the new encoding is $\varphi(s)$ + '→' + $\lambda(p,t)$, and otherwise it is φ + $\lambda(p,t)$ (Fig. 10.7 [6] & [7]). For example, from Fig. 10.4, a strip with *sid* 0, extending A_0 (A at position 0) with A_2 (A at position 2) generates $A{\rightarrow}A$ since $2 > \Psi(0,0)$ for the given strip. We know that $\Psi(0,0)$ from Fig. 10.4 is 1. Although a constraint represents an important concept, we do not particularly focus on it in this chapter. However, the current scheme can be extended easily to incorporate the *gap constraint* described in the previous section, for example, by restricting the extension using the event-id delta.

10.6.4 Frequency Counting

We utilize a hybrid strategy for enumerating subsequences by extension and counting by the vertical join approach to make use of the desired property from each technique. Enumeration by extension makes sure that the candidates generated are all valid. What we mean by a valid candidate is a candidate whose support is greater than zero. The vertical counting approach is used to help us achieve a space efficient and less expensive full pruning strategy.

We define two operators *inter-join* and *intra-join* (Fig. 10.7 [10] & [11]) to tackle both sub problems of mining sequence of itemsets and items. The inter-join operator is used whenever the tail items of two joinable (*k-1*)-subsequences are from different events, whereas *intra-join* operator is used if they are from the same event. *Intra-join* will join any two occurrence coordinates whose *sid* and *eid* are equal. Inter-join will join any two occurrence coordinates whose *sid* are equal but the prior *eid* is less than the preceeding *eid*. If the *gap constraint* is to be introduced, it will be to define the *inter-join* operator by constraining the gap between the two *eid*s. Space efficiency is achieved through the vertical join approach by freeing the memory sooner than the horizontal counting approach does. The details of how the less expensive full pruning is achieved through the vertical join approach are discussed below.

10.6.5 Pruning

According to Apriori theory, a pattern is frequent if and only if all its subsets are frequent. To make sure that no generated subsequences contain infrequent subsequences, full (*k-1*) pruning must be performed. Full (*k-1*) pruning or *full pruning* implies that at most (*k-1*) numbers of (*k-1*)-subsequences need to be generated from the currently expanding *k*-subsequences. This process is expensive. Using the join approach principle, any two (*k-1*)-subsequences that are obtained from a *k*-subsequence by removing one node at a time can be joined to form the *k*-subsequence back. This implies that we could accelerate the full pruning by only doing root and tail pruning (Fig. 10.7 [8] & [9]). Root pruning is done by removing the root node, i.e. the first node of the sequence and checking if the pattern has been generated previously. Similarly, tail pruning is done by removing the tail node. Since we use a hashtable to perform frequency counting this can be done by simply checking if the pattern exists in the *k*-1 hashtable. If it does not exist, we can safely prune the generated *k*-subsequences. However, it is not completely safe to assume that if both root-pruned and tail-pruned (*k-1*)-subsequences exist means the generated *k*-subsequences must be frequent. At least, the root and tail pruning will do the partial check and the final check is done by joining the occurrence coordinates of the root-pruned and tail-pruned (*k-1*)-subsequences which is done as part of the support counting. This scheme can only be done using vertical support counting since it counts the support of a subsequence vertically, i.e. the result is known immediately. Horizontal counting, on the other hand, increments the support of a subsequence one at a time and the final support of a subsequence is not known until the database has been completely traversed.

10.6.6 SEQUEST Pseudo-Code

Overall, the SEQUEST algorithm (Tan et al. 2006b) is described by the pseudo-code in Fig. 10.7.

```
Sdb:Sequence Database
Lk:Frequent k-subsequences

L1= Scan-Sdb                                              [1]
Sdb'= intersect(Sdb ∩ L1)                                 [2]
DMA-STRIPS = Construct-DMA-STRIPS(Sdb')                   [3]
L2 = Generate-2-Subsequences(DMA-STRIPS)                  [4]
k = 3
while(Lk > 0)
  Lk= Generate-k-Subsequences(Lk-1)                       [5]
k++

Generate-k-Subsequences(Lk-1):
foreach (k-1)-sequence s in Lk-1{
  foreach occurrence-coordinate oc(sid, r) in s {
    for t = r + 1 to |DMA-STRIPS[sid]| {
      if(t > scope(sid, r)) {
        join = inter-join
        φ(s') = φ(s) + → + label(sid, t)                  [6]
      } else {
        join = intra-join
        φ(s') = φ(s) + label(sid, t)                      [7]
      }

      root-pruned = remove-root(φ(s'))                    [8]
      tail-pruned = φ(s')                                 [9]
      if(join == intra-join)
        Intra-Join(root-pruned, tail-pruned)              [10]
      else
        Inter-Join(root-pruned, tail-pruned)              [11]

      if(InFrequent(φ(s')))
          Prune(φ(s'))
      else
          Insert(φ(s'),Lk)
    }
  }
}
```

Fig. 10.7 Pseudo-code of SEQUEST

10.7 Experimental Results and Discussions

This section explores the effectiveness of the proposed SEQUEST algorithm, by comparing it with the PLWAP algorithm that processes sequences of items. The formatting of the datasets used by PLWAP can be found in (Ezeife & Lu, 2005). SEQUEST, on the other hand, uses the datasets format that is used by the IBM data generator (Agrawal & Srikant, 1995; Srikant & Agrawal, 1996). The minimum support σ is denoted as (Sxx), where xx is shown as the absolute support count. Experiments were run on Dell Rack Optimized Slim Servers (dual 2.8Ghz CPU, 4Gb RAM) running Fedora 4.0 ES Linux. The source code for each algorithm was compiled using GNU g++ version 3.4.3 with the –O3 parameter.

The datasets chosen for this experiment were previously used in (Ezeife & Lu, 2005), in particular data.ncust_25 (200K) (Fig. 10.8), data.100K, data.600K, data.ncust_125 (1000K) (Fig. 10.9 & Fig. 10.10), and data.ncust_1000 (7000K) (Fig. 10.11). To obtain the 500K dataset for experiments in Fig. 10.9, we omitted the last 100K sequences from the data.600K dataset.

10.7.1 Performance Test

Fig. 10.8(a) shows the comparison of the algorithms with respect to time, for varying support thresholds. The experiment uses the 200K dataset. In this experiment, we try two schemes for SEQUEST, namely SH (refers to SEQUEST-hybrid) and S (refers to SEQUEST). SH uses a vertical join counting approach only for generating frequent *k*-sequences where $k \geq 3$, whereas S uses a horizontal counting approach using a hashtable for all *k*. The results in Fig. 10.8(a) show that SH performs better than S and PLWAP. The performance of SH remains nearly constant for a very small support (σ:2 or 0.001%). The difference between SH and PLWAP performance at σ:2 is in the order of ~800 times.

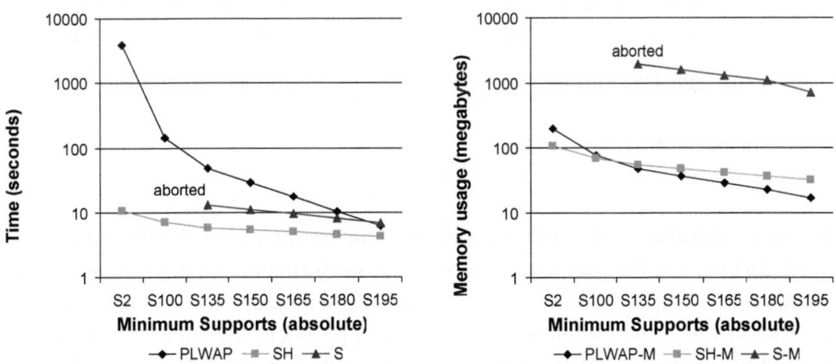

(a) time performance (b) memory profiles

Fig. 10.8 200K dataset

Fig. 10.8(b) shows comparison of memory usage between PLWAP, SH and S. We can see that when the horizontal counting approach is used S suffers from memory blow-up and this contributes to S aborting at $\sigma = 135$. SH memory profiles, on the other hand, show a superior overall performance over the other two, especially when the minimum support is lowered.

10.7.2 Scalability Test

We perform a scalability test by varying the datasets of different sizes while σ is fixed at 0.005%. The experiment uses 100K, 500K, and 1000K dataset. In Fig. 10.9(a), SH outperformed PLWAP with respect to time in the order of ~250 times faster. In this experiment, we try two different methods for generating frequent 2-subsequences. We modify SH such that it uses the structure-guided enumeration approach as described in the previous section for enumerating 2-subsequences, and S-L2J uses a join approach for enumerating 2-subsequences.

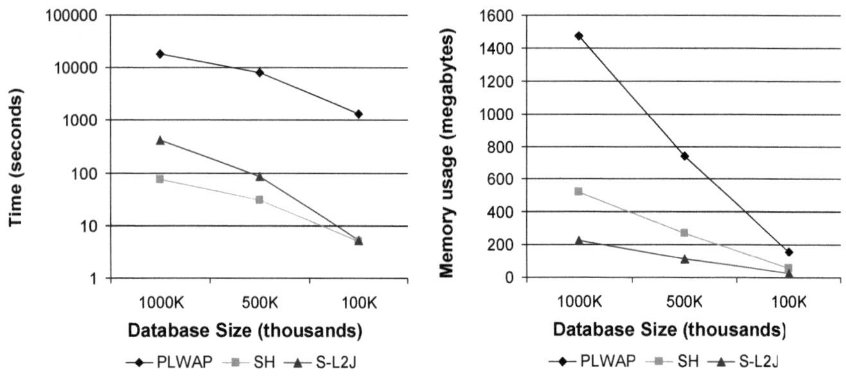

(a) time performance (b) memory profiles

Fig. 10.9 100K, 500K, 1000K dataset

By implementing horizontal enumeration for enumerating 2-subsequences, in Fig. 10.9(a) we see that SH has a more linear performance than S-L2J which implements the join approach for enumerating 2-subsequences. We observe that if the enumeration for 2-subsequences uses the join approach, the time performance is increased by roughly n^2 if the datasets size is scaled up by n times. On the other hand, it is interesting to see that whenever the join approach is used, the memory usage is slashed by almost half as evident in Fig. 10.9(b). Fig. 10.9(b) shows that SH memory usage is 2x greater than S-L2J. Here, we observe that by using the horizontal enumeration approach for generating 2-subsequences, the overall performance of the SEQUEST is improved with a trade-off that the memory usage will be higher. When using the join approach with numerous candidates, we can see that the performance degrades.

From this, we can infer that optimal performance can be obtained whenever it is known when to switch from the horizontal to vertical join counting approach. The horizontal counting approach suffers from memory blow-up but is faster whenever the numbers of candidates generated are very large. Although the vertical counting join approach is space efficient, the performance degrades when the number of candidates to be generated is very large.

PLWAP performs well whenever the average length of the extracted subsequences is relatively short. In other words, PLWAP performs well for large sparse databases but when applied to dense databases with long patterns, the performance is significantly reduced. Overall, we witness that all SEQUEST variants, SH and S-L2J have outperformed PLWAP in terms of speed and space performance.

10.7.3 Frequency Distribution Test

For the experiment in Fig. 10.10, we use 100K, 500K, 1000K datasets. Fig. 10.10 shows the distribution of candidate subsequences and frequent subsequences generated over different dataset sizes with σ fixed at 0.005%.

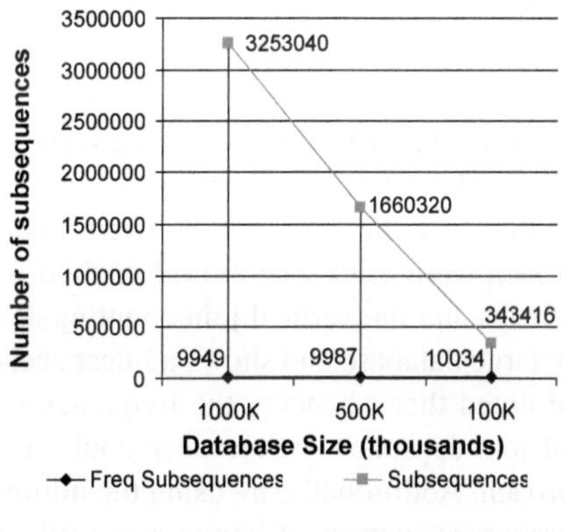

Fig. 10.10 100K, 500K, 1000K dataset frequency distribution

10.7.4 Large Database Test

Fig. 10.11 shows the time performance comparison between PLWAP and SH for a dataset of 7.8 million sequences (7000K dataset). For this particular dataset, we see that PLWAP outperforms SH by a small margin. However, at σ:2500 (0.03%), PLWAP aborted due to a memory blow-up problem. At σ:10000 and σ:5000, the number of extracted frequent subsequences is relatively low: 130 and 1211 respectively.

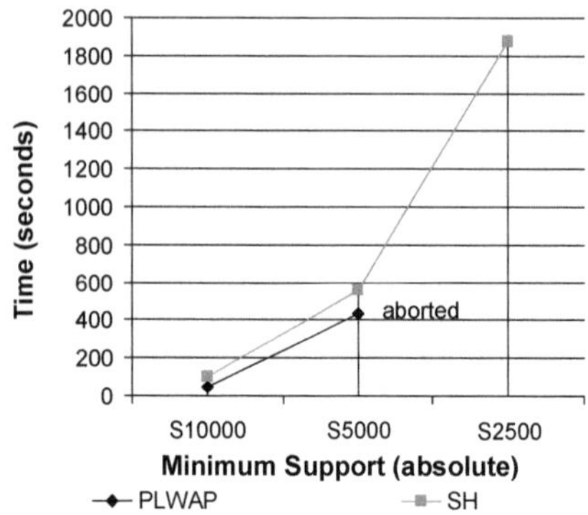

Fig. 10.11 7000K time performance

10.7.5 Overall Conclusions

PLWAP is a space efficient technique but it suffers when the support is decreased and the numbers of frequent subsequences extracted are high. On the other hand, when the horizontal counting approach is used, SEQUEST suffers from a memory blow-up problem due to the BFS approach for generating candidate subsequences. It uses two hash-tables and additional space is needed when the length of the sequences to be enumerated increases. We also show that if SEQUEST uses a vertical join counting approach, it performs extremely well for both speed and space performance. If the vertical join counting approach is used, the space can be freed much sooner than if the horizontal counting approach were used. The hybrid method of structure-guided enumeration using DMA-Strips and the vertical join counting approach enables SEQUEST to process a very large database and shows a linear scalability performance. However, it should be noted that whenever the frequency of extracted subsequences is high, the vertical join approach performance could degrade due to the complexity of the join approach. Additionally, by using the notion of *scope* in DMA-Strips, SEQUEST can process sequences of itemsets as well as sequences of items.

10.8 Conclusion

In this chapter, we have presented the algorithm for mining frequent subsequences from a database of sequences as another extension of the TMG framework to mine frequent subtrees from a database of trees. The aim of the extension work is to show how the TMG framework previously used for mining frequent ordered subtrees can also be used to address other problems that are in similar domains.

There are two different sub-problems in mining frequent subsequences, i.e. problems of addressing (1) sequences of items and (2) sequences of itemsets. Addressing sequences of itemsets, i.e. sequences whose elements consist of multi items, is a more difficult problem and poses more challenges. We have shown how SEQUEST can reuse the TMG enumeration approach utilized for mining frequent ordered subtrees by modelling the sequences of itemsets as the vertical trees. By modelling sequences of itemsets as vertical trees, SEQUEST is designed to tackle not only the problem of mining frequent subsequences of items, but also subsequences of itemsets.

We have also demonstrated how a hybrid approach, which implements horizontal enumeration and vertical counting through the join approach, enables SEQUEST to process a very large database and shows a linear scalability performance. It should be noted, however, that with vertical counting through the join approach, performance could degrade due to the complexity of this approach.

Experiments utilizing very large databases of sequences, which compare our technique with the existing technique PLWAP (Ezeife & Lu, 2005) have demonstrated a degree of effectiveness and the ability of the proposed framework to tackle the problem of mining frequent subsequences in addition to addressing the problem of mining frequent ordered subtrees. In future, we aim to compare SEQUEST with other techniques also capable of addressing the problem of mining sequences of itemsets. Further optimizations are to take place, after revealing the strengths and weaknesses of the technique through extensive experimental evaluation and comparison.

References

1. Agrawal, R., Srikant, R.: Mining sequential patterns. Paper presented at the Proceedings of the 11th International Conference on Data Engineering, Taipei, Taiwan, March 6-10 (1995)
2. Antunes, C., Oliveira, L.: Sequential Pattern Mining Algorithms: Trade-offs between Speed and memory. Paper presented at the Proceedings of the 2nd ECML PKDD Workshop on Mining Graphs, Trees and Sequences (MGTS 2004), Pisa, Italy, September 20-24 (2004)
3. Arimura, H., Uno, T.: Mining Maximal Flexible Patterns in a Sequence. Paper presented at the Proceedings of the JSAI 5th Workshop on Learning with Logics and Logics for Learning, Miyazaki, Japan, June 18-22 (2007)
4. Ayres, J., Flannick, J., Gehrke, J., Yiu, T.: Sequential Pattern Mining Using Bitmap Representation. Paper presented at the Proceedings of the 8th ACM SIGKDD International Conference on Knowledge Discovery and Data Mining, Edmonton, Alberta, Canada, July 23-26 (2002)
5. Capelle, M., Masson, C., Boulicaut, J.-F.: Mining Frequent Sequential Patterns under a Similarity Constraint. Paper presented at the Proceedings of the 3rd International Conference on Intelligent Data Engineering and Automated Learning (IDEAL), Manchester, UK, August 12-14 (2002)
6. Chen, Y.L., Chiang, M.C., Ko, M.-T.: Discovering time-interval sequential patterns in sequence databases. Expert Systems with Applications 25(3), 343–354 (2003)
7. Chen, G., Wu, X., Zhu, X.: Sequential Pattern Mining in Multiple Streams. Paper presented at the Proceedings of the 5th IEEE International Conference on Data Mining (ICDM 2005, Houston, Texas, USA, Novemeber 27-30 (2005)

8. Chi, Y., Muntz, R.R., Nijssen, S., Kok, J.N.: Frequent Subtree Mining - An Overview. Fundamenta Informaticae, Special Issue on Graph and Tree Mining 66(1-2), 161–198 (2005)
9. Cule, B., Goethals, B., Robardet, C.: A new constraint for mining sets in sequences. Paper presented at the Proceedings of the SIAM International Conference on Data Mining (SDM, Sparks, Nevada, USA, April 30 - May 2 (2009)
10. Cule, B., Goethals, B.: Mining Association Rules in Long Sequences. In: Zaki, M.J., Yu, J.X., Ravindran, B., Pudi, V. (eds.) PAKDD 2010. LNCS, vol. 6118, pp. 300–309. Springer, Heidelberg (2010)
11. Esposito, F., Mauro, N.D., Basile, T.M.A., Ferilli, S.: Multi-Dimensional Relational Sequence Mining. Fundamenta Informaticae 89(1), 23–43 (2009)
12. Ezeife, C.I., Chen, M.: Incremental Mining of Web Sequential Patterns Using PLWAP Tree on Tolerance MinSupport. Paper presented at the Proceedings of the 8th International Database Engineering and Applications Symposium (IDEAS 2004), Coimbra, Portugal, July 7-9 (2004)
13. Ezeife, C.I., Lu, Y.: Mining Web Log sequential Patterns with Position Coded Pre-Order Linked WAP-tree. Data Mining and Knowledge Discovery 10(1), 5–38 (2005)
14. Feng, L., Dillon, T.S., Liu, J.: Inter-transactional association rules for multi-dimensional contexts for prediction and their application to studying meteorological data. Data and Knowledge Engineering 37(1), 85–115 (2001)
15. Fredkin, E.: Trie Memory. Communications of the ACM 3(9), 490–499 (1960)
16. Gomathi, C., Morthi, M.: Coded Web Access Pattern Tree in Education Domain. International Education Studies 1 (4):28-32 (2008)
17. Gouda, K., Hassaan, M., Zaki, M.J.: PRISM: A Primal-Encoding Approach for Frequent Sequence Mining. Paper presented at the Proceedings of the 7th IEEE International Conference on Data Mining (ICDM 2007), Omaha, Nebraska, USA, October 28-31 (2007)
18. Guil, F., Bosch, A., Marin, R.: TSETMAX: an algorithm for mining frequent maximal temporal patterns. Paper presented at the Proceedings of the IEEE ICDM Workshop on Temporal Data Mining: Algorithms, Theory and Applications (TDM 2004), Brighton, UK, Novemebr 1-4 (2004)
19. Guralnik, V., Karypis, G.: Parallel tree-projection-based sequence mining algorithms. Parallel Computing 30(4), 443–472 (2004)
20. Han, J., Pei, J., Mortazavi-Asl, B., Chen, Q., Dayal, U., Hsu, M.: FreeSpan: frequent pattern-projected sequential pattern mining. Paper presented at the Proceedings of the 6th International Conference on Knowledge Discovery and Data Mining (KDD 2000), Boston, MA, USA, August 20-23 (2000)
21. Han, J., Pei, J., Yin, Y., Mao, R.: Mining Frequent Patterns without Candidate Generation: A Frequent-Pattern Tree Approach. Data Mining and Knowledge Discovery 8(1), 53–87 (2004)
22. Han, J., Kamber, M.: Data Mining: Concepts and Techniques, 2nd edn. Elsevier, Morgan Kaufmann Publishers, San Francisco, CA, USA (2006)
23. Joshi, M., Karypis, G., Kumar, V.: A universal formulation of sequential patterns. Paper presented at the Proceedings of the KDD 2001 workshop on Temporal Data Mining, San Francisco, California, USA, August 26-29 (2001)
24. Kum, H.C., Pei, J., Wang, W., Duncan, D.: ApproxMAP: Approximate mining of consensus sequential patterns. Paper presented at the Proceedings of the 3rd SIAM International Conference on Data Mining (SDM 2003), San Francisco, CA, USA, May 1-3 (2003)
25. Kum, H.C., Paulsen, S., Wang, W.: Comparative Study of Sequential Pattern Mining Models. Studies in Computational Intelligence: Foundations of Data Mining and Knowledge Discovery 6, 45–71 (2005)

26. Kum, H.-C., Chang, J.-H., Wang, W.: Sequential Pattern Mining in Multi-Databases via Multiple Alignment. Data Mining and Knowledge Discovery 12(2-3), 151–180 (2006)
27. Laird, P.: Identifying and using patterns in sequential data. Paper presented at the Proceedings of the 4th International Workshop on Algorithmic Theory (ALT 1993), Tokyo, Japan, November 8-10 (1993)
28. Leleu, M., Rigotti, C., Boulicaut, J.-F., Euvrard : GO_SPADE: Mining Sequential Patterns over Datasets with Consecutive Repetitions. In: Paper presented at the Proceedings of the 3rd International Conference on Machine Learning and Data Mining in Pattern Recognition (MLDM 2003), Leipzig, Germany, July 5-7 (2003)
29. Lin, M., Lee, S.: Incremental update on sequential patterns in large databases. In: Paper presented at the Proceedings of the 10th IEEE International Conference on Tools with Artificial Intelligence (ICTAI 1998), Taipei, Taiwan, Novemeber 10-12 (1998)
30. Lin, M., Lee, S.: Fast Discovery of Sequential Patterns by Memory Indexing. Paper presented at the Proceedings of the 4th International Conference on Data Warehousing and Knowledge Discovery (DaWak 2002), Aix-en-Provence, France, September 4-6 (2002)
31. Luo, C., Chung, S.M.: A Scalable Algorithm for Mining Maximal Frequent Sequences Using Sampling. Paper presented at the Proceedings of the 16th IEEE International Conference on Tools with Artificial Intelligence (ICTAI 2004), Boca Raton, FL, USA, Novemeber 15-17 (2004)
32. Luo, C., Chung, S.M.: A scalable algorithm for mining maximal frequent sequences using a sample. Knowledge and Information Systems 15(2), 149–179 (2008)
33. Mannila, H., Toivonen, H., Verkamo, A.I.: Efficient algorithms for discovering association rules. Paper presented at the Proceedings of AAAI 1994, Workshop on Knowledge Discovery in Databases Seattle, WA, USA, July 31- August 4 (1994)
34. Mannila, H., Toivonen, H.: Discovering generalized episodes using minimal occurrences. Paper presented at the Proceedings of the 2nd International Conference on Knowledge Discovery and Data Mining (KDD 1996), Portland, Oregon, USA, August 2-4 (1996)
35. Mannila, H., Toivonen, H., Verkamo, A.I.: Discovery of frequent episodes in event sequences. Data Mining and Knowledge Discovery 1(3), 259–289 (1997)
36. Masseglia, F., Cathala, F., Poncelet, P.: The PSP Approach for Mining Sequential Patterns. In: Żytkow, J.M. (ed.) PKDD 1998. LNCS, vol. 1510, Springer, Heidelberg (1998)
37. Masseglia, F., Poncelet, P., Teisseire, M.: Incremental mining of sequential patterns in large databases. Data and Knwoledge Engineering 46(1), 97–121 (2003)
38. Masseglia, F., Teisseire, M., Poncelet, P.: Sequential Pattern Mining: A Survey on Issues and Approaches. Encyclopedia of Data Warehousing and Mining, 3–29 (2005)
39. Oates, T., Schmill, M.D., Jensen, D., Cohen, P.R.: A family of algorithms for finding temporal structure in data. Paper presented at the Proceedings of the 6th International Workshop on AI and Statistics (March 1997)
40. Parthasarathy, S., Zaki, M.J., Ogihara, M., Dwarkadas, S.: Incremental and interactive sequence mining. Paper presented at the Proceedings of the 8th International Conference on Information and Knowledge Management (CIKM 1999), Kansas City, Missouri, USA, November 2-6 (1999)
41. Pei, J., Han, J., Mortazavi-asl, B., Zhu, H.: Mining Access Patterns Efficiently from Web Logs. In: Terano, T., Chen, A.L.P. (eds.) PAKDD 2000. LNCS, vol. 1805, Springer, Heidelberg (2000)
42. Pei, J., Han, J., Mortazavi-Asl, B., Pinto, H., Chen, Q., Dayal, U., Hsu, M.: PrefixSpan: Mining Sequential Patterns by Prefix-Projected Pattern Growth. Paper presented at the Proceedings of the 17th International Conference on Data Engineering (ICDE), Heidelberg, Germany, April 2-6 (2001)

43. Pei, J., Han, J., Wang, W.: Mining Sequential Patterns with Constraints in Large Databases. Paper presented at the Proceedings of the 2002 ACM International Conference on Information and Knowledge Management (CIKM 2002), McLean, Virginia, USA, November 4-9 (2002)
44. Pei, J., Han, J., Wang, W.: Constraint-based sequential pattern mining: the pattern-growth methods. Journal of Intelligent Information Systems 28(2), 133–160 (2007)
45. Pinto, H., Han, J., Pei, J., Wang, K., Chen, Q., Dayal, U.: Multi-dimensional Sequential Pattern Mining. Paper presented at the Proceedings of the 2001 ACM CIKM International Conference on Information and Knowledge Management (CIKM 2001), Atlanta, Georgia, USA, November 5-10 (2001)
46. Plantevit, M., Choong, Y.W., Laurent, A., Laurent, D., Teisseire, M.: M2SP: Mining sequential patterns among several dimensions. Paper presented at the Proceedings of the 9th European Conference on the Principles and Practice of Knowledge Discovery in Databases (PKDD), Porto, Portugal, October 3-7 (2005)
47. Plantevit, M., Laurent, A., Laurent, D., Teisseire, M., Choong, Y.W.: Mining multidimensional and multilevel sequential patterns. ACM Transactions on Knowledge Discovery from Data 4(1), 1–37 (2010)
48. Seno, M., Karypis, G.: SLPMiner: an algorithm for finding frequent sequential patterns using lengthdecreasing support constraint. Paper presented at the Proceedings of the 2nd IEEE International Conference on Data Mining (ICDM 2002), Maebashi City, Japan, December 9-12 (2002)
49. Shintani, T., Kisuregawa, M.: Mining Algorithms for Sequential Patterns in Parallel: Hash Based Approach. In: Wu, X., Kotagiri, R., Korb, K.B. (eds.) PAKDD 1998. LNCS, vol. 1394, Springer, Heidelberg (1998)
50. Srikant, R., Agrawal, R.: Mining Sequential Patterns: Generalizations and Performance Improvements. In: Apers, P.M.G., Bouzeghoub, M., Gardarin, G. (eds.) EDBT 1996. LNCS, vol. 1057, Springer, Heidelberg (1996)
51. Srikant, R., Vu, Q., Agrawal, R.: Mining Association Rules with Item Constraints. In: Proceedings of the 3rd International Conference on Knowledge Discovery and Data Mining (KDD 1997), Newport Beach, CA, USA, August 14-17, pp. 67–73 (1997)
52. Stefanowski, J., Ziembinski, R.: Mining context based sequential patterns. In: Szczepaniak, P.S., Kacprzyk, J., Niewiadomski, A. (eds.) AWIC 2005. LNCS (LNAI), vol. 3528, pp. 401–407. Springer, Heidelberg (2005)
53. Tan, H., Dillon, T.S., Feng, L., Chang, E., Hadzic, F.: X3-Miner: Mining Patterns from XML Database. Paper presented at the Proceedings of the 6th International Conference on Data Mining, Text Mining and their Business Applications, Skiathos, Greece, May 25 (2005)
54. Tan, H., Dillon, T.S., Hadzic, F., Chang, E.: SEQUEST: Mining Frequent Subsequences using DMA Strips. Paper presented at the Proceeding of the 7th International Conference on Data Mining and Information Engineering, Prague, Czech Republic, July 11-13 (2006)
55. Tan, H., Dillon, T.S., Hadzic, F., Chang, E., Feng, L.: IMB3-Miner: Mining Induced/Embedded Subtrees by Constraining the Level of Embedding. In: Ng, W.-K., Kitsuregawa, M., Li, J., Chang, K. (eds.) PAKDD 2006. LNCS (LNAI), vol. 3918, pp. 450–461. Springer, Heidelberg (2006)
56. Tan, X., Yao, M., Zhang, J.: Mining Maximal Frequent Access Sequences Based on Improved WAP-tree. Paper presented at the Proceedings of the 6th International Conference on Intelligent Systems Design and Applications (ISDA 2006), Jinan, China, October 16-18 (2006)

57. Tan, H.: Tree Model Guided (TMG) enumeration as the basis for mining frequent patterns from XML documents. University of Technology Sydney, Sydney (2008)
58. Tan, H., Hadzic, F., Dillon, T.S., Feng, L., Chang, E.: Tree Model Guided Candidate Generation for Mining Frequent Subtrees from XML. ACM Transactions on Knowledge Discovery from Data 2(2) (2008)
59. Trasarti, R., Bonchi, F., Goethals, B.: Sequence Mining Automata: A New Technique for Mining Sequences under Regular Expressions. Paper presented at the Proceedings of the 8th IEEE International Conference on Data Mining (ICDM 2008), Pisa, Italy, December 15-19 (2008)
60. Tzvetkov, P., Yan, X., Han, J.: TSP: Mining top-k closed sequential patterns. Knowledge and Information Systems 7(4), 438–457 (2005)
61. Vasumathi, D., Govardhan, A.: BC-WASPT: Web Access Sequential Pattern Tree Mining. International Journal of Computer Science and Network Security (IJCSNS) 9(6), 288–293 (2009)
62. Wang, J.T.-L., Chirn, G.-W., Marr, T.G., Shapiro, B., Shasha, D., Zhang, K.: Combinatorial pattern discovery for scientific data: Some preliminary results. Paper presented at the Proceedings of the ACM SIGMOD Conference on Management of Data, Minneapolis, Minnesota, USA, May 24-27 (1994)
63. Wang, K., Tan, J.: Incremental Discovery of Sequential Patterns. Paper presented at the Proceedings of the ACM SIGMOD Data Mining Workshop (DMKD 1996), Montreal, Canada, USA, June 4-6 (1996)
64. Wang, K.: Discovering patterns from large and dynamic sequential data. Journal of Intelligent Information Systems 9(1), 33–56 (1997)
65. Wang, J., Han, J.: BIDE: Efficient Mining of Frequent Closed Sequences. Paper presented at the Proceedings of the International Conference on Data Engineering (ICDE 2004), Boston, USA, March 30 - April 2 (2004)
66. Wang, K., Xu, Y., Yu, J.X.: Scalable Sequential Pattern Mining for Biological Sequences. Paper presented at the Proceedings of the ACM 13th Conference on Information and Knowledge Management (CIKM 2004), Washington, DC, USA, November 8-13 (2004)
67. Yan, X., Zhou, X.J., Han, J.: Mining Closed Relational Graphs with Connectivity Constraints. Paper presented at the Proceedings of the 11th ACM SIGKDD International Cofnerence on Knowledge Discovery and Data Mining (KDD 2005), Chicago, Illinois, USA, August 21-24 (2005)
68. Yang, J., Wang, W., Yu, P.S., Han, J.: Mining Long Sequential Patterns in a Noisy Environment. Paper presented at the Proceedings of the 2002 ACM SIGMOD International Conference on Management of Data, Madison, Wisconsin, USA, June 3-6 (2002)
69. Yang, J., Wang, W.: CLUSEQ: Efficient and effective sequence clustering. Paper presented at the Proceedings of the 19th International Conference on Data Engineering (ICDE 2003), Bangalore, India, March 5-8 (2003)
70. Yu, C.-C., Chen, Y.-L.: Mining sequential patterns from multidimensional sequence data. IEEE Transactions on Knowledge and Data Engineering 17(1), 136–140 (2005)
71. Yusheng, X., Lanhui, Z., Zhixin, M., Lian, L., Chen, X., Dillon, T.S.: Mining Sequential Pattern Using DF2Ls. Paper presented at the Proceedings of the 5th International Conference on Fuzzy Systems and Knowledge Discovery (FSKD), Shanding, China, October 18-20 (2008)
72. Yusheng, X., Zhixin, M., Lian, L., Dillon, T.S.: Effective pruning strategies for sequential pattern mining. Paper presented at the Proceedings of International Workshop on Knowledge Discovery and Data Mining (WKDD1) in Conjunction with E-Forensics Conference, Adelaide, Australia, January 21-22 (2008)

73. Zaki, M.J.: Fast mining of sequential patterns in very large databases. University of Rochester Computer Science Department, New York (1997)
74. Zaki, M.J.: Parallel Sequence Mining on Shared-Memory Machines. Paper presented at the Proceedings of the ACM SIGKDD Workshop on Large-Scale Parallel KDD Systems, San Diego, CA, USA, August 15 (1999)
75. Zaki, M.J.: SPADE: An Efficient Algorithm for Mining Frequent Sequences. Machine Learning 42(1/2), 31–60 (2001)
76. Zaki, M.J.: Sequence Mining in Categorical Domains: Incorporating Constraints. Paper presented at the Proceedings of the 9th ACM CIKM International Conference on Information and Knwoledge Management, McLean, VA, USA, Novemeber 6-11 (2002)
77. Zaki, M.J.: Efficiently Mining Frequent Trees in a Forest: Algorithms and Applications. IEEE Transactions on Knowledge and Data Engineering 17(8), 1021–1035 (2005)
78. Zeitouni, K.: From Sequence Mining to Multidimensional Sequence Mining. In: Mining Complex Data. SCI, vol. 165, pp. 133–152. Springer, Heidelberg (2009)
79. Zhang, M., Kao, B., Yip, C., Cheung, D.: A GSP-based efficient algorithm for mining frequent sequences. Paper presented at the Proceedings of the, International Conference on Artificial Intelligence (IC-AI 2001), Las Vegas, Nevada, USA (June 2001)
80. Zhang, C., Hu, K., Chen, Z., Chen, L., Dong, Y.: ApproxMGMSP: A Scalable Method of Mining Approximate Multidimensional Sequential Patterns on Distributed System. Paper presented at the Proceedings of the 4th International Conference on Fuzzy Systems and Knowledge Discovery (FSKD), Haikou, Hainan, China, August 24-27 (2007)
81. Zhao, Q., Bhowmick, S.S.: Sequential Pattern Mining: A Survey. Nanyang Technological University, Singapore (2003)

Chapter 11
Graph Mining

11.1 Introduction

The contents of the book have focused so far on the mining of data where the underlying structure is characterized by special types of graphs where cycles are not allowed, i.e. acyclic graphs or trees. The focus of this chapter is on the frequent pattern mining problem where the underlying structure of the data can be of general graph type where cycles are allowed. These kinds of representations allow one to model complex aspects of the domain such as chemical compounds, networks, the Web, bioinformatics, etc. Generally speaking, graphs have many undesirable theoretical properties with respect to algorithmic complexity. In the graph mining problem, the common requirement is the systematic enumeration of sub-graphs from a given graph, known as the frequent subgraph mining problem. From the available graph analysis methods, we will narrow our focus to this problem as it is the prerequisite for the detection of interesting associations among graph-structured data objects, and has many important applications. For an extensive overview of graph mining in a general context, including different laws, data generators and algorithms, please refer to (Chakrabati & Faloutsos 2006; Washio & Motoda 2003, Han & Kamber 2006). Due to the existence of cycles in a graph, the frequent subgraph mining problem is much more complex than the frequent subtree mining problem. Even though theoretically it is an NP complete problem, in practice, a number of approaches are very applicable to the analysis of real-world graph data. We will look at a number of different approaches to the frequent subgraph mining problem and a number of approaches for the analysis of graph data in general.

The rest of the chapter is organized as follows. The necessary concepts related to the graph mining problem are discussed in Section 11.2. The graph isomorphism problem is addressed in Section 11.3, which is an important aspect of the frequent subgraph mining process. In Section 11.4, an overview is given of some existing graph mining methods where they have been categorized according to the main underlying approach to the problem. The chapter is concluded in Section 11.5.

F. Hadzic, H. Tan, and T.S. Dillon: Mining of Data with Complex Structures, SCI 333, pp. 287–300.
springerlink.com

11.2 General Graph Concepts and Definitions

A graph is a set of nodes where each node can be connected to another node including itself. There are many types of graphs, such as directed/undirected, weighted, finite/infinite, regular and complete graphs. Often, these types exist for specialized applications where the specific relationships or constraint hold, or are to be enforced to hold. However, most semi-structured documents where the underlying structure is a graph, can be modeled as a general graph with undirected and unlabeled edges. In that sense, a graph can be denoted as $G = (V,L,E)$, where (1) V is the set of vertices or nodes; (2) L is a labelling function that assigns a label $L(v)$ to every vertex $v \in V$; and (3) $E = \{(v_1,v_2) | v_1,v_2 \in V\}$ is the set of edges in G. The number of nodes in G (i.e. $|V|$) is called the *order* of G while the number of edges ($|E|$) is referred to as the *size* of G. Each edge usually has two nodes associated with it, but it is also possible that only a single node is associated with it in the case when there is an edge from one node to itself (i.e. a cycle). If two vertices v_1 and v_2 are connected by an edge, then they are said to be *adjacent* to one another, and *nonadjacent* or *independent*, otherwise. Two edges that meet at a vertex are said to be *adjacent*, and *nonadjacent* otherwise. In other words, if two edges (v_{a1}, v_{a2}) and (v_{b1}, v_{b2}) are adjacent and $(v_{a1} \neq v_{a2})$ and $(v_{b1} \neq v_{b2})$, then one of the following holds: $(v_{a1} = v_{b1})$ or $(v_{a1} = v_{b2})$ or $(v_{a2} = v_{b1})$ or $(v_{a2} = v_{b2})$. The set of all vertices adjacent to a vertex v corresponds to the neighbourhood of v. A *path* is defined as a finite sequence of edges between any two nodes and, as opposed to a tree where there is a single unique path between any two nodes, in a graph there could be multiple paths. The *length of a path p* is the number of edges in p. The fan-out/degree of a node is the number of edges emanating from that node.

A graph *G'* is a subgraph of another graph G if there exists a subgraph isomorphism between *G'* and G. Isomorphism is a one-to-one and onto mapping from the node set of one graph to the node set of another where the adjacency and non-adjacency is preserved.

As is the case with trees, there are a number of different types of subgraphs and some of the most common ones in the context of frequent subgraph mining are discussed next together with a formulization of the subgraph isomorphism problem.

11.3 Graph Isomorphism Problem

Two graphs $G_1(V_1,L_1,E_1)$ and $G_2(V_2,L_2,E_2)$ are said to be isomorphic to one another if there exists a structure preserving vertex bijection $f{:}V_1 \rightarrow V_2$ such that $(v_1, v_2) \in E_1$ iff $(f(v_1), f(v_2)) \in E_2$. Hence, two labeled graphs $G_1(V_1,L_1,E_1)$ and $G_2(V_2,L_2,E_2)$ are isomorphic to each other if there is one-to-one mapping from V_1 to V_2 that preserves vertices, labels of vertices and adjacency and non-adjacency of the vertices.

As mentioned earlier, a graph is a subset of another graph if there exists a subgraph isomorphism between them. More formally, this can be stated as follows:

A graph G_S (V_S,L_S,E_S) is a *subgraph* of graph G (V,L,E) iff $V_S \subseteq V$ and $E_S \subseteq E$.

The frequent subgraph mining problem can be generally stated as: Given a database of graphs G_{DB} and a minimum support threshold (σ), extract all subgraphs that occur at least σ times in G_{DB}.

In addition to the general subgraph definition provided above, other common subgraph types are spanning, connected and induced subgraphs.

A graph G_S (V_S,L_S,E_S) is a *spanning* subgraph of graph G (V,L,E) iff $V_S = V$ and $E_S \subseteq E$.

A graph G_S (V_S,L_S,E_S) is a *connected* subgraph of graph G (V,L,E) iff $V_S \subseteq V$ and $E_S \subseteq E$, and all vertices in V_S are mutually reachable through some edges in E_S.

A graph G_S (V_S,L_S,E_S) is an *induced* subgraph of graph G (V,L,E) iff $V_S \subseteq V$, $E_S \subseteq E$, and there exists a mapping $f{:}V_S \rightarrow V$ such that for any pair of vertices v_x and $v_y \in V_S$ if there is an edge $(f(v_x), f(v_y)) \in E \Leftrightarrow (v_x, v_y) \in E_S$.

To illustrate these aspects, please consider Fig. 11.1 where we show an example graph database G_{DB} consisting of three transactions. The different types of subgraphs of transaction 1 are shown in Fig. 11.2, and some frequent (general) subgraphs for support 2 and 3 are shown in Figure. 11.3.

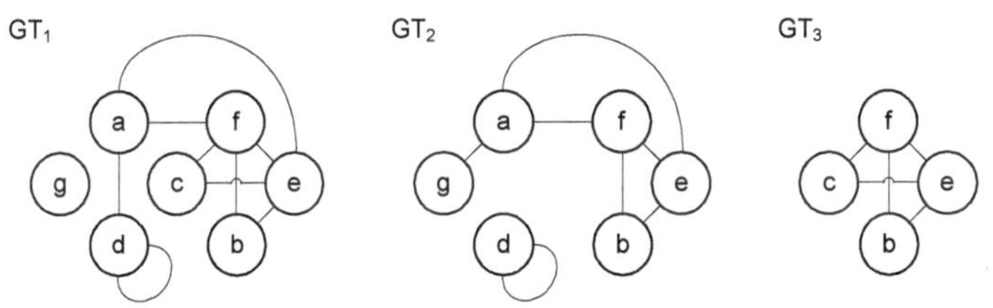

Fig. 11.1 Example graph database G_{DB} consisting of 3 transactions GT_1, GT_2 and GT_3

In this section, we narrowed our focus to the most commonly used subgraph definitions within the frequent pattern mining problem from graph-structured data in the data mining field. There are quite a few different variations of graphs considered in the large research problem of analysis of graph-structured data, as well as different types of measures and constraints that can be imposed on the analysis task. For details about these aspects and other related issues, we refer the interested reader to (Chakrabati & Faloutsos 2006, Washio & Motoda 2003). In the next section, we consider a variety of methods developed for the frequent graph mining problem. In some of these methods, constraints have been imposed to either orient the analysis better toward a specific application aim, or to decrease the number of patterns enumerated to alleviate the complexity problem introduced by mining graph-structured data.

11.4 Existing Graph Mining Methods

A number of graph-based data mining methods have been developed. In this section, we will focus mainly on the methods developed for solving the frequent subgraph

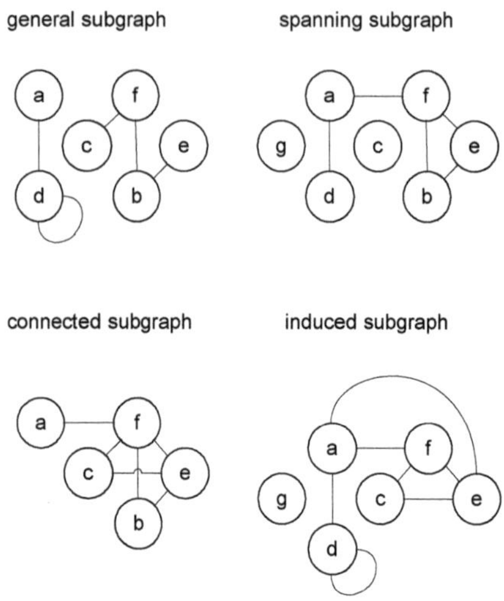

Fig. 11.2 Example subgraphs of graph representing transaction 1 (GT_1) from graph database G_{DB} of Fig. 11.1

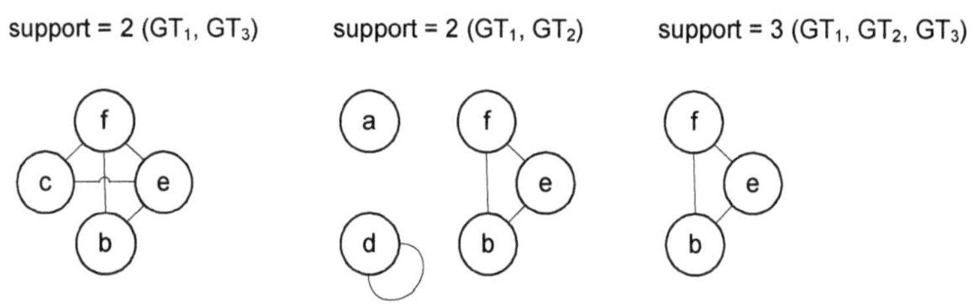

Fig. 11.3 Some frequent general subgraphs from graph database G_{DB} of Fig. 11.1

mining task explained earlier in Section 11.3. At the end of the section we overview a number of methods developed for solving related problems to frequent subgraph mining and the analysis of graph data in general.

11.4.1 Apriori-Like Methods

This section focuses on the graph mining methods which were developed using the same principle of the Apriori-based frequent itemset/sequence/subtree mining. The candidate enumeration process is performed in a bottom-up manner, starting with subgraphs consisting of one node (i.e. 1-subgraphs). At each iteration, the candidate k-subgraphs are checked for frequency and only the frequent patterns are used for generating (k+1)-subgraphs. There are a number of variations among the Apriori-based algorithms in regards to the way the candidates are generated. As was discussed earlier in the book, the join approach that works so well for the frequent itemset mining may not be so suitable for application to cases where the structural properties of data patterns need to be taken into account. Many candidates that are invalid are unnecessarily generated and pruned. In frequent subgraph mining, this is even more so the case, due to the many ways in which the join operation on two substructures can be performed.

The AGM algorithm (Inokuchi, Washio & Motoda, 2000) adopts the level-wise Apriori approach where graphs are represented using an adjacency matrix. Starting from the smallest subgraphs, the candidates are enumerated by performing the join operation on the adjacency matrices representing candidate subgraphs. A canonical form for induced subgraphs is defined according to which each adjacency matrix needs to be sorted so that the frequency can be correctly determined. The FSG algorithm (Kuramochi & Karypis 2001), was developed to extract frequent connected undirected subgraphs. To minimize storage and computational processing, FSG uses a sparse graph representation to effectively store input transactions, candidate and frequent subgraphs. They perform canonical form labelling of adjacency matrix and then convert it to adjacency-list representation. To generate candidate (k+1)-subgraphs from the frequent k-subgraphs, the join operation is performed on the frequent k-subgraphs that contain the same (k-1)-subgraph. The frequency of the newly generated (k+1)-subgraph is determined as the size of the intersection of the transactional identifier lists (TID lists) of the joined k-subgraphs. The use of the TID lists simplifies the candidate enumeration and counting process in FSG, but the storage requirements of all the TID lists for large graph data may cause memory problems. The same authors (Kuramochi & Karypis 2002), have proposed the gFSG algorithm which is an extension to the problem of finding frequently occurring geometric patterns in geometric graphs. These geometric graphs are graphs whose vertices have two or three dimensional coordinates associated with them. The gFSG also uses a level-wise approach extending frequent subgraphs by one edge at a time, and a number of algorithms are integrated for computing the isomorphism between geometric subgraphs, that are rotation, scaling and translation invariant.

Another Apriori-based algorithm has been proposed in (Vanetik, Gudes & Shimony, 2002). It uses canonical representations of paths and path sequences, and a lexicographical ordering over path pairs is defined based on node labels and degrees of nodes within paths. A graph is represented as a union of *edge-disjoint paths*, where two paths are edge-disjoint if they do not share any common edge. A bottom-up approach is used where firstly, all frequent subgraphs consisting of single path are found and these are combined, where possible, to construct subgraphs consisting of two paths. The algorithm progressively enumerates subgraphs consisting of k paths, by joining frequent subgraphs with k-1 paths.

11.4.2 Pattern-Growth Methods

The commonality among the algorithms that adopt an Apriori-like approach is that subgraphs are systematically enumerated in a bottom-up manner where problems can occur when the data is too large or the structural relationships are fairly complex. This is especially the case when the candidates are generated using the join approach. There are many possibilities where two subgraphs can be joined, and the structural variations of a candidate subgraph may not always exist in the database. The subgraph isomorphism is an expensive test, and hence, generating candidates which then need to be pruned is wasteful. These problems have motivated the development of a number of algorithms which adopt more of a graph-structure-guided approach to minimize the unnecessary candidate subgraphs generated.

The gSpan algorithm (Yan & Han, 2002) was the first approach that adopts a depth-first based search for frequent subgraphs. The search is supported by a novel canonical labelling system. Each graph is mapped to a sequential code and all codes are sorted according to the ascending lexicographical order. A depth-first search is then applied to the trees matching the first nodes of the codes and progressively adding more frequent descendants. This subgraph increase stops when the support of a graph becomes less than the minimum support or when it has already been discovered earlier in the method. The gSpan algorithm is very efficient in terms of the time taken and the memory required.

The algorithm presented in (Borgelt & Berthold 2002), focuses on the discovery of frequent substructures in molecular compounds and also uses a depth-first search strategy to find frequent subgraphs. The occurrence of a fragment within all molecules is kept in parallel during the pattern (fragment) growth process. The local order of the atoms and bonds is used to prune the unnecessary fragment extensions which speeds up the search and avoids the generation of redundant patterns. The FFSM algorithm (Huan, Wang & Prins 2003) uses an algebraic graph framework for an unambiguous enumeration of frequent subgraphs. Each graph is represented using an adjacency matrix and the subgraph isomorphism is avoided by exploiting the embeddings of each frequent subgraph. The GASTON (GrAph/Sequence/Tree extractiON) algorithm (Nijssen & Kok 2004) has been developed with the underlying idea of a quick start for search of frequent graph structures by integrating the search for frequent paths, trees and graphs into one approach. A level-wise approach is used that first enumerates simple paths, followed by tree structures and the most complex graph structures at the end. This kind of approach is motivated by the observation that in practice, most graphs are not actually so complex and the number of cycles is not too large. Hence, in the enumeration phase, sequences/paths are enumerated first and then tree structures are enumerated by repeatedly adding one node and an incident edge to the nodes in the path. Graph structures are then enumerated by adding edges in between the nodes of the tree structure.

11.4.3 Inductive Logic Programming (ILP) Methods

The approaches that fall within this category are characterized by the use of the ILP to represent graph data using horn clauses. They come from the multi relational data mining field, which is essentially the mining of data that is organized in multiple interlinked tables of a relational database.

The WARMR algorithm (Dehaspe & Toivonen 1999) has been developed for frequent query extraction from a DATALOG database. Compared with standard frequent pattern mining methods, an extra parameter (*key*) is used, which is essentially an attribute that must be contained in all of the patterns extracted. It allows the user to restrict the search space to only those patterns that contain the item of interest. The algorithm is flexible in the types of patterns that can be extracted. A declarative language bias (WRMODE) is formulized that restricts the search space to only the admissible and potentially interesting queries. The main processes of the algorithm are candidate generation and candidate evaluation. The candidate evaluation step is used to determine the frequency of the generated candidates. In the candidate generation step, the patterns that contain the specified key are found and extended incrementally, i.e. starting from the smallest patterns. At each step, the frequent and infrequent patterns are determined and the frequent patterns are extended by adding one node at a time. The patterns are pruned if they are either already contained in the frequent pattern set, or if they are a specialization of a pattern contained in the infrequent pattern set. The method was applied to the analysis of telecommunication alarm and chemical toxicology. The major shortcoming of the algorithm is its efficiency since the equivalence test between first-order clauses is very complex. This has motivated Nijssen & Kok (2001) to propose an alternative solution for this expensive step. Their algorithm FARMER still uses the first order logic notation and is by and large very similar to WARMR, with a substantial time performance gain. The main difference is that instead of the expensive equivalence tests, the FARMER utilizes a tree data structure to both count and generate candidate queries in an efficient manner. De Raedt & Kramer (2001) developed a method to find frequent molecular fragments from a molecular compound database. A molecular fragment is defined as a sequence of linearly connected atoms, and the database contains information of the elements of the atoms of a molecule and bond orders. The approach allows the specification of a generality constraint so that all the found molecular fragments satisfy a conjunction of primitive constraints. In addition to the traditionally used minimum frequency parameter, their approach allows the use of the maximum frequency constraint on patterns. These constraints allow more flexibility in the types of queries that can be answered through the patterns, and in the molecular fragment finding domain, more interesting patterns could be found. Another ILP-based approach has been presented in (Lisi & Malerba 2004) and a hybrid language is proposed to retain information regarding the relational as well as structural properties in the multi-level association rules mined from multiple relations. Query subsumption relationships are utilized to define a generality order and a downward refinement operator.

11.4.4 Greedy Search Methods

The characteristics of the greedy search-based methods are that they avoid excessive computation required for the search of all frequent subgraphs, and hence use a greedy approach to reduce the number of subgraphs considered. The high complexity of the graph isomorphism problem is avoided at the cost of missing some frequent subgraphs.

One of the first graph mining approaches is known as the SUBDUE system (Cook & Holder 1993) and many refinements and optimizations of this system were made (Cook & Holder 2000; Cook et al. 2001; Jonyer, Holder & Cook 2002; Noble

& Cook 2003; Holder et al. 2003; Ketkar, Holder & Cook 2005). The SUBDUE system is based on the minimum description length (MDL) principle, measured as the number of bits necessary to completely describe the graph. Instances of discovered subgraphs are replaced by concept definitions. This compresses the original dataset and provides a basis for discovering hierarchically-defined structures. The motivation of this kind of approach is to identify conceptually interesting substructures that enhance the interpretation of the data. Besides using the MDL principle, the SUBDUE system allows for the incorporation of other background knowledge to focus the search on more suitable subgraphs. A greedy beam search approach is used to discover candidate subgraphs from the data. It starts from single nodes and iteratively expands the instances of substructures with one neighboring edge at a time to cover all possible expansions. Each newly generated candidate subgraph is then expanded and the algorithm considers all possible substructures in the search for the best substructures according to the minimum description length. The algorithm stops when either all possible substructures have been considered or the computational limit has been reached. An optional pruning technique is used to avoid expansions of those substructures whose description length would increase. In its search for matching substructures, SUBDUE utilizes an inexact graph match algorithm to allow for slight variations since interesting substructures may often show up in a slightly different form throughout the data.

Similarly, the Graph Based Induction (GBI) method (Yoshida, Motoda & Indurkhya 1994) compresses the graph data in order to arrive at interesting subgraphs with minimal size. It was initially developed to find interesting concepts from frequent patterns found in the inference trace. The compression technique used is the so called pair-wise chunking, where two nodes are paired (chunked) into one node. The links between paired nodes are removed and if necessary the links to other nodes in the graph are 'remembered' so that at any time during the search, the original graph can be reconstructed. The pair-wise chunking can be nested and the graph can be repeatedly compressed. A minimal size is chosen to limit the amount of compression which reflects the sizes of extracted patterns and the compressed graph. The search for local subgraph candidates is performed using an opportunistic beam search.

11.4.5 Other Methods

The methods discussed in the previous sections all belong to the family of frequent substructure (subgraph) mining methods. The amount of research that has gone into automatic analysis of graph data in general is enormous and for a broad overview of the different graph mining laws, constraints, specialized applications and algorithms please refer to (Chakrabarti & Faloutsos, 2006; Washio & Motoda, 2003, Han & Kamber 2006). This section will mention some alternative approaches to graph data mining, which are usually motivated by the different data analysis aims or specific application needs or constraints.

Clustering of graph-structured data is generally important in many applications, as discovered clusters often provided the basis for analysis of similarities and differences, and can be used to classify graph-structured data objects based on the characteristics of the formed clusters. For example, the graph clustering technique based on the network flow theory has been applied to the tissue segmentation of magnetic resonance images of the human brain in (Wu & Leahy 1993). The data is represented as an undirected adjacency edge where each edge is assigned a flow capacity indicating the similarity of the nodes connected by the edge. Clusters are formed by the continuous removal of edges from the graph until mutually exclusive subgraphs are formed. The edges are removed based on their flow capacity with the aim of minimizing the largest maximum flow (similarity) among the formed subgraphs (clusters). Mancoridis et al. (1998) have developed a number of clustering techniques for the automatic recovery of the modular structure of a software system from its source code. The source code is represented as a module dependency graph and a combination of clustering, hill-climbing and genetic algorithms are used to discover the high level structure of the systems organization.

A number of graph-clustering algorithms are focused on optimizing specific clustering criteria that occur in the context of graph clustering, and often borrow methods from the more general graph optimization problems. For example, a graph theory-based cluster analysis approach, has been presented in (Hartuv & Shamir 2000). A similarity graph is defined and clusters are defined as subgraphs with connectivity that is above half the number of vertices, and the algorithm has low polynomial complexity. The graph-clustering algorithm presented in (Flake, Tarjan and Tsioutsiouliklis 2004) is based on the general idea of maximum flow techniques to maximize intra-cluster similarity and minimize inter-cluster similarity. An artificial sink is placed in the graph that is connected to all the nodes, and the maximum flows are calculated between all the nodes and the sink. A parameter is chosen to limit the number of edges used in connecting the artificial sink to the other nodes of the graph. The clustering is based on minimum cut trees. A minimal cut tree is a subgraph of the original graph where the path between two given nodes in the minimal cut tree, is guaranteed to be the smallest path between those two nodes in the original graph. Hence, by selecting minimum cut trees from the expanded graphs, the algorithm discovers quality clusters and heavily connected components. A kernel function between two graphs has been proposed in (Kashima, Tsuda & Inokuchi 2003). The graphs are represented as a feature vector by counting the *label paths* that appear in the graph. The label paths are produced by random walk functions on the graph, and the kernel is defined as the inner product of the feature vectors averaged over all possible label paths. The computation of the kernel becomes the problem of finding the stationary state of a discrete-time linear system, and the solution becomes that of solving simultaneous linear equations with a sparse coefficient matrix.

Most of the clustering approaches just described will not completely solve the frequent subgraph mining problem as defined in Section 11.3, as it is not guaranteed that the discovered subgraphs using clustering approaches will contain all the frequent subgraphs for a given support. Hence, they provide an approximate solution, and can often overcome some of the complexity limitations that occur when all frequent subgraphs need to be completely enumerated. This approximation can be useful in certain applications, as often, discovering all the frequent subgraphs for a given support will result in too many patterns that are not of interest for the application at hand,

Another approximate frequent subgraph mining method based on spanning trees has been proposed in (Zhang, Yang & Cheedella 2007). As mentioned in Section 11.3, a spanning subgraph of a graph must contain all its vertices. A spanning tree is a spanning subgraph that is a tree (i.e. no cycles). In other words, a spanning tree from an undirected, connected graph *G*, is a selection of edges from *G* that span every vertex from *G*, with no cycles. An example spanning tree is shown in Fig. 11.4. The Monkey algorithm (Zhang, Yang & Cheedella 2007), is motivated by the observation that in cases of edge distortion, one needs to allow for some flexibility in the graph isomorphism checking in order to detect these frequent patterns in spite of the variation on edges. This flexibility in the isomorphism checking is introduced by searching for frequent spanning trees since, in a spanning subtree, all the vertices must be present while some of the edges can be absent as long as the subgraph is still a tree. Monkey is based on depth-first search and two constraints are integrated to allow one to control the unrelated edges in affecting the frequency of corresponding patterns.

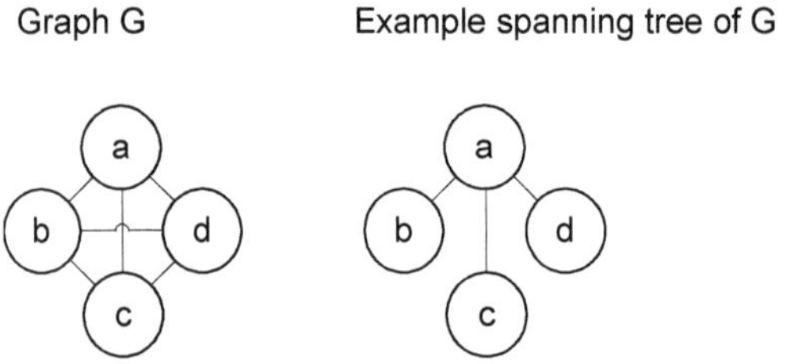

Fig. 11.4 An example spanning tree of a graph *G*

The problem of mining large disk-based graph databases has been addressed in (Wang et al. 2004). The main contribution is the *ADI* (adjacency index) structure to support major tasks in graph mining when the graph databases cannot be held in main memory. The previously discussed gSpan algorithm (Yan & Han, 2002) was adapted to use the *ADI* structure, and the resulting algorithm *ADI-Mine* is shown to outperform gSpan when large disk-based graph databases are in question. The ADI-Mine algorithm and a number of constraint-based mining techniques have been integrated together to form the GraphMiner system (Wang et al. 2005). The system provides a graphical user interface so that a number of constraints can be selected, such as: pattern size constraints, the edges/vertices that must appear or not appear in the patterns, graphs of which the patterns must be super- or sub-patterns and aggregate constraints. Furthermore, the system allows for easy browsing and analysis of patterns and querying capabilities to focus on the application-specific or interesting patterns.

Saigo & Tsuda (2008) have proposed an iterative subgraph mining method for principal component analysis, where salient patterns are progressively collected by separate graph mining calls. At each graph mining call, real-value weights are assigned to graph transactions, and patterns satisfying the weighted support threshold are enumerated. The search strategy is based on a branch-and-bound algorithm that uses the depth-first search code tree as the canonical search space. The patterns are organized as trees, such that a child node has a supergraph of the pattern in its parent node. Patterns are enumerated by systematically generating patterns from the root to the leaves in a recursive manner using right-most-node expansion.

11.4.6 Mining Closed/Maximal Subgraph Patterns

Similar to other frequent pattern mining problems, the mining of frequent closed and maximal subgraphs is an important area of research as it is one of the ways to reduce the complexity problem introduced by enumerating all of the frequent subgraphs.

The CloseGraph algorithm (Yan & Han 2003), mines closed frequent subgraphs and its overall framework is similar to the gSpan algorithm (Yan & Han 2002) with the addition of suitable graph pruning techniques to effectively reduce the search space and enumerate only closed subgraphs. The algorithm starts by pruning all infrequent nodes and edges, and then generates all frequent graphs consisting of a single node. It then extends the set of frequent graphs recursively using the depth-first search and right-most extension. The depth-first search is based on the DFS lexicographical order, i.e the novel canonical labelling system first introduced in their gSpan algorithm. The authors define an early termination condition based upon the equivalence of occurrence of frequent subgraphs in the original graph, so that not all the graphs need to be extended and checked for satisfying the closed subgraph property. This early termination condition is not valid for all subgraphs and another condition is imposed to detect such cases in order to avoid loss of information regarding potentially frequent close subgraphs. These conditions allow the CloseGraph algorithm to detect conditions in which all of descendant subgraphs of a frequent graph cannot be closed and need not be checked.

The problem of mining frequent closed subgraphs with connectivity constraints in relational graphs, has been addressed in (Yan, Zhou & Han 2005), and two algorithms, CloseCut and Splat, have been proposed. The CloseCut algorithm is a pattern-growth approach, i.e. small frequent subgraphs are first enumerated and extended by adding new edges. The candidate subgraphs generated by this process must satisfy the defined connectivity constraints and the minimum frequency threshold. The candidate graphs are extended by adding new edges until the resulting graph after edge addition is no longer frequent. The Splat algorithm, on the other hand, is a pattern-reduction based approach, whereby the relational graphs are intersected and decomposed in order to obtain highly connected graphs. At each step, the algorithm checks whether a newly generated graph exists in the results set, in which case it needs to no longer be processed, since no closed highly connected graphs can be enumerated from that candidate graph.

The MARGIN algorithm (Thomas, Valluri & Karlapalem 2006) mines frequent maximal subgraphs in a top-down manner. For each graph record of a database of graphs, the algorithm repetitively drops one edge at a time, without generating disconnected graphs, until a frequent representative of that graph record is found. This frequent representative is likely to be a frequent maximal pattern once a number of infrequent edges and nodes have been removed. Hence, a recursive procedure is called on the generated frequent connected subgraphs, which progressively cuts infrequent edges and nodes from the given subgraph until it satisfies the properties of a frequent maximal subgraph.

11.5 Conclusion

This chapter has given a brief introduction to the problem of graph mining, which is not restricted to acyclic graphs which is the main focus of this book. Some formal definitions were given with an illustrative example to allow the reader to gain some understanding of the complexity of the problem and some of the undesirable properties causing this. This complexity also explains the reason behind so many different approaches to solving the graph mining problem as was discussed in Section 11.4. We can see that the novelty often comes when the problem is alleviated to some extent such as is the case in GASTON algorithm (Nijssen & Kok 2004) which gets a quick start by first discovering simpler patterns and increasing the complexity by moving from sequence, trees and finally to cyclic graphs. However, it is worth mentioning that even though the graph mining problem appears theoretically infeasible, in practical situations the existing algorithms can often complete the task within a reasonable amount of time, on real-world graph-structured data. Applications of graph mining are manifold and in general they are useful for analysis of data in any domain, given that the data objects are organized in a graph structure. When the algorithms were discussed previously, we mentioned a few application areas, and we can generally state that the applications are somewhat similar to the many applications of tree mining discussed in Chapter 9, with the main difference being that in graph mining the application data can contain cycles among the data objects. Some other example applications are chemical compound analysis, social network analysis, web structure mining, ontology mining, and in general graph mining techniques are useful for many web and social-media applications.

References

1. Bern, M., Eppstein, D.: Approximation Algorithms For Geometric Problems. In: Hochbaum, D.S. (ed.) Approximation Algorithms for NP-Hard Problems, pp. 296–345. PWS Publishing Company (1996)
2. Borgelt, C., Berthold, M.R.: Mining Molecular Fragments: Finding Relevant Substructures of Molecules. Paper presented at the Proceedings of the 2002 IEEE International Conference on Data Mining (ICDM), Maebashi City, Japan, December 9-12 (2002)
3. Chakrabarti, D., Faloutsos, C.: Graph mining: Laws, Generators and Algorithms. ACM Computing Surveys 38(1), 2-es (2006)

4. Cook, D.J., Holder, L.B.: Substructure Discovery Using Minimum Description Length and Background Knowledge. Journal of Artificial Intelligence Research 1(1), 231–255 (1993)
5. Cook, D.J., Holder, L.B.: Graph-Based Data Mining. IEEE Transactions on Intelligent Systems 15(2), 32–41 (2000)
6. Cook, D.J., Holder, L.B., Galal, G., Maglothin, R.: Approaches to Parallel Graph-Based Knowledge Discovery. Journal of Parallel and Distributed Computing 61(3), 427–446 (2001)
7. De Raedt, L., Kramer, S.: The levelwise version space algorithm and its application to molecular fragment finding. Paper presented at the Proceedings of the 17th International Joint Conference on Artificial intelligence, Seattle, WA, USA, August 4-10 (2001)
8. Dehaspe, L., Toivonen, H.: Discovery of frequent DATALOG patterns. Data Mining and Knowledge Discovery 3(1), 7–36 (1999)
9. Flake, G.W., Tarjan, R.E., Tsioutsiouliklis, K.: Graph Clustering and Minimum Cut Trees. Internet Mathematics 1(4), 385–408 (2004)
10. Han, J., Kamber, M.: Data Mining: Concepts and Techniques, 2nd edn. Elsevier, Morgan Kaufmann Publishers, San Francisco, CA, USA (2006)
11. Hartuv, E., Shamir, R.: A Clustering Algorithm Based on Graph Connectivity. Information Processing Letters 76(4-6), 175–181 (2000)
12. Holder, L.B., Cook, D.J., Djoko, S.: Substructure Discovery in the SUBDUE System. Paper presented at the Proceedings of the AAAI Workshop on Knowledge Discovery in Databases, Seattle, Washington, USA, July 31- August 4 (1994)
13. Holder, L., Cook, D., Gonzalez, J., Jonyer, I.: Structural Pattern Recognition in Graphs. In: Chen, D., Chen, X. (eds.) Pattern Recognition and String Matching, pp. 255–279. Kluwer Academic Publishers, Dordrecht (2003)
14. Huan, J., Wang, W., Prins, J.: Efficient mining of frequent subgraph in the presence of isomorphism. Paper presented at the Proceedings of the 3rd IEEE International Conference on Data Mining (ICDM 2003), Melbourne, Florida, USA, December 19-22 (2003)
15. Inokuchi, A., Washio, T., Motoda, H.: An apriori-based algorithm for mining frequent substructures from graph data. Paper presented at the Proceedings of the 4th European Conference on Principles and Practice of Knowledge Discovery in Databases, Lyon, France, September 13-16 (2000)
16. Jonyer, I., Holder, L.B., Cook, D.J.: Graph-based hierarchical conceptual clustering. Journal of Machine Learning Research 2, 19–43 (2002)
17. Kashima, H., Tsuda, K., Inokuchi, A.: Marginalized kernels between labeled graphs. Paper presented at the Proceedings of the 20th International Conference on Machine Learning (ICML 2003), Washington, DC, USA, August 21-24 (2003)
18. Ketkar, N.S., Holder, L.B., Cook, D.J.: Subdue: compression-based frequent pattern discovery in graph data. Paper presented at the Proceedings of the ACM SIGKDD 1st International Workshop on Open source Data Mining, Chicago, Illinois, USA, August 21-24 (2005)
19. Kuramochi, M., Karypic, G.: Frequent Subgraph Discovery. Paper presented at the Proceedings of the IEEE International Conference on Data Mining (ICDM 2001), San Jose, California, USA, November 29 - December 2 (2001)
20. Kuramochi, M., Karypis, G.: Discovering Frequent Geometric Subgraphs. Paper presented at the Proceedings of the 2nd IEEE International Conference on Data Mining (ICDM 2002), Maebashi City, Japan, December 9-12 (2002)
21. Lisi, F.A., Malerba, D.: Inducing Multi-Level Association Rules from Multiple Relations. Machine Learning 55(2), 175–210 (2004)

22. Mancoridis, S., Mitchell, B., Rorres, C., Chen, Y., Gansner, E.: Using Automatic Clustering to Produce High-Level System Organizations of Source Code. Paper presented at the Proceedings of the 6th International Workshop on Program Comprehension (IWPC 1998), Los Alamitos, CA, USA, June 26 (1998)
23. Nijssen, S., Kok, J.N.: A Quickstart in Frequent Structure Mining Can Make a Difference. Paper presented at the Proceedings of the, International Conference on Knowledge Discovery and Data Mining (KDD 2004), Seattle, WA, USA, August 22-25 (2004)
24. Noble, C.C., Cook, D.J.: Graph-based anomaly detection. Paper presented at the Proceedings of the 9th ACM SIGKDD International Conference on Knowledge Discovery and Data Mining, Washington, DC, USA, August 24-27 (2003)
25. Saigo, H., Tsuda, K.: Iterative Subgraph Mining for Principal Component Analysis. Paper presented at the Proceedings of the 8th IEEE International Conference on Data Mining (ICDM 2008), Pisa, Italy, December 15-19 (2008)
26. Thomas, S., Sarawagi, S.: Mining Generalized Association Rules and Sequential Patterns using SQL Queries. In: Proc. 4th Intl. Conf. on Knowledge Discovery and Data Mining (KDD 1998), pp. 344–348 (1998)
27. Vanetik, N., Gudes, E., Shimony, S.E.: Computing Frequent Graph Patterns from Semistructured Data. Paper presented at the Proceedings of the 2002 IEEE International Conference on Data Mining (ICDM 2002), Maebashi City, Japan, December 9-12 (2002)
28. Wang, C.W., Pei, J., Zhu, Y., Shi, B.: Scalable Mining of Large Disk-Based Graph Databases. Paper presented at the Proceedings of the 10th ACM SIGKDD International Conference on Knowledge Discovery and Data Mining, Seattle, WA, USA, August 22-25 (2004)
29. Wang, W., Wang, C., Zhu, Y., Shi, B., Pei, J., Yan, X., Han, J.: GraphMiner: a structural pattern-mining system for large disk-based graph databases and its applications. Paper presented at the Proceedings of the, ACM SIGMOD International Conference on Management of Data, Baltimore, Maryland, USA, June 14-16 (2005)
30. Washio, T., Motoda, H.: State of the art of graph-based data mining. ACM SIGKDD Explorations Newsletter 5(1), 59–68 (2003)
31. Wilson, R., Hancock, E., Luo, B.: Pattern vectors from algebraic graph theory. IEEE Transactions on Pattern Analysis and Machine Intelligence 27(7), 1112–1124 (2005)
32. Yan, X., Han, J.: gSpan: Graph-based substructure pattern mining. Paper presented at the Proceedings of the, IEEE International Conference on Data Mining (ICDM), Maebashi City, Japan, December 9-12 (2002)
33. Yan, X., Han, J.: CloseGraph: mining closed frequent graph patterns. In: Proceedings of the 9th ACM SIGKDD International Conference on Knowledge Discovery and Data Mining, Washington, DC, USA, August 24-27, pp. 286-295 (2003)
34. Yan, X., Zhou, X.J., Han, J.: Mining Closed Relational Graphs with Connectivity Constraints. Paper presented at the Proceedings of the 11th ACM SIGKDD International Cofnerence on Knowledge Discovery and Data Mining (KDD 2005), Chicago, Illinois, USA, August 21-24 (2005)
35. Yoshida, K., Motoda, H., Indurkhya, N.: Graph-based induction as a unified learning framework. Journal of Applied Intelligence 4(3), 297–316 (1994)
36. Zhang, S., Yang, J., Cheedella, V.: Monkey: Approximate Graph Mining Based on Spanning Trees. Paper presented at the Proceedings of the IEEE 23rd International Conference on Data Engineering (ICDE 2007), Istanbul, Turkey, April 15-20 (2007)

Chapter 12
New Research Directions

12.1 Introduction

This chapter will discuss some new research directions in the frequent subtree mining field. This will be discussed from both the application and technical perspectives. Since frequent subtree mining (FSM) is a relatively new field compared with frequent itemset/sequence mining, many lessons can be learned form the more mature research in frequent itemset/sequence mining. A drawback of frequent pattern mining in general is that often, for a set support threshold, the number of frequent patterns becomes quite large due to some characteristics of the database. This may cause not only algorithm complexity problems, but also significant delays in the analysis and interpretation of the results. Many of the patterns may not be useful for the application at hand and/or are redundant, or not of interest to the user. Furthermore, it is also not always clear what support threshold is satisfactory for obtaining reasonable results. These are all important research areas, with some significant achievements in complexity reduction from the algorithmic and application perspectives. Some of these or similar ideas can, to a certain extent, already be applied in the FSM field, but others will need refinements and extensions to be flexible enough to cope with the additional structural properties of the data. In Section 12.2, we highlight some of the work in frequent itemset/sequence mining where the same or similar idea can be applied and prove useful in the FSM field. At the end of Section 12.2, we look at some work that has already been initiated in frequent pattern filtering and the incorporation of application-oriented constraints.

This book has focused mainly on the traditional frequent pattern mining problem based on the support and confidence association rule mining framework. The approaches we have examined in this book tackle the problem using a systematic enumeration approach whereby all candidates satisfying the chosen thresholds will be extracted, and in this sense these can be viewed as exact approaches to the problem. In Section 12.3, we will explore the idea of enumerating frequent subtrees in a more approximate approach using a neural network clustering technique. One of the main difficulties is how to effectively represent the label and structural information contained in the tree-structured data for a network and how this additional structural

F. Hadzic, H. Tan, and T.S. Dillon: Mining of Data with Complex Structures, SCI 333, pp. 301–326.
springerlink.com

information is to be preserved in the learned knowledge. We therefore give a brief overview of some existing approaches for mining graph-structured data using neural networks, and discuss some promising properties of a Self-Organizing Map neural network to address the frequent pattern mining problem. Addressing the frequent subtree mining problem using a clustering technique will have some important differences and similarities with the traditional systematic enumeration approach, but it is important here to realize that they are governed by different laws for frequent pattern generation.

A tree, being a special type of a graph, causes the frequent subtree and frequent subgraph mining problems to be strongly related. Section 12.4 looks at future work in extending the tree-model guided framework to graph-model guided framework for mining frequent subgraphs. A potential way to apply the frequent subtree mining algorithms to analyze graph-structured data, is indicated.

This book has so far discussed data mining from its traditional perspective, where a method is applied to analyze only one type of data at a time. With the different types of data being used historically and in the present, related data and complementary information about a domain is found not only in one data source, but can often exist in part in relational, semi-structured or unstructured data. In such scenarios, it may not be sufficient to mine each of the data sources separately as some significant associations that exist when the information from different data sources is combined, may be missed. One needs to combine the information and mine it in a conjoint manner in order to arrive at this additional knowledge. In Section 12.5, we will look at some initiatives in this direction, and describe a conceptual framework for the conjoint mining of relational and semi-structured data.

In the last section, we describe an important application which we believe can make good use of the automated tree structured data analysis. We have already looked at a number of existing and well studied FSM applications in Chapter 9. In Section 12.6, we focus on ontology learning, where to the best of our knowledge, the use of FSM has not been explored. The areas within the broad application where FSM can advance the analysis task will be highlighted and a possible approach indicated. The chapter concludes with Section 12.7.

12.2 Frequent Pattern Reduction

As discussed in the introduction, there are a few important reasons for reducing the frequent pattern set. A number of constraints are often defined which need to be adhered to either during the pattern generation phase or in the post-processing phase where a rule is considered only when it satisfies some constraint or measure. This section starts with a brief overview of existing studies and techniques for frequent itemset/sequence mining in Section 12.2.1, which are believed to be useful for frequent subtree mining. In Section 12.2.2, we look at some recently proposed methods for reducing the frequent patterns set and incorporating constraints for frequent subtree mining.

12.2.1 Frequent Pattern Reduction and Rule Evaluation

As already discussed throughout the book, one way for reducing the frequent pattern set is by mining for maximal and/or closed frequent patterns, for which many methods have been developed for the different data types. In general, approaches to pattern reduction can be roughly split into pattern generation constraints and pattern/rule interestingness constraints. The pattern generation constraints are incorporated during the pattern generation phase and can lead to the saving of significant memory and execution time. Hence, it is one of the ways of handling a situation where the data set has such characteristics that the generation of the complete set of frequent patterns would be unfeasible. A discussion and detailed explanation of existing constraints is beyond the scope of this book, and we refer the interested reader to a more general data mining book (Han & Kamber 2006), where the pattern evaluation phase of data mining has been discussed in great detail and common techniques explained and illustrated with examples. The second half of this section is concerned with the various statistical and heuristic-based measures for evaluating the interestingness of rules.

12.2.1.1 Pattern Pruning, Focusing and Compression

One line of work that focuses on more interesting association rules, inserts meta-queries during the mining process which allows one to focus on patterns of interest to the application at hand (Shen et al. 1996). For example, the work described in (Klemettinen et al. 1994) first classifies the attributes of the original dataset to an inheritance hierarchy, and then uses the terms in that hierarchy to define rule templates (domain-oriented constraints) according to which the rule set is pruned. More recently, the problem of meta-rule guided mining of multiple-level association rules has been addressed in (Fu & Han 2005). The rules allow for syntactic constraints to be imposed and a top-down deepening technique is developed that guides the generation of multiple-level association rules with the pre-defined rules.

An *item constraint* has been incorporated in frequent itemset mining in (Srikant, Vu & Agrawal 1997) where algorithms were presented that integrate constraints as Boolean expressions that indicate the items that must be present or absent in the discovered patterns. The usefulness of incorporating *anti-monotone* and *succinct* constraints on the patterns during the candidate generation phase was first explored in (Ng. et al. 1998). Using anti-monotone properties, a user can specify SQL-like queries that need to be satisfied by all the patterns (e.g. avg_bidding_amount = 2000), and the algorithm will prune any itemsets at the earliest k-level (i.e. generating itemsets with k items) to avoid any generation of its supersets, since they can never satisfy the constraint. A succinct constraint works in the other direction whereby the algorithm will directly enumerate only those patterns that satisfy the user-supplied constraint, before the support counting even begins. Hence, integrating these constraints can often result in substantial reduction in memory requirements and execution time. This work has been extended in (Lakshmanan et al. 1999) and a *quasi-succinctness* constraint is defined which allows one to simultaneously impose constraints on the antecedent and precedent of a rule. Hence, the constraint is considered as a 2-variable where each variable constraint is enforced in a singular succinct constraint which are jointly used for pruning optimizations. Correlation among itemsets has been explored as a constraint in (Grahne, Lakshmanan & Wang 2000). DualMiner (Bucila, Gherke & Kifer 2003) combine both monotone and anti-monotone constraints simultaneously to effectively prune the search space. The convertible constraints have been defined in (Pei, Han & Lakshmanan 2001), with a few different categorizations, *convertible monotone*, *convertible anti-monotone* and *strongly convertible*. These convertible constraints are defined to allow one to incorporate additional constraints such as median function and other algebraic functions. These types of constraints do not exhibit properties such as monotonocity, and succinctness to be expressed by such constraints. The authors propose a strategy that imposes an appropriate order of items so that the convertible constraints can be converted into the ones exhibiting monotone behaviour, and the strategy is integrated inside a pattern-growth-based approach for frequent itemset mining. The Data-Peeler algorithm (Cerf et al. 2008) mines n-ary relations to extract all closed n-sets using a new class of constraints which generalizes both monotonic and anti-monotonic properties of a pattern. A new enumeration strategy is defined which allows for the easy checking of the closeness of a pattern. The framework for finding highly correlated association patterns, referred to as the *hyperclique patterns*, has been proposed in (Xiong, Tan & Kumar 2006). These hyperclique patterns are discovered using a *h-confidence* measure which is the minimum probability of an item from a pattern in one transaction implying the presence of all other items in the same transaction. Hence, hyperclique patterns will satisfy the minimum *h-confidence* threshold. In addition, the authors define a cross-support property of the *h-confidence* measure in order to avoid the generation of patterns with substantially different frequency, which have low correlation.

Some recent work has focused more on compressing the frequent pattern sets without losing significant amounts of information. The work presented in (Afrati, Gionis & Mannila 2004) is concerned with the problem of finding the k sets of frequent items that best approximate the information conveyed by the full pattern set. The measure used in their approach is defined based on the size of the collection covered by the k set and a bound on the extra sets allowed to be covered. Xin et al. (2005) incorporate two greedy measures inside a pattern-growth-based algorithm to find patterns that are close to minimal representatives of information in the set of frequent closed patterns. An efficient algorithm, TFP, was proposed in (Wang et al. 2005) for mining top k frequent closed itemsets with length larger or equal to a user specified threshold. Siebes, Vreeken & Leeuwen (2006) propose a minimum description length (MDL) -based approach to discard any redundant patterns. The compression is actually performed on the database itself, where the MDL principle is used to search for the subset of all frequent item sets that best compress the database. For a detailed overview of frequent pattern compression in general, we refer the interested reader to (Xin, Han, Yan & Cheng 2006).

In regards to the frequent pattern mining from sequential data, a number of important works exist for reducing the number of sequential patterns. The SPIRIT algorithm (Garofalakis, Rastogi & Shim 1999) is a sequence mining algorithm that allows the user to specify regular expression constraints, which are integrated with the sequence mining process for effective pruning. The PlanMine approach (Zaki, Lesh & Ogihara 1999) has been specifically developed for the mining of sequences of plan executions for patterns that predict plan failures. The pruning strategy consists of three phases to arrive at the most predictive rules. In the first phase, rules consistent with background knowledge are pruned as they do not contribute any novel information. Redundant patterns are then eliminated which are defined as those that have the same frequency as at least one of their super-sequences. In the last pruning phase, only the dominant patterns are retained which are defined as more predictive than any of their proper sub-sequences. A method for clustering of sequential data, CLUSEQ, has been proposed in (Yang & Wang 2003). It explores significant statistical properties of the sequences and calculates conditional probability distribution of the next occurring symbol given its preceding segment. This information is used to characterize the sequences and to support the similarity measure so that high quality clusters are obtained. An extensive study of the constraint-based sequence pattern mining has been provided in (Pei, Han & Wang 2007), where the different constraints that can be incorporated inside a pattern-growth-based approach have been compared and evaluated.

12.2.1.2 Rule Evaluation

Various objective interestingness criteria have been used to limit the nature of rules extracted, such as support and confidence (Agrawal, Imieliski & Swami 1993), collective strength (Aggarwal & Yu 1998), lift/interest (Brin et al. 1997), chi-squared test (Silverstein, Brin & Motwani 1998), correlation coefficient and log linear analysis (Brijs, Vanhoof & Wets 2003), leverage (Piatetsky-Shapiro 1991; Webb 2007), empirical Bayes correlation (Brijs, Vanhoof & Wets 2003), three alternative interest measures: any-confidence, all confidence, and bond (Omiecinski 2003), etc. Both (Bing, Wynne & Yiming 1999) and (Yun et al. 2003) proposed and successfully developed two approaches, namely multiple support and relative support for generating rules for significant rare data that appear infrequent in the database but are highly associated with specific data. Mutual information and J-Measures are common information theory approaches in objective interestingness measures (Tan, Kumar & Srivastava 2002).

A number of researchers have anticipated an assessment of pattern discovery by applying a statistical significance test. For example, in (Zhang, Padmanabhan & Tuzhilin 2004) focus is on significant statistical quantitative rules, in (Brin, Motwani & Silverstein 1997) correlation of rules is used, Liu, Hsu & Ma (1999) proposed a pruning and summarizing approach, in (Bay & Pazzani 2001) a statistical test with a correction for multiple comparison by using a Bonferroni adjustment is applied, Meggido & Srikant (1998) proposed an alternative approach of encountering rules by change and applying hypothesis testing, and (Webb 2003, 2007) summarizes holdout evaluation techniques. More recently, a unified framework was proposed in (Shaharee, Hadzic & Dillon 2009) which combines a statistical-heuristic feature selection criterion and several objective metrics to measure the interestingness and relevance of the extracted association rules, and remove any redundant, misleading and irrelevant rules. An overview of the latest development in discovering significant rules was provided in (Webb 2007), and some areas worthy of further exploration were indicated, which involve: issues concerning the optimal split between the subset of data used for training and testing, selection of an appropriate statistical test, and assessment of the rules with more than one itemset in the consequent (Webb 2007).

Another problem in association rule mining is that the discovered rules may reflect only those aspects of the database being observed, and not the general aspects of the domain. Assessing whether a rule satisfies a particular constraint is accompanied by a risk that the rule will satisfy the constraint with respect to the sample data but not with respect to the whole data distribution (Webb 2007). As such, the rules may not reflect the "real" association between the underlying attributes. The hypotheses reflected in the generated rules must be validated by a statistical methodology for them to be useful in practice, because the nature of data mining techniques is data-driven (Goodman, Kamath & Kumar 2008). However, even if the rules satisfy appropriate statistical tests, it can still be the case that the underlying association is caused purely by a statistical coincidence (Aumann & Lindell 2003).

To date, there has been limited work done on the rule evaluation phase of tree-structured rules. Many of the developed statistical and objective measures for relational data have had great success in evaluating rule interestingness. The applicability of these interestingness measures needs to be explored in the context of frequent subtree mining, where necessary adjustments and extensions need to be made to ascertain the validity of the methods in the presence of more complex structural aspects of the data, which often need to be preserved in the rules.

12.2.2 *Reducing Frequent Subtrees*

Many lessons on compression, pattern reduction and useful constraints learned in the frequent pattern mining of relational and sequential data can, to a certain extent, be applied in the frequent subtree mining field, or can be applied after some major extensions and/or refinements to take into account the structural aspects. For example, the work presented in (Nakamura & Kudo 2005) extends the idea of the item constraint (Srikant, Vu & Agrawal, 1997) to that of node-inclusion constraint in subtrees. The work done in (Bathoorn, Kopman & Siebes 2006) uses the method proposed for database compression in regards to item set mining in (Siebes, Vreeken & Leeuwen, 2006), to demonstrate how the same minimum description length principle can yield good results for sequential and tree-structured data. The use of monotone constraints in frequent subtree mining has been explored in (Knijf & Feelders, 2005). The types of constraints investigated are anti-monotone, monotone, convertible and succinct, which have been discussed in Section 12.1.1. An opportunistic pruning strategy is used to mine frequent subtrees under the defined constraints. A fuzzy approach to the matching of subtrees has been proposed in (Lopez et al. 2007), where the notion of subtree/supertree inclusion is extended to that of fuzzy inclusion. Hence, in their approach, the degree to which a tree is included in another tree is measured as the number of matching nodes between the two trees. An algorithm using this definition of tree inclusion has been presented in (Lopez et al. 2009), but due to the fuzzy measure used for tree comparison, more subtrees are considered as frequent in comparison with the traditional approach. A constraint-based approach on the smallest supertrees was explored in (Knijf 2008) who presents the MASS algorithm for discovering almost the smallest supertrees. A smallest supertree is defined as the one of which every tree in the database is a subset, and there is no other tree with fewer nodes that is also a superset of all the trees. The motivation for this work is that the smallest supertree can be seen as the shortest description of the database and the best according to the minimum description length principle. The problem of deriving the smallest supertree is considered as a special case of tree alignment and hence, the almost smallest supertree is arrived at through incremental alignment of candidate tree patterns. The work in (Murakami, Doi & Yanamoto 2008) extends the FREQT (Asai et al. 2002) algorithm so that it can be applied to a compressed format of data using the data compression technique proposed in (Onuma, Doi & Yamamoto 2006). A reduction in execution time is achieved by counting the occurrence of candidate subtrees directly from compressed data without expansion. The problem of mining mutually dependent ordered subtrees has been addressed in (Ozaki & Ohkawa 2009). The proposed algorithm utilizes the hyperclique method (Xiong, Tan, Kumar 2006) (see Section 12.1.1) in the tree mining context so that all the components of a subtree are highly correlated. From the application perspective, a method for mining significant tree patterns in carbohydrate sugar chains was proposed in (Hashimoto et al. 2008). They defined the concept of *α-closed* frequent subtree, as a subtree *st* where non of its frequent supertrees has a support $\geq \alpha$ * support(*st*). Such a subtree has a number of properties that are effectively exploited in the candidate enumeration phase. Once the frequent *α-closed* frequent subtrees have been extracted, they are ranked using statistical hypothesis testing to measure their significance. The experiments demonstrated that the top 10 ranked subtrees, after some parameter setting, were consistent with significant motifs or cores in glycobiology.

Some alternative tree mining methods have been developed, not necessarily for compressing the frequent pattern set, but rather with specific applications in mind. For example, the FAT-miner algorithm (Knijf 2007) discovers frequent attribute trees from a labeled rooted ordered tree, where in addition each node is assigned a non-empty attribute set. The algorithm is applied for exploring the properties of good and bad loans from a multi-relational financial database. A different means of frequency counting is explored in (Tosaka, Nakamura & Kudo 2007), in their approach to mining subtrees with frequent occurrence of similar subtrees. Hence, as opposed to the traditional approach, an occurrence of a subtree is counted for not only equivalent, but similar subtrees. The similarity measure used is based on the tree edit/alignment distance.

One can already see a parallel here to the methods that have been adopted from itemset mining to sequence mining and more recently to subtree mining. It is expected that many of the remaining techniques hitherto successfully applied to itemset/sequence mining for reducing the frequent pattern set and obtaining more focused results, will be explored for their use in frequent subtree mining. There has been limited work performed on feature selection from XML data, i.e. ways of selecting the most useful attributes (nodes) for the task at hand. As usual, the feature selection needs to be performed prior to the mining process and at times during and after the mining process in order to discard irrelevant attributes from the rules. A difficulty in subtree mining is that, even though some nodes are irrelevant for the informative content of the patterns, they are crucial for preserving the structural context of a pattern. A specialized feature selection criterion may be needed for XML data, whereby we can measure attribute relevance for both informational and/or predictive properties as well as its capability of preserving context in patterns with relevant nodes. This would allow one to incorporate such measures during the pattern generation to avoid the expensive generation of patterns that will result in many candidates that are irrelevant as well as expensive to generate.

12.3 Top-Down Approach for Frequent Subtree Mining

The frequent subtree mining approaches/algorithms described in this book generally differ in the candidate enumeration and counting strategy, tree representation and representative structures. However, a commonality that remains is that all candidate subtrees are first enumerated and then counted by employing a bottom-up lattice search. Even though the existing algorithms are well scalable for large datasets, a difficulty that remains is that if the instances in the tree database are characterized by complex tree structures, the approach will result in an enormous number of candidate subtrees that need to be enumerated. For example, the mathematical analysis provided in Chapter 5 indicates a potentially unfeasible situation, because of enormous number of candidates that would need to be generated for a tree of maximum depth 3 where all the nodes have degree 3. To alleviate this problem, some work has shifted toward the mining of maximal and closed subtrees as explained in Chapter 8. The question still remains whether the problem of enormous candidates can be alleviated using this approach, since if the database is characterized by complex tree structures it is likely to remain computationally unfeasible. While most of the maximal/closed subtree mining methods actually reduce the number of patterns, it may be the case that during the top-down enumeration of the patterns, one can still get stuck with too many candidates, since the approach is still done in a systematic manner. It may be useful to explore a different approach altogether where the frequent subtree patterns are obtained in a more human-like manner, i.e. where we mimic that the patterns that more frequently influence the learning space are "remembered".

In (Hadzic et al. 2007), it was discussed how a clustering approach using the Self-Organizing Map (SOM) (Kohonen 1990) can be used for extracting frequent patterns from relational data. When the aim is to use a similar technique for frequent subtree mining, the main difficulty is how to represent an instance from a tree-structured database to the input layer of the SOM, and how to preserve the structural relationships in the extracted patterns. There has been limited work in this area. The Contextual SOM method (Voegtlin 2000) is based on context quantization. It develops near-optimal representations of context where recurrent connections between the nodes in the map are used. In (Hagenbuchner & Sperduti 2003), an unfolding procedure is adopted in recurrent and recursive neural networks. Each leaf node in an input tree is treated as a standard input for a SOM, while the coordinates of the winning neuron are a pointer within a parent node. Hence, the work is based on the assumption that the information transferred between the parent and the children is via the coordinates of the winning neuron in a child node. This method has been further extended in (Hagenbuchner, Sperduti & Tsoi 2005) to enable better discrimination of all directed trees in data, by incorporating the idea of a recursive cascade-correlation model (Micheli, Sona, & Sperduti 2004). More recently, a contextual constructive approach has been presented for neural networks in general (Micheli 2009). It is based on a constructive, feedforward architecture with state variables using neurons with no feedback connection. A traversal process is applied to the input graphs and the contextual information of state variables for each node. The abovementioned approaches are developed mainly for classification and regression tasks, and their suitability for the mining of frequent maximal subtrees has not been explored.

There is a known relationship between clustering and frequent pattern mining, as it has often been used as an approximation approach to some frequent pattern mining problems in general. The Self-Organizing map has been reported as the most biologically plausible neural network (Sestito & Dillon 1994), as it is based on the functions of cognitive maps found in the brain. SOM (Kohonen 1990) is an unsupervised neural network that effectively creates spatially organized "internal representations" of the features and abstractions detected in the input space. It consists of an input layer and an output layer in the form of a map (see Fig. 12.1). SOM is based on the competition among the cells in the map for the best match against a presented input pattern. Each node in the map has a weight vector associated with it, which are the weights on the links emanating from the input layer to that particular node. When an input pattern is imposed on the network, a node is selected from among all the output nodes as having the best response according to some criterion. This output node is declared the 'winner' and is usually the cell having the smallest Euclidean distance between its weight vector and the presented input vector. The 'winner' and its neighboring cells are then updated to match the presented input pattern more closely. Neighborhood size and the magnitude of update shrink as the training proceeds. After the learning phase, cells that respond in similar manner to the presented input patterns are located close to each other, and so clusters can be formed in the map. Existing similarities in the input space are revealed through the ordered or topology preserving mapping of high dimensional input patterns into a lower-dimensional set of output cluster (Sestito & Dillon 1994).

SOM groups frequently-occurring data object characteristics together in the form of clusters. As it is based on competitive learning, the data object characteristics will

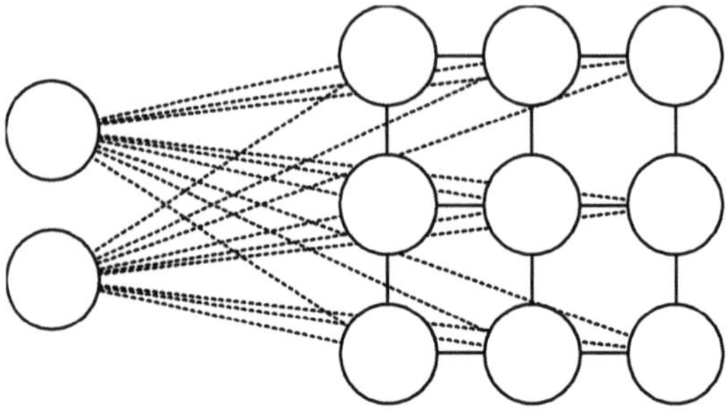

Fig. 12.1 Example SOM consisting of 2 input nodes and 3*3 map

compete to be projected onto the limited output space. After the training of the SOM is complete, the output map will be organized according to the data object characteristics that were most frequently presented to the SOM during the training phase. The map size is the determining factor in the number of distinct patterns that will be projected onto the output space. With smaller space, only the most frequently occurring patterns will be projected, while with a larger space, there will be enough resolution to project some of the not-so-frequently occurring characteristics. Another contributing factor is the intensity of the competition that occurs within the limited output space.

In (Hadzic et al. 2007), the similarity between the type of patterns extracted by the SOM-based approach and those by the traditional approach, was shown for the frequent pattern mining problem from structured or relational data. One could argue that a clustering-based approach to frequent pattern learning is more human-like as the patterns are not enumerated in a systematic bottom-up lattice search, but rather are 'remembered' due to their frequent influence on the learning space. The main motivation behind the future extensions for tree-structured data is that, for datasets with complex tree structures, the traditional approach leads to an enormous number of candidates that need to be generated and counted, while a top-down clustering approach will at least be able to provide a (partial) solution. The set of frequent subtrees may not be complete, but at least a good approximation is provided in cases where the traditional bottom-up approaches struggle due to inherent complexity. The learning parameters of SOM are dependent on each other, and at this stage it would be hard to provide the exact guidelines for mimicking support. More investigation is needed into finding the exact parameters so that the total set of patterns would be guaranteed. However, it is worth noting that if SOM is to be used effectively for frequent pattern discovery, the support threshold would be replaced by a threshold where major decisive factors are the output space size and the intensity of competition that occurs within that limited space. In other words, the traditional and SOM-based techniques for frequent pattern discovery are parameterized differently, and whereas support and confidence framework parameterize the traditional approach, output space dimension and intensity of competition parameterize the SOM approach.

12.4 Model Guided Approach for Graph Mining

From the frequent subtree mining approaches overviewed, we see that when generating only valid frequent subtrees by following a model of the underlying structure of the database, this results in many fewer candidate patterns being generated that need to be pruned afterwards anyway. It will be interesting to see whether a tree-model-guided approach can be generalized to a graph-model-guided approach for graph mining. Since the difficulty in frequent subgraph mining exists because of the existence of cycles, rather than mining subgraphs, another option would be to convert the graph structure into the tree structure where the information regarding cycles is preserved in some manner. For example, the work presented in (Feng, Chang & Dillon 2002), had to deal with the transformation from semantic networks (graph with cycles) to XML Schema (tree). To enable the transformation, the approach taken is to break the cycles, and turn a cyclic directed graph into an equivalent acyclic directed graph (i.e. tree) without the loss of semantics. Whenever a cycle occurs (n_1, n_2, n_3, ..., n_m), where node n_1 points to n_2, n_2 to n_3, and so on until n_m points to n_1 again, a new leaf node n_x is introduced so that n_m points to n_x. The referential integrity constraint is applied on n_x so that its key refers to node n_1 (Feng, Chang & Dillon 2002). This ensures that the information contained in the original graph structure is preserved in the transformed tree structure. Hence, if a similar method could be applied for transforming graph-structured to tree-structured data, the applicability of frequent subtree mining would extend to data that is originally in graph-structured form, and the complexity would be significantly reduced in many cases as the cyclic properties have been removed.

12.5 Conjoint Mining of Different Data Types

Digital information within an enterprise consists of (1) *relational data* (2) *semistructured data and (3) unstructured content*. The structured data includes enterprise and business data such as sales, customer and product information, accounts, inventory, etc. while the content includes contracts, emails, customer opinions, transcribed calls, etc. The structured data is managed by the relational database system (RDB), the semi/unstructured content is managed by the content manager, and these two exist as silos. Similarly, in the biomedical area, resources can be structured data in, for example, Swiss-Prot or unstructured text information in journal articles stored in content repositories such as PubMed. This leads to information of different types being in their own silos being managed and queried separately. This is undesirable since the information content of these sources is conceptually related and complementary. Due to this separation, an application (like business intelligence (BI) and deep knowledge discovery and data mining (DKDD)) would need to straddle the two disparate information sources, querying one and then the other. Effective knowledge management and market intelligence requires seamless access to both kinds of information. Understanding the need to bridge this separation, enterprises are demanding integrated retrieval, management and intelligent analysis of both data and content sources in order to gain critical insights into customer and market trends and optimize the decision-making process. This issue has recently become of major importance, as the relative proportion of structured to unstructured information has shifted from 50-50 in the 1960s to 5-95 today (Blumberg & Atre 2003; Brodie 2007). So, unless we can effectively utilize the unstructured/semistructured content conjointly with the structured data, we will obtain only very limited BI or limited and shallow knowledge discovery from an increasingly narrow slice of information.

Consider the following scenario. A credit card provider wants to find out the common profile characteristics of the clients likely to discontinue the use of the credit card service after three complaints have been received about transactions involving amounts greater than $200. Customer profiles are usually stored in a Relational Database, containing values for attributes such as 'date of birth', 'nationality, 'salary', 'address' and so on. The reasons for the discontinuation of the particular bank service are more likely to be found in other data sources that allow more flexibility in the information being stored, but are still structured and organized according to domain requirements. For this purpose, semi-structured documents such as XML are commonly used. It allows one to collect information about customers lodging complaints through customer contact centers, sending emails, filling survey forms or giving their opinions through website forums for customers' feedback. It is hard to extract meaningful information from plain text unless it has been organized following some domain-defined structure. Hence, this type of information is usually processed into XML format to facilitate analysis. The partial structure involves attributes such as email, complaints, or survey entries, but the value content is more flexible to embed unstructured text within those well-defined tags. To access this information, crucial for answering the query of the motivating scenario, we need to link the complaint to the related records or information in the RDB. Once the complete information has been integrated, there is enough information to answer the example query. To the best of our knowledge, the existing data mining methods can mine only structured or semi-structured data sources separately, and hence there is a need for a method or a system which enables both data sources to be mined in a conjoint manner. In the existing work on schema matching and data integration of structured and semi-structured data, the focus has been on querying and other knowledge management related tasks, rather than data mining of the merged data (Beeri & Milo 1999; Do & Rahm 2002; McBrien & Poulvassilis 2005).

12.5.1 A Framework for Conjoint Mining of Relational and Semi-structured Data

In this section, we focus our attention on the problem of combining the complementary information from structured and semi-structured data sources, and mining it in a conjoint manner. The described framework was initially introduced in (Pan, Hadzic & Dillon 2008), and its general steps are categorized into Data Pre-process, Data Linkage and Data Mining models as shown in Fig. 12.2.

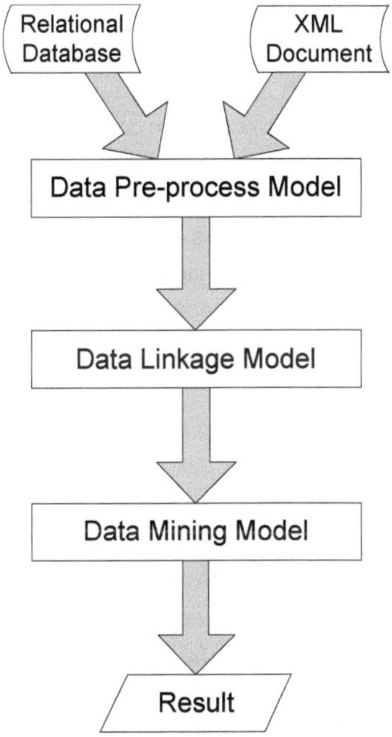

Fig. 12.2 General steps of a framework for mining relational and semi-structured data conjointly

Each of these models is likely to consist of many subtasks as there are usually many important issues to be considered when detecting, combining and conjointly mining information from different types of data sources (relational and semi-structured in this case). For example, some main issues causing difficulty in the process are:

1. Being able to identify that a particular document instance in the XML repository corresponds (relates) to a particular record in the relational database
2. Once we are able to link these instances to develop a mining methodology capable of mining these joined data sources.

Hence, a large number of sub-problems exist, each of which often comprises a research area on its own, some of which have still not reached a mature stage. There are still many more advances and refinements of existing knowledge matching, data linking and merging techniques to be made, and the problem of conjoint mining of data sources of different types is still in its very early stages of research.

12.6 Ontology Learning

The existing methods for ontology learning are currently, to a large extent, based on the experience of ontology building for a particular domain. As such, many of them can be seen as guidelines and may not be applicable to more general domains. Hence, at the present time there is still not a single widely accepted methodology since the approach chosen often depends on the application and extensions that the developer has in mind. However, the main trend should be towards an automation of the ontology learning process since a manual effort, besides being more prone to errors, is also very costly in terms of time and resources. The ontology development process has often been considered as more of an art than a science (Jones, Bench-Capon & Visser 1998). While some semi-automatic approaches have been proposed, there is still a lack of automation in the process and the core development is still mainly achieved through a manual effort. The next section provides some general ontology concepts and definitions.

12.6.1 Ontology Definition and Formulations

'Ontology' is a term that was first used in philosophical studies where it refers to a 'systematic account of existence'. It is a study of the essence of objects and their relationships. In the Artificial Intelligence field, ontology is defined as a formal, explicit specification of a shared conceptualization (Gruber 1993a). 'Formal' corresponds to the fact that the ontology should be machine readable, 'explicit' means that the concepts and their constraints should be explicitly defined, and 'conceptualization' refers to the definition of concepts and their relationships that occur in a particular domain (Gruber 1993a, Gruber 1993b). The important factor that distinguishes ontologies from other formal and explicit domain knowledge conceptualizations is that it captures consensual knowledge, i.e. the conceptualization is accepted by a large community of users. This enables the sharing and reuse of the consensual knowledge across applications and groups of people. Ontologies have been developed for a variety of domains serving a wide range of purposes and the amount of semantic constraints may vary among the domains. However, it needs to contain a vocabulary of terms and some specification of their meaning and interrelationships in order to limit possible interpretations (Uschold & Gruninger 1996). There are a few other aspects that can be included in an ontology specification and these will be explained in more detail later. At this stage, a more general notation will be used to represent the core aspects of an ontology. An ontology can be populated with instances but for the sake of simplicity we do not consider these in the general description of an ontology that is given next. An ontology can be described as a knowledge structure that is a four tuple ($<C, A, R, F>$) where:

$C - \{c_1, c_2, \ldots, c_n\}$ set of classes or concepts

$A - \{a_1, a_2, \ldots, a_n\}$ set of attributes

$R - \{r_1, r_2, \ldots, r_n\}$ relationships among the ontology concepts

$F - \{f_1, f_2, \ldots, f_n\}$ set of formal axioms modelling knowledge that is always true

Classes – In the set theory field of mathematics, a class (or equivalence class) corresponds to a collection of sets that are unambiguously defined by a property or relation that all its members share. In the ontology field, the term 'class' is taken to correspond to a concept in a broad sense. Concept is a unit of thought and may have abstract and physical meanings. In other words, it can refer to abstract groups, sets or collections of objects, as well as more specific concepts (people, locations, books). A 'concept term' (or name of a concept) is a lexical representation of a concept. Depending on the purpose of a particular ontology, classes can subsume or be subsumed by other classes. The term 'class' is often being replaced by the term 'concept' when referring to an ontology in the Artificial Intelligence field, and from this point forward, the same will be done in this chapter.

Attributes – Attributes are properties that are assigned to a concept for describing the information specific to that particular concept. When an ontology is populated with instances, then the attributes will be assigned values that can be singular or of a complex type.

Relations - Relations are used to represent different types of associations or relationships among the concepts in the domain. The relations within an ontology are commonly of the binary type and the most important relation is the subsumption relation. This relation is commonly implied by the way that concepts are organized in the concept hierarchy or taxonomy, so that relations such as sub-class-of are implied by the lower level concepts, and super-class-of by the concepts at the higher level in the structure. It is analogous to the child-parent and ancestor-descendant relationships in a tree structure that were discussed earlier. Another common type of relation is a composite relationship that describes the way that concepts can combine to form composite concepts.

Axioms – An axiom corresponds to underlying assumptions used during theory building and are considered as either obvious or have previously been proven. In an ontology, formal axioms are used to represent knowledge that cannot be formally defined by the other components. They are useful for verifying the consistency of the represented knowledge and can be used as the basis for inference of new knowledge.

The set of concept terms and their relationships is organized in a representational structure so that the inheritance relationships can be applied. It is often referred to as a 'concept hierarchy' or 'taxonomy' and the underlying structures that are commonly used are graphs or trees. Graphs are commonly used where there is a need to connect concepts situated in different parts of the hierarchy, which thereby introduces a cycle and hence the basic tree hierarchy becomes a graph. Except for very complex domains where many concepts are interrelated, there may not be so many cycles in the graph representation of the ontology and the core knowledge would still mainly be represented in a hierarchy. Since an ontology describes consensual knowledge, matching of heterogeneous knowledge representations describing the same domain and discovering the generic knowledge can, to some extent, be viewed as an automated ontology building process. While many graph-structured knowledge representations and ontologies exist, the focus in this section is on matching tree-structured knowledge representations to obtain a tree-structured ontology.

The main differences among knowledge representation occur in: vocabularies, naming of knowledge concepts (i.e. concept terms), knowledge representation, concept granularity and level of generalization/specification. These differences make the knowledge matching a challenging task and to perform it manually would be too time consuming and error-prone. Before the knowledge representations can be matched, the main problem is to find the semantically correct matches among the concept terms. This is known as the concept matching problem and is discussed in the next section.

12.6.2 Concept Term Matching Problem

The concept term *matching problem* is the problem of finding a mapping between the concept terms of two (or more) knowledge representations (KR). Since knowledge representations commonly differ in the amount of specific/general knowledge stored about some aspect of the domain, the number of concept terms will differ among them and the mappings will not be restricted just to one-to-one (1:1) but to many-to-one (M:1), one-to-many (1:M) and many-to-many (M:M). The last case does not occur so often because it can usually be considered as two mappings. Representing it as a 1:M and M:1 match, gives a better indication of the point in the structure (particular concept) where the difference in the knowledge level of detail occurs. A complex match generally indicates that, while one KR uses a singular concept term to represent some aspect of the domain, the second KR uses multiple concept terms or stores more detailed information about that aspect of the domain, i.e. the granularity of the representation in the two cases is different. As a simple example, consider Fig. 12.3 where the concept with label 'DeliverTo' in KR_2 corresponds to two concepts in KR_1 namely 'DeliverTo' and 'Address' (the same holds for the concept 'BillTo').

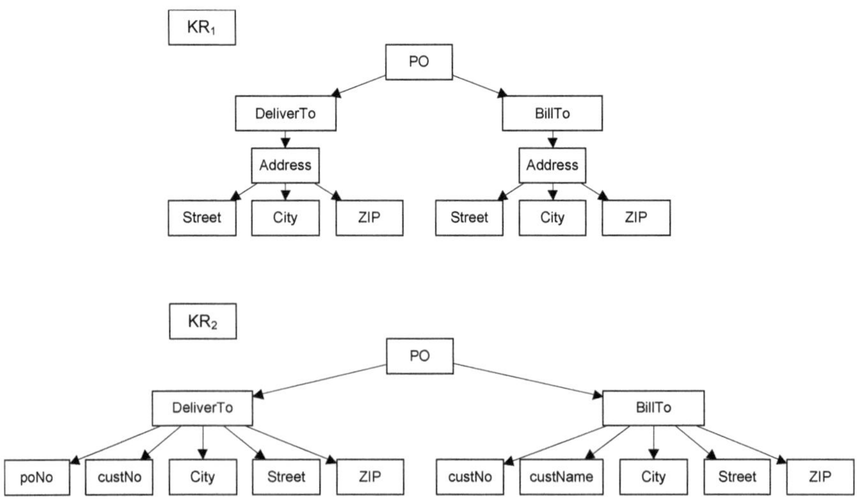

Fig. 12.3 XML schema structures describing different post order documents (KR_1 and KR_2) (obtained from www.ontologymatching.org)

Complex matches are very hard to detect since many possible combinations exist and to date not many approaches exist that can detect complex matches. The problem of concept matching can be generally stated as follows:

Definition: Let KR_1 and KR_2 correspond to two knowledge representations, $C_1 = \{c_{11}, c_{12}, \ldots, c_{1n}\}$ (where $n = |C_1|$) and $C_2 = \{c_{21}, c_{22}, \ldots, c_{2m}\}$ (where $m = |C_2|$) are complete sets of concept terms used in KR_1 and KR_2, respectively. The task of concept term matching is to find a set of mappings $M = \{m_1, m_2, \ldots, m_p\}$ (where $p = |M|$), where each $m \in M$ is a 2-tuple denoted as (e,f) such that set $e \subseteq C_1$ and set $f \subseteq C_2$.

Most of the sets e and f from the mappings m will consist of a single concept, but in cases of a complex match, a number of concept terms will be present in one of the sets. In ideal cases, the set M would contain all concept terms from C_1 and C_2 in all the sets e and f from its elements, respectively. However, due to the many different ways in which knowledge is represented by different organizations, there may remain some elements from C_1 and/or C_2 which are not present in the respective element sets (e and f) of the found mappings (m). This usually occurs in cases where the extent to which the domain knowledge is covered by one organization is much larger, and there are some aspects of the domain totally unaccounted for by the other organization. This is in contrast to complex matches where it can be detected that for the same aspect of the domain (concept), additional information is stored in one of the matched knowledge representations in the form of additional concept terms. For example, consider Fig. 12.3 again where an example of a complex match was given before (i.e. 'DeliverTo' vs ('DeliverTo', 'Address'). The 'DeliverTo' and ('DeliverTo', 'Address') concept terms both describe one concept or aspect of the domain, which is where to deliver the goods. An example of concept terms which are unaccounted for and do not have a corresponding match are concepts with labels 'poNo' and 'custNo' in the left subtree of KR_2. The same holds for concepts with labels 'custNo' and 'custName' in the right subtree of KR_2. The aspect of the domain represented by these concept terms is unaccounted for in KR_1 and a valid mapping cannot be formed with these concept terms. If we let C_1 correspond to the concept terms from KR_1, and C_2 to concept terms from KR_2 the set of correct mappings found for the example structures are M = {({'PO'},{'PO'}), ({'DeliverTo', 'Address'}, {'DeliverTo'}), ({'BillTo', 'Address'}, {'BillTo'}), ({'City'},{'City'}), ({'Street'},{'Street'}), ({'Zip'},{'Zip'})}. In this case all the concept terms from set C_1 are contained in the concept terms of sets e in all $m \in M$, while the set C_2 is not contained completely in the concepts of sets f in all $m \in M$.

While the problem has been stated here for two knowledge representations, it is easily extended for three or more knowledge representations by performing the task for each pair separately.

There exist a number of criteria and ways of forming the mappings; many are based on some type of string matching and compare the concept terms using existing online dictionaries and thesauruses. In this case, the actual string label used for describing a concept of the domain is used as the basis for comparison. In the example given in Fig. 12.3, the concept terms describing the same aspect of the domain were labeled the same to make the illustration easier, but in reality, when two heterogeneous knowledge structures are matched, the concept terms representing the same domain concept may often have different labels.

In our future work, the aim is to avoid initially making comparisons based on string labels, but rather analyze the structure in which the concept terms are presented. String matching approaches can work well for some cases, but at the same time they are not always reliable since different naming conventions are used among knowledge representations and the same name may refer to different things at times. Furthermore, this type of matching that relies on labels has been done by many works and the aim is to approach the problem in a different manner and see how close one can get to the correct set of mappings without considering the labels. Hence, the focus will be purely on semantic matching rather than the syntactic approach. Once the mappings have been formed, there is also a need for the capability of verifying whether the mappings found are correct. Up to date, this has been mostly done through manual effort which is very time consuming. It would beneficial if this task could also be automated so that the found mappings are verified automatically, and if incorrect, the process could immediately continue in its search for a correct set of mappings. Automatic verification of the formed mappings is another problem that needs to be investigated in future work. Next, the problem of knowledge representation matching at the structural level is discussed.

12.6.3 Structural Representation Matching Problem

At this stage of the problem, the assumption is that the concept term matching has been performed and the same concept terms are used for representing the same domain concepts. The aim is to find the largest common knowledge structure among the knowledge representations since this would correspond to the shared conceptualization of the domain. Each of the knowledge representations can be modeled as a rooted ordered labeled tree. When matching knowledge structures, if the sibling nodes are ordered differently, the knowledge represented is still considered the same. Furthermore, the same concept terms can occur at different levels in the tree. Therefore, the task of finding the common structure is equivalent to finding the largest common unordered embedded subtree among the available trees. If tree T_S is the largest unordered embedded subtree of a tree T_K, the following notation will be used $T_S \cap_{UE} T_K$. The problem of knowledge structure matching among n number of tree-structured knowledge representations can then be more formally stated as follows:

Definition: Given a set of trees $T = \{T_1, T_2, \ldots, T_n\}$ find a tree T_S such that $\forall\ T_i \in T$, where $i = (1, \ldots, n)$, $T_S \cap_{UE} T_i$.

While the focus has been on ontology learning through matching of knowledge representations, the progress in this area will be highly applicable to the problem of ontology matching where ontologies are of tree-structured form. The capability to automatically match knowledge representations at both the structural and conceptual levels is one of the hardest tasks in the ontology matching problem which is still, to a large extent, accomplished through a manual or semi-automatic effort.

12.6.4 Ontology Learning Method Aims

In future work, the aim will be to apply the developed tree mining algorithms to enable ontology building by matching the existing knowledge representations from the same domain. The main problem that will need to be addressed in this process is to find semantically correct matches among the concept terms in heterogeneous knowledge representations. We will initially avoid considering concept labels as a guide for the formation of candidate mappings, but rather use the structural information in which concepts occur in a particular knowledge representation. Taking into account the structural position of the concept term nodes is, to a certain extent, a promising approach for considering the context in which the concept terms are used. Taking context into consideration is one of the main difficulties in existing approaches. As opposed to matching concepts based upon label comparison, taking the structural aspects into account will indicate possible complex matches (i.e. cases where a concept term in one knowledge representation maps to multiple concept terms in another knowledge representation).

The relations considered are limited to the subsumption relations implied by the concept hierarchy or taxonomy. In this respect, the two main problems that are considered are the matching of knowledge representations at the conceptual and structural levels.

12.6.4.1 Conceptual Level Matching

To detect the concept term mappings between two (or more) knowledge representations, the main aim is to use the structural information in which concept terms occur in a particular knowledge representation. Taking into account the position of the concept terms in the representational structure, is to some extent a promising approach for taking the context in which the concepts are used into account. Furthermore, there is usually some indication of the possible complex matches which is not so easily obtained by label comparison (unless performed manually).

The basic idea is to use one of the developed frequent subtree mining algorithms to extract all possible subtrees from the knowledge representations (KR_1 and KR_2). The general idea of this process is illustrated in Fig. 12.4

A similarity matrix will be set up where all the concept terms from KR_1 and KR_2 are organized into rows and columns in pre-order traversal of the underlying tree structures. There will be a corresponding entry for all combinations of possible one-to-one concept term mappings. The process starts by taking in an initial match that is preferably supplied by the user. Otherwise, the root nodes of the knowledge representations are good candidates for the initial match. The process continues by traversing the subtree sets from each knowledge representation in which the matched concepts occur. Whenever a pair of k-subtrees (one from KR_1 and one from KR_2) have the same structure and the matched concepts occur at the same position, the similarity value between the remaining concept terms in those subtrees is increased in the corresponding entries in the similarity matrix. Whenever a similarity

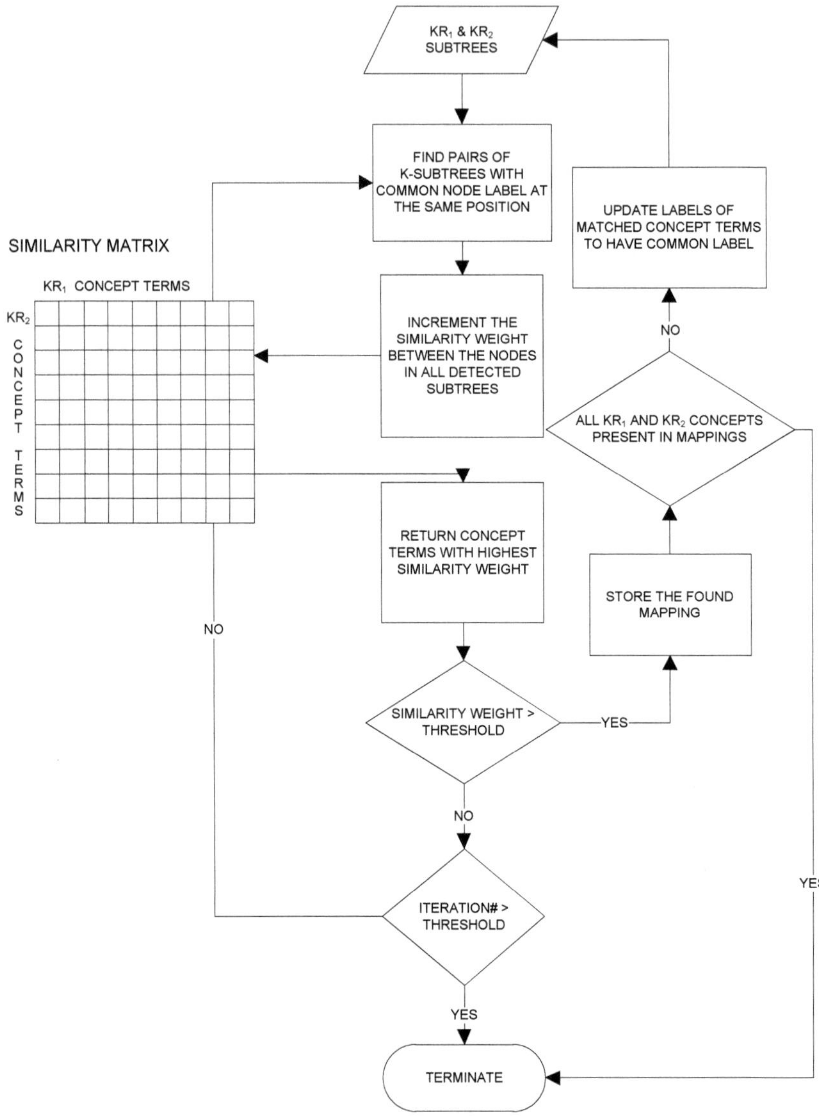

Fig. 12.4 Proposed Concept Term Matching Process

value between two concept terms exceeds a predetermined threshold value, it becomes a new match and the process is repeated using the newly matched concepts. The criteria for complex matches will also be checked during the process, and the discovered mappings will be verified utilizing instance information to validate the found mapping by structural comparison and statistical measures. However, if no instances are available, the validation will be done manually, or some of the available dictionaries, thesauruses, and syntactic string matchers may be used to check whether the concept terms in the mappings do correspond to the same aspect of the domain.

The general idea behind this approach is to enable the ontology learning process to occur through a more constructive approach where the instances used to derive heterogeneous knowledge models are available. With this information, validation of formed mappings can automatically occur using some powerful machine learning and statistical techniques. The explanation above is given for two knowledge representations, but if there are more available, then the process can be repeated for matching the concept terms of each pair separately.

12.6.4.2 Structural Level Matching

Once the correct concept term mappings are found between the available knowledge representations, the next task is to find the largest common structure. This will represent the shared representation of the domain knowledge for that domain. The common concept terms among the available knowledge representations are updated so that they have a common label.

With these assumptions, the way in which the problem is approached is equal to the way in which the knowledge matching problem is approached as described in Chapter 9. Hence, if there are n transactions, then one will be interested in finding the largest unordered embedded k-subtree that occurs in all n transactions (i.e. minimum transaction-based support = n). This subtree which in fact corresponds to the maximal frequent unordered embedded subtree, can then be taken to represent the shared knowledge of the domain at hand. This detected shared knowledge structure (KS) could be less specific than the KS from a particular organization, but it is therefore valid for all the organizations. Furthermore, each of the different organizations could have their own specific part of knowledge which is valid only from their perspective, and which can be added to the shared KS so that every aspect relevant for that organization is covered. Hence, the shared KS can be used as the basis for structuring the knowledge for that particular domain and different communities of users can extend this model when required for their own organization's specific purposes. In other words, it can be considered as an ontology for that domain since it represents a formal and shared conceptualization of the domain. It is shared within the organizations from which the knowledge representations were matched.

In the above discussion, the ontologies will be developed by matching heterogeneous knowledge representations where the underlying structure is a tree. Hence, the end result is also limited to the ontologies where all the concepts are organized in a hierarchy. Having a graph mining algorithm, the same idea of ontology learning can be applied, except that in this case we are matching knowledge representations where the underlying structure is a graph. Hence, in this sense one would be matching concepts by exploiting the similarity of structural properties of the concepts in the graph structure, and the largest subgraph that occurs in all of the knowledge structures compared, can be considered as the ontology for that domain. In this future approach of applying frequent subtree mining to ontology learning, we aim to indicate the many aspects where it can prove useful, and automate the process to a large extent. We must note here that in some domains, to obtain a reliable and easy to use and application oriented ontology without human intervention is almost impossible, and in some cases even undesirable as humans want to make it more intelligible. In such domains, the planned approach may still be of great use to automate a large portion of the similarity analysis task, where one needs to find shared aspects of heterogeneous knowledge representations (understanding) of the domain.

12.7 Conclusion

This chapter has looked at some future directions in the field of frequent subtree mining, which as evident from the current status, deals with reducing the complexity requirements and focusing the pattern extraction to only the most relevant and important patterns for the application at hand. The ways that the complexity and pattern irrelevance issues have been approached for frequent itemset/sequence mining have been overviewed. A number of such approaches have already been explored for frequent subtree mining, and we have looked at a number of existing methods for pattern reduction and rule evaluation in the frequent subtree mining field. The results were promising, and are evidence that further exploration in this area needs to take place. A more human-like approach to frequent subtree mining using the Self-organizing map, was then discussed. The important issues of combining different information from related data sources of different type has also been explored. To conclude, we have looked at the potential application of frequent subtree mining to aid in automating a part of the ontology construction phase, when heterogeneous knowledge representations of the same domain are to be investigated in order to detect shared conceptualizations of the domain.

References

1. Afrati, F., Gionis, A., Mannila, H.: Approximating a collection of frequent sets. Paper presented at the Proceedings of the 10th ACM SIGKDD International Conference on Knowledge Discovery and Data Mining, Seattle, Washington, USA, August 22-25 (2004)
2. Aggarwal, C.C., Yu, P.S.: A new framework for itemset generation. Paper presented at the Proceedings of the 17th ACM SIGACT-SIGMOD-SIGART Symposium on Principles of database systems, Seattle, Washington, USA, June 1-3 (1998)
3. Agrawal, R., Imieliski, T., Swami, A.: Mining Association Rules between Sets of Items in Large Databases. In: Proceedings of the ACM SIGMOD International Conference on Management of Data, Washington D.C., USA, May 26-28, pp. 207–216. ACM, New York (1993)
4. Asai, T., Abe, K., Kawasoe, S., Arimura, H., Sakamato, H., Arikawa, S.: Efficient substructure discovery from large semi-structured data. Paper presented at the Proceedings of the 2nd SIAM International Conference on Data Mining (SIAM 2002), Arlington, VA, USA, April 11-13 (2002)

5. Bathorn, R., Kopman, A., Siebes, A.: Reducing the Frequent Pattern Set. Paper presented at the Proceedings of the 6th IEEE international Conference on Data Mining – Workshops (ICDMW 2006), Hong Kong, China, December 18-22 (2006)
6. Bay, S.D., Pazzani, M.J.: Detecting Group Differences: Mining Contrast Sets. Data Mining and Knowledge Discovery 5(3), 213–246 (2001)
7. Beeri, C., Milo, T.: Schemas for integration and translation of structured and semi-structured data. In: Beeri, C., Bruneman, P. (eds.) ICDT 1999. LNCS, vol. 1540, pp. 296–313. Springer, Heidelberg (1998)
8. Blumberg, R., Atre, S.: The Problem with Unstructured Data. Information Management Magazine (2003)
9. Brijs, T., Vanhoof, K., Wets, G.: Defining interestingness for association rules. International Journal of Information Theories and Applications 10(4), 370–376 (2003)
10. Brin, S., Motwani, R., Silverstein, C.: Beyond Market Baskets: Generalizing Association Rules to Correlations. In: Proceedings of the ACM SIGMOD International Conference on Management of Data, Tucson, Arizona, USA, May 13-15, pp. 265–276. ACM, New York (1997)
11. Brin, S., Motwani, R., Ullman, J., Tsur, S.: Dynamic Itemset Counting and Implication Rules for Market Basket Data. Paper presented at the Proceedings of the 1997 ACM SIGMOD International Conference on Management of Data, Tucson, Arizona, USA, May 13-15 (1997)
12. Brodie, M.L.: Computer Science 2.0: A New World of Data Management. In: Proceedings of the 33rd International Conference on Very Large Data Bases (VLDB 2007), Vienna, Austria, September 23-27, p. 1161 (2007)
13. Bucila, C., Gehrke, J., Kifer, D., White, W.: DualMiner: A Dual-Pruning Algorithm for Itemsets with Constraints. Data Mining and Knowledge Discovery 7(3), 241–272 (2003)
14. Cerf, L., Besson, J., Robardet, C., Boulicaut, J.-F.: Data-Peeler: Constraint Based Closed Pattern Mining in n-ary Relations. Paper presented at the Proceedings of the SIAM International Conference on Data Mining (SDM 2008), Atlanta, Georgia, USA, April 24-26 (2008)
15. Do, H., Rahm, E.: COMA: a system for flexible combination of schema matching approaches. Paper presented at the Proceedings of 28th International Conference on Very Large Data Bases (VLDB 2002), Hong Kong, China, August 20-23 (2002)
16. Doan, A., Halevy, A.Y.: Semantic-Integration Research in the Database Community: A Brief Survey AI Magazine, 26 (2005)
17. Feng, L., Chang, E., Dillon, T.S.: A semantic network-based design methodology for XML documents. ACM Transactions on Information Systems 20(4), 390–421 (2002)
18. Fu, Y., Han, J.: Meta-rule guided mining of association rules in relational databases. Paper presented at the Proceedings of the 1st International Workshop on Knowledge Discovery in Databases with Deductive and Object-Oriented Databases (KDOOD 1995), Singapore, December 8 (1995)
19. Garofalakis, M.N., Rastogi, R., Shim, K.: SPIRIT: Sequential Pattern Mining with Regular Expression Constraints. Paper presented at the Proceedings of the 25th International Conference on Very Large Databases (VLDB 1999), Edinburgh, Scotland, UK, September 7-10 (1999)
20. Goodman, A., Kamath, C., Kumar, V.: Data Analysis in the 21st Century. Statistical Analysis and Data Mining 1(1), 1–3 (2008)
21. Grahne, G., Lakshmanan, L.V.S., Wang, X.: Efficient mining of constrained correlated sets. Paper presented at the Proceedings of the 16th International Conference on Data Engineering (ICDE 2000), San Diego, CA, USA, February 28 - March 3 (2000)

22. Gruber, T.R.: A translation approach to portable ontology specifications. Knowledge Acquisition 5(2), 199–220 (1993a)
23. Gruber, T.R.: Towards Principles for the Design of Ontologies Used for Knowledge Sharing. International Journal of Human and Computer Studies 43(5/6), 907–928 (1993b)
24. Hadzic, F., Dillon, T.S., Tan, H., Feng, L., Chang, E.: Mining Frequent Patterns using Self-Organizing Map. In: Taniar, D. (ed.) Advances in Data Warehousing and Mining Series, pp. 121–142. Idea Group Inc., USA (2007)
25. Hagenbuchner, M., Sperduti, A.: A Self-Organizing Map for Adaptive Processing of Structured Data. IEEE Transactions on Neural Networks 14(3), 491–505 (2003)
26. Hagenbuchner, M., Sperduti, A., Tsoi, A.: Contextual Processing of Graphs using Self-Organizing Maps. Paper presented at the Proceedings of the 13th European Symposium on Artificial Neural Networks, Bruges, Belgium, April 27-29 (2005)
27. Han, J., Kamber, M.: Data Mining: Concepts and Techniques, 2nd edn. Elsevier, Morgan Kaufmann Publishers, San Francisco, CA, USA (2006)
28. Hashimoto, K., Takigawa, I., Shiga, M., Kanehisa, M., Mamitsuka, H.: Mining significant tree patterns in carbohydrate sugar chains. Bioinformatics 24(16), 167–173 (2008)
29. He, B., Chang, K.C.C.: Statistical Schema Matching across Web Query Interfaces. Paper presented at the Proceedings of the 2003 ACM SIGMOD International Conference on Management of Data, San Diego, CA, USA, June 9-12 (2003)
30. Jones, D.M., Bench-Capon, T.J.M., Visser, P.R.S.: Methodologies for Ontology Development. Paper presented at the Proceedings of the IT&KNOWS Conference of the 15th IFIP World Computer Congress, Budapest, Hungary, August 31 - Septmeber 4 (1998)
31. Kappel, G., Kapsammer, E., Retschitzegger, W.: Integrating XML and Relational Database Systems. World Wide Web 7(4), 343–384 (2004)
32. Klemettinen, M., Mannila, H., Ronkainen, P., Toivonen, H., Verkamo, A.I.: Finding interesting rules from large sets of discovered association rules. Paper presented at the Proceedings of the 3rd International Conference on Information and Knowledge Management (CIKM 1994), Gaithersburg, Maryland, USA, November 29 - December 2 (1994)
33. Kniif, J.D., Feelders, A.: Monotone Constraints in Frequent Tree Mining. Paper presented at the Proceedings of the 14th Annual Machine Learning Conference of Belgium and the Netherlands (BENELEARN 2005), Enschede, Netherlands, Februrary 17-18 (2005)
34. Knijf, J.D.: FAT-miner: Mining Frequent Attribute Trees. Paper presented at the Proceedings of the 22nd Annual ACM Symposium on Applied Computing, Seoul, Korea, March 11-15 (2007)
35. Knijf, J.D.: Mining Tree Patterns with Almost Smallest Supertrees. Paper presented at the Proceedings of the 2008 SIAM International Conference on Data Mining (SDM), Atlanta, Georgia, USA., April 24-26 (2008)
36. Kohonen, T.: The Self-Organizing Map. Proceedings of the IEEE 78(9), 1460–1480 (1990)
37. Lakshmanan, L.V.S., Ng, R.T., Han, J., Pang, A.: Optimization of Constrained Frequent Set Queries with 2-variable Constraints. In: Proceedings ACM SIGMOD International Conference on Management of Data, Philadelphia, Pennsylvania, USA, June 1-3, pp. 157–168. ACM, New York (1999)
38. Liu, B., Hsu, W., Ma, Y.: Mining Association Rules with Multiple Minimum Supports. In: Proceedings of the 5th ACM SIGKDD International Conference on Knowledge Discovery and Data Mining, San Diego, CA, USA, August 15-18, pp. 337–341. ACM, New York (1999)

39. Liu, B., Hsu, W., Ma, Y.: Pruning and summarizing the discovered associations. Paper presented at the Proceedings of the 5th ACM SIGKDD International Conference on Knowledge Discovery and Data Mining, San Diego, CA, USA August 15-18 (1999)
40. Lopez, F.R., Laurent, A., Poncalet, P., Teisseire, M.: Fuzzy Tree Mining: Go Soft on Your Nodes. In: Melin, P., Castillo, O., Aguilar, L.T., Kacprzyk, J., Pedrycz, W. (eds.) IFSA 2007. LNCS (LNAI), vol. 4529, pp. 145–154. Springer, Heidelberg (2007)
41. Lopez, F.R., Laurent, A., Poncalet, P., Teisseire, M.: FTMnodes: Fuzzy tree mining based on partial inclusion. Fuzzy Sets and Systems 160(15), 2224–2240 (2009)
42. McBrien, P., Poulovassilis, A.: A Semantic Approach to Integrating XML and Structured Data Sources. In: Dittrich, K.R., Geppert, A., Norrie, M.C. (eds.) CAiSE 2001. LNCS, vol. 2068, p. 330. Springer, Heidelberg (2001)
43. Meggido, N., Srikant, R.: Discovering Predictive Association Rules. Paper presented at the Proceedings of the 4th International Conference on Knowledge Discovery and Data Mining (KDD 1998), New York City, New York, USA, August 27-31 (1998)
44. Micheli, A., Sona, D., Sperduti, A.: Contextual processing of structured data by recursive cascade correlation. IEEE Transactions on Neural Networks 15(6), 1396–1410 (2004)
45. Micheli, A.: Neural network for graphs: a contextual constructive approach. IEEE Transactions on Neural Networks 20(3), 498–511 (2009)
46. Murakami, S., Doi, K., Yamamoto, A.: Finding Frequent Patterns from Compressed Tree-Structured Data. Paper presented at the Proceedings of the 11th International Conference on Discovery Science, Budapest, Hungary, October 13-16 (2004)
47. Nakamura, A., Kudo, M.: Mining Frequent Trees with Node-Inclusion Constraints. In: Ho, T.-B., Cheung, D., Liu, H. (eds.) PAKDD 2005. LNCS (LNAI), vol. 3518, pp. 850–860. Springer, Heidelberg (2005)
48. Ng, R.T., Lakshmanan, L.V.S., Han, J., Pang, A.: Exploratory Mining and Pruning Optimizations of Constrained Associations Rules. In: Proceedings of the ACM-SIGMOD International Conference on Management of data, SIGMOD 1998, Seattle, WA, USA, June 2-4, pp. 13–24. ACM, New York (1998)
49. Onuma, J., Doi, K., Yamamoto, A.: Data compression and anti-unification for semistructured documents with tree grammars. In: IEIC Technical Report, Artificial intelligence and knowledge-based processing, vol. 106(38), Institute of Electronics, Information and Communication Engineer, Kyoto (2006) (in Japanese)
50. Ozaki, T., Ohkawa, T.: Mining Mutually Dependent Ordered Subtrees in Tree Databases. Paper presented at the Proceedings of the PAKDD 2009, Wrokshop on New Frontiers in Applied Data Mining, Osaka, Japan, May 20-23 (2009)
51. Pan, Q.H., Hadzic, F., Dillon, T.S.: Conjoint Data Mining of Structured and Semistructured Data. In: Proceedings of the 4th International Conference on the Semantics, Knowledge and Grid (SKG 2008), Beijing, China, December 3-5, pp. 87-94 (2008)
52. Pei, J., Han, J., Lakshmanan, L.V.S.: Mining frequent itemsets with convertible constraints. Paper presented at the Proceedings of the 17th International Conference on Data Engineering, Heidelberg, Germany, April 2-6 (2001)
53. Pei, J., Han, J., Wang, W.: Constraint-based sequential pattern mining: the pattern-growth methods. Journal of Intelligent Information Systems 28(2), 133–160 (2007)
54. Piatetsky-Shapiro, G.: Discovery, analysis, and presentation of strong rules. In: Piatetsky-Shapiro, G., Frawley, W.J. (eds.) Knowledge Discovery in Databases, pp. 229–238. AAAI/MIT Press (1991)
55. Sestito, S., Dillon, T.S.: Automated Knowledge Acquisition. Prentice Hall, Sydney (1994)
56. Mohd Shaharanee, I.N., Hadzic, F., Dillon, T.S.: Interestingness of Association Rules using Symmetrical Tau and Logistic Regression. In: Nicholson, A., Li, X. (eds.) AI 2009. LNCS, vol. 5866, pp. 422–431. Springer, Heidelberg (2009)

57. Shen, W.-M., Ong, K., Mitbander, B., Zaniolo, C.: Metaqueries for Data Mining. In: Fayyad, U.M., Piatetsky-Shapiro, G., Smyth, P., Uthurusamy, R. (eds.) Advances in Knowledge Discovery and Data Mining, pp. 375–398. AAAI/MIT Press (1996)
58. Siebes, A., Vreeken, J., Leeuwen, M.V.: Itemsets that compress. Paper presented at the Proceedings of the 6th SIAM International Conference on Data Mining (SDM 2006), Bethesda, MD, USA, April 20-22 (2006)
59. Silverstein, C., Brin, S., Motwani, R.: Beyond Market Baskets: Generalizing Association Rules to Dependence Rules. Data Mining and Knowledge Discovery 2(1), 39–68 (1998)
60. Srikant, R., Vu, Q., Agrawal, R.: Mining Association Rules with Item Constraints. In: Proceedings of the 3rd International Conference on Knowledge Discovery and Data Mining (KDD 1997), Newport Beach, CA, USA, August 14-17, pp. 67-73 (1997)
61. Tan, P.N., Kumar, V., Srivastava, J.: Selecting the right interestingness measure for association patterns. Paper presented at the Proceedings of the 8th ACM SIGKDD International Conference on Knowledge Discovery and Data Mining (KDD 2002), Edmonton, Alberta, Canada, July 23-26 (2002)
62. Tosaka, H., Nakamura, A., Kudo, M.: Mining Subtrees with Frequent Occurrence of Similar Subtrees. Paper presented at the Proceedings of the 10th International Conference on Discovery Science, Sendai, Japan, October 1-4 (2007)
63. Uschold, M., Grninger, M.: Ontologies: Principles, Methods and Applications. Knowledge Engineering Review 11(2), 93–136 (1996)
64. Voegtlin, T, Context quantization and contextual self-organizing maps. Paper presented at the Proceedings of the IEEE-INNS-ENNS International Joint Conference on Neural Networks (IJCNN 2000), Como, Italy, July 24-27 (2000)
65. Wang, J., Han, J., Lu, Y., Tzvetkov, P.: TFP: An efficient algorithm for mining top-k frequent closed itemsets. IEEE Transactions on Knowledge and Data Engineering 17(5), 652–664 (2005)
66. Webb, G.I.: Preliminary investigations into statistically valid exploratory rule discovery. Paper presented at the Proceedings of the Australasian Data Mining Workshop (AusDM 2003), Canberra, Australia, December 8 (2003)
67. Webb, G.I.: Discovering Significant Patterns. Machine Learning 68(1), 1–33 (2007)
68. Xin, D., Han, J., Yan, X., Cheng, H.: Mining compressed frequent-pattern sets. Paper presented at the Proceedings of the 31st International Conference on Very Large Databases (VLDB 2005), Trondheim, Norway, August 30 - September 2 (2005)
69. Xin, D., Han, J., Yan, X., Cheng, H.: On compressing frequent patterns. Data and Knowledge Engineering 60(1), 5–29 (2006)
70. Xiong, H., Tan, P.-N., Kumar, V.: Hyperclique pattern discovery. Data Mining and Knowledge Discovery 13(2), 219–242 (2006)
71. Yang, J., Wang, W.: CLUSEQ: Efficient and effective sequence clustering. Paper presented at the Proceedings of the 19th International Conference on Data Engineering (ICDE 2003), Bangalore, India, March 5-8 (2003)
72. Yun, H., Ha, D., Hwang, B., Ryu, K.H.: Mining association rules on significant rare data using relative support. Journal of Systems and Software 67(3), 181–191 (2003)
73. Zaki, M.J., Lesh, N., Ogihara, M.: PlanMine: Predicting Plan Failures Using Sequence Mining. Artificial Intelligence Review 14(6), 421–446 (2000)
74. Zhang, H., Padmanabhan, B., Tuzhilin, A.: On the discovery of significant statistical quantitative rules. Paper presented at the Proceedings of the 10th ACM SIGKDD International Conference on Knowledge Discovery and Data Mining (KDD 2004), Seattle, WA, USA, August 22-25 (2004)

MIX
Papier aus verantwortungsvollen Quellen
Paper from responsible sources
FSC® C105338

If you have any concerns about our products,
you can contact us on
ProductSafety@springernature.com

In case Publisher is established outside the EU,
the EU authorized representative is:
Springer Nature Customer Service Center GmbH
Europaplatz 3, 69115 Heidelberg, Germany

Printed by Libri Plureos GmbH
in Hamburg, Germany